PROJECT MANAGEMENT METRICS, KPIs, AND DASHBOARDS

T0312614

PROJECT MANAGEMENT METRICS, KPIs, AND DASHBOARDS

A Guide to Measuring and Monitoring Project Performance

Fourth Edition

Harold Kerzner, PhD

Sr. Executive Director for Project Management,
The International Institute for Learning

INTERNATIONAL
Institute for Learning, Inc.

Published by John Wiley & Sons, Inc., Hoboken, New Jersey
Published simultaneously in Canada

For general information about our other products and services, please contact our Customer Care Department within the United States at (800) 762-2974, outside the United States at (317) 572-3993 or fax (317) 572-4002.

Wiley publishes in a variety of print and electronic formats and by print-on-demand. Some material included with standard print versions of this book may not be included in e-books or in print-on-demand. If this book refers to media such as a CD or DVD that is not included in the version you purchased, you may download this material at http://booksupport.wiley.com. For more information about Wiley products, visit www.wiley.com.

Library of Congress Cataloging-in-Publication Data:
Names: Kerzner, Harold, author.
Title: Project management metrics, KPIs, and dashboards : a guide to measuring and monitoring project performance / Harold Kerzner, Ph.D., Sr. Executive Director for Project Management The International Institute for Learning.
Description: Fourth Edition. | Hoboken, New Jersey : John Wiley & Sons, [2023] | Revised edition of the author's Project management metrics, KPIs, and dashboards, [2017] | Includes bibliographical references and index.
Identifiers: LCCN 2021062269 (print) | LCCN 2021062270 (ebook) | ISBN 9781119851554 (paperback) | ISBN 9781119851578 (pdf) | ISBN 9781119851561 (epub) | ISBN 9781119851592 (obook)
Subjects: LCSH: Project management. | Project management–Quality control. | Performance standards. | Work measurement.
Classification: LCC HD69.P75 K492 2022 (print) | LCC HD69.P75 (ebook) | DDC 658.4/04–dc23/eng/20220131
LC record available at https://lccn.loc.gov/2021062269
LC ebook record available at https://lccn.loc.gov/2021062270

Cover image: Screenshots © Dundas Data Visualization, Inc. Desktop Computer SC2D40 © Cover Action Pro 3.
Cover design: Wiley

Set in 10/12pt ITC Giovanni Std by Integra Software Services Pvt. Ltd, Pondicherry, India

SKY10042305_020123

CONTENTS

v

6 DASHBOARDS 247

PREFACE

The ultimate purpose of metrics and dashboards is not to provide more information but to provide the right information to the right person at the right time, using the correct media and in a cost-effective manner. This is certainly a challenge. As computer technology has grown, so has the ease with which information can be generated and presented to management and stakeholders. Today, everyone seems concerned about information overload. Unfortunately, the real issue is non-information overload. In other words, there are too many useless reports that cannot easily be read and that provide readers with too much information, much of which may have no relevance. This information simply distracts us from the real issues and accurate performance reporting. Furthermore, the growth in metric measurement techniques has encouraged us to measure everything regardless of its value as part of performance reporting.

The purpose of status reporting is to show us what actions the viewer must consider. Insufficient or ineffective metrics prevent us from understanding what decisions really need to be made. In traditional project review meetings, emphasis is placed on a detailed schedule analysis and a lengthy review of the cost baseline versus actual expenditures. The resulting discussion and explanation of the variances are most frequently pure guesswork. Managers who are upset about the questioning by senior management then make adjustments that do not fix the problems but limit the time they will be grilled by senior management at the next review meeting. They then end up taking actions that may be counterproductive to the timely completion of the project, and real issues are hidden.

You cannot correct or improve something that cannot be effectively identified and measured. Without effective metrics, managers will not respond to situations correctly and will end up reinforcing undesirable actions by the project team. Keeping the project team headed in the right direction cannot be done easily without effective identification and measurement of metrics.

When all is said and done, we wonder why we have studies like the Chaos Report, which has shown us that over a period of 20 years only about 30% of the IT projects were completed successfully. We then identify hundreds of causes as to why projects fail but neglect what is now being recognized as perhaps the single most important cause: a failure in metrics management.

Metrics management should be addressed in all of the areas of knowledge in the *PMBOK® Guide*,[1] especially communications management. We are now struggling to find better ways of communicating on projects. This will become increasingly important as companies compete in a global marketplace. Our focus today is on the unique needs of the receiver of the information. The need to make faster and better decisions mandates better information. Human beings can absorb information in a variety of ways. We must address all of these ways in the selection of the metrics and the design of the dashboards that convey this information. Dashboards and data visualization techniques are now part of information warehouses and business intelligence systems.

The three most important words in a stakeholder's vocabulary are "making informed decisions." This is usually the intent of effective stakeholder relations management. Unfortunately, this cannot be accomplished without an effective information system based on meaningful and informative metrics and key performance indicators (KPIs).

All too often, we purchase project management software and reluctantly rely on the report generators, charts, and graphs to provide the necessary information, even when we realize that this information either is not sufficient or has limited value. Even those companies that create their own project management methodologies neglect to consider the metrics and KPIs that are needed for effective stakeholder relations management. Informed decisions require effective information. We all seem to understand this, yet it has only been in recent years that we have tried to do something about it.

For decades we believed that the only information that needed to be passed on to the client and the stakeholders was information related to time and cost. Today we realize that the true project status cannot be determined from time and cost alone. Each project may require its own unique metrics and KPIs. The future of project management may very well be metric-driven project management.

Information design has finally come of age. Effective communications is the essence of information design. Today we have many small companies that are specialists in business information design. Larger companies may maintain their own specialist team and call these people graphic designers, information architects, or interaction designers. These people maintain expertise in the visual display of both quantitative and qualitative information necessary for informed decision making.

Traditional communications and information flow has always been based on tables, charts, and indexes that were, it is hoped, organized properly by the designer. Today information or data graphics combines points, lines, charts, symbols, images, words, numbers,

---

1 PMBOK is a registered mark of the Project Management Institute, Inc.

shades, and a symphony of colors necessary to convey the right message easily. What we know with certainty is that dashboards and metrics are never an end in themselves. They go through continuous improvement and are constantly updated. In a project management environment, each receiver of information can have different requirements and may request different information during the life cycle of the project.

With this in mind, the book is structured as follows:

- Chapters 1 and 2 identify how project management has changed over the last few years and how more pressure is being placed on organizations for effective metrics management.
- Chapter 3 provides an understanding of what metrics are and how they can be used.
- Chapter 4 discusses key performance indications and explains the difference between metrics and KPIs.
- Chapter 5 focuses on the value-driven metrics and value-driven KPIs. Stakeholders are asking for more metrics related to the project's ultimate value. The identification and measurement of value-driven metrics can be difficult.
- Chapter 6 describes how dashboards can be used to present the metrics and KPIs to stakeholders. Examples of dashboards are included together with some rules for dashboard design.
- Chapter 7 identifies dashboards that are being used by companies.
- Chapter 8 provides various business-related metrics that are currently used by portfolio management project management offices to ensure that the business portfolio is delivering the business value expected.

HAROLD KERZNER, PhD
Sr. Executive Director for Project Management
The International Institute for Learning

ABOUT THE COMPANION WEBSITE

This book is accompanied by a companion website which includes a number of resources created by author that you will find helpful.

www.wiley.com/go/kerznermetrics4e

This is an Instructor website.

Please note that the resources in instructor website are password protected and can only be accessed by instructors who register with the site.

1

THE CHANGING LANDSCAPE OF PROJECT MANAGEMENT

CHAPTER OVERVIEW

The way project managers managed projects in the past will not suffice for many of the projects being managed now or for the projects of the future. The complexity of these projects will place pressure on organizations to better understand how to identify, select, measure, and report project metrics, especially metrics showing value creation. The future of project management may very well be metric-driven project management. In addition, new approaches to project management, such as those with agile and Scrum, have brought with them new sets of metrics.

CHAPTER OBJECTIVES

- To understand how project management has changed
- To understand the need for project management metrics
- To understand the need for better, more complex project management metrics

KEY WORDS

- Certification boards
- Complex projects
- Engagement project management
- Frameworks
- Governance
- Project management methodologies
- Project success

1.0 INTRODUCTION

For more than 50 years, project management has been in use but perhaps not on a worldwide basis. What differentiated companies in the early years was whether they used project management or not, not how well they used it. Today, almost every company uses project management, and the differentiation is whether they are simply good at project management or whether they truly excel at project management. The difference between using project management and being good at it is relatively small, and most companies can become good at project management in a relatively short time,

Project Management Metrics, KPIs, and Dashboards: A Guide to Measuring and Monitoring Project Performance, Fourth Edition. Harold Kerzner.
© 2023 John Wiley & Sons, Inc. Published 2023 by John Wiley & Sons, Inc.
Companion Website: www.wiley.com/go/kerznermetrics4e

especially if they have executive-level support. A well-organized project management office (PMO) can also accelerate the maturation process. The difference, however, between being good and excelling at project management is quite large. One of the critical differences is that excellence in project management on a continuous basis requires more metrics than just time and cost. The success of a project cannot be determined just from the time and cost metrics, yet many companies persist in the belief that this is possible.

The growth of project management applications to nontraditional projects such as those involving strategic issues, innovation, and long-term business investment opportunities have forced companies to rethink how project management can be better utilized. Companies have come to the realization that they must excel at project management rather than just being good at it. This requires the use of flexible methodologies rather than the one-size-fits-all approach, new tools, specialized metrics, creation of information warehouses, new data visualization programs, and packaging all of this into business intelligence systems.

Companies such as IBM, Microsoft, Siemens, Hewlett-Packard (HP), and Deloitte, to name just a few, have come to the realization that they must excel at project management. Doing this requires additional tools and metrics to support project management. IBM has more than 300,000 employees, more than 70 percent of whom are outside of the United States. This includes some 30,000 project managers. HP has more than 8000 project managers and 3500 PMP® credential holders. HP's goal is 8000 project managers and 8000 PMP® credential holders. These numbers are now much larger with HP's acquisition of Electronic Data Systems (EDS).

1.1 EXECUTIVE VIEW OF PROJECT MANAGEMENT

Companies today perform strategic planning for project management and are focusing heavily on the future. Several of the things that these companies are doing will be discussed in this chapter, beginning with senior management's vision of the future. Years ago, senior management paid lip service to project management, reluctantly supporting it to placate the customers. Today, senior management appears to have recognized the value in using project management effectively and maintains a different view of project management, as shown in Table 1.1.

Project management is no longer regarded as a part-time occupation or even a career path position. It is now viewed as a strategic competency needed for the survival of the firm. Superior project management capability can make the difference between winning and losing a contract.

For more than 30 years, becoming a PMP® credential holder was seen as the light at the end of the tunnel. Today, that has changed. Becoming a PMP® credential holder is the light at the *entryway* to the tunnel. The light at the end of the tunnel may require multiple

TABLE 1.1 Executive View of Project Management

OLD VIEW	NEW VIEW
Project management is a career path.	Project management is a strategic or core competency necessary for the growth and survival of the company.
We need our people to receive Project Management Professional certifications.	We need our people to undergo multiple certifications and, at a minimum, to be certified in both project management and corporate business processes.
Project managers will be used for project execution only.	Project managers will participate in strategic planning, the portfolio selection of projects, and capacity-planning activities.
Business strategy and project execution are separate activities.	Part of the project manager's job is to bridge strategy and execution.
Project managers just make project-based decisions.	Project managers make both project and business decisions.

certifications. As an example, after becoming a PMP® credential holder, a project manager may desire to become certified in

- Business Analyst Skills or Business Management
- Program Management
- Business Processes
- Managing Complex Projects
- Six Sigma
- Risk Management
- Agile Project Management

Some companies have certification boards that meet frequently and discuss what certification programs would be of value for their project managers. Certification programs that require specific knowledge of company processes or company intellectual property may be internally developed and taught by the company's own employees.

Executives have come to realize that there is a return on investment in project management education. Therefore, executives are now investing heavily in customized project management training, especially in behavioral courses. As an example, one executive commented that he felt that presentation skills training was the highest priority for his project managers. If a project manager makes a highly polished presentation before a client, the client believes that the project is being managed the same way. If the project manager makes a poor presentation, then the client might believe the project is managed the same way. Other training programs that executives feel would be beneficial for the future include:

- Establishing metrics and key performance indicators (KPIs)
- Dashboard design
- Managing complex projects
- How to perform feasibility studies and cost–benefit analyses

- Business analysis
- Business case development
- How to validate and revalidate project assumptions
- How to establish effective project governance
- How to manage multiple stakeholders many of whom may be multinational
- How to design and implement "fluid" or adaptive enterprise project management (EPM) methodologies
- How to develop coping skills and stress management skills

Project managers are now being brought on board projects at the beginning of the initiation phase rather than at its end. To understand the reason for this, consider the following situation:

SITUATION: A project team is assembled at the end of the initiation phase of a project to develop a new product for the company. The project manager is given the business case for the project together with a listing of the assumptions and constraints. Eventually the project is completed, somewhat late and significantly over budget. When asked by marketing and sales why the project costs were so large, the project manager responds, "According to my team's interpretation of the requirements and the business case, we had to add in more features than we originally thought."

Marketing then replies, "The added functionality is more than what our customers actually need. The manufacturing costs for what you developed will be significantly higher than anticipated, and that will force us to raise the selling price. We may no longer be competitive in the market segment we were targeting."

"That's not our problem," responds the project manager. "Our definition of project success is the eventual commercialization of the product. Finding customers is your problem, not our problem."

Needless to say, we could argue about what the real issues were in this project that created the problems. For the purpose of this book, three issues stand out. First and foremost, project managers today are paid to make business decisions as well as project decisions. Making merely project-type decisions could result in the development of a product that is either too costly to build or overpriced for the market at hand. Second, the traditional metrics used by project managers over the past several decades were designed for project rather than business decision making. Project managers must recognize that, with the added responsibilities of making business decisions, a new set of metrics may need to be included as part of their responsibilities. Likewise, we could argue that marketing was remiss in not establishing and tracking business-related metrics throughout the project and simply waited until the project was completed to see the results. Third, with the growth in the number of projects that companies must work on, executives do not have the time to act as heavily involved sponsors on all the projects without sacrificing some of their required day-to-day responsibilities. Data visualization systems and dashboards will ease some of the pain in knowing when a project requires immediate executive attention. The growth in metric measurement techniques and dashboard designs will allow executives to have customized metrics accompanied by real time updates so that decisions can be made based upon facts and evidence rather than guesses.

1.2 COMPLEX PROJECTS

TIP Today's project managers see themselves as managing part of a business rather than simply managing a project. Therefore, they may require additional metrics for informed decision making.

For four decades, project management has been used to support traditional projects. Traditional projects are heavily based on linear thinking; there exist well-structured life cycle phases and templates, forms, guidelines, and checklists for each phase. As long as the scope is reasonably well defined, traditional project management works well.

Unfortunately, only a small percentage of all of the projects in a company fall into this category. Most nontraditional or complex projects use seat-of-the-pants management because they are largely based on business scenarios where the outcome or expectations can change from day to day. Project management techniques were neither required nor used on these complex projects that were more business oriented and aligned to 5-year or 10-year strategic plans that were constantly updated.

Project managers have finally realized that project management can be used on these complex projects, but the traditional processes may be inappropriate or must be modified. This includes looking at project management metrics and KPIs in a different light. The leadership style for complex projects may not be the same as that for traditional projects. Risk management is significantly more difficult on complex projects, and the involvement of more participants and stakeholders is necessary.

Now that companies have become good at traditional projects, we are focusing our attention on the nontraditional or complex projects. Unfortunately, there is no clear-cut definition of a complex project. Some of the major differences between traditional and nontraditional or complex projects, in the author's opinion, are shown in Table 1.2.

Comparing Traditional and Nontraditional Projects

The traditional project that most people manage usually lasts less than 18 months. In some companies, the traditional project might last six months or less. The length of the project usually depends on the industry. In the auto industry, for example, a traditional project lasts three years.

Section 1.2 is adapted from Harold Kerzner and Carl Belack, *Managing Complex Projects* (Hoboken, NJ: John Wiley & Sons, 2010), Chapter 1.

TABLE 1.2 Traditional versus Nontraditional Projects

TRADITIONAL PROJECTS	NONTRADITIONAL PROJECTS
Time duration is 6–18 months.	Time duration can be several years.
Assumptions are not expected to change over the project's duration.	Assumptions can and will change over the project's duration.
Technology is known and will not change over the project's duration.	Technology will most certainly change.
People who started on the project will remain through to completion (the team and the project sponsor).	People who approved the project and are part of the governance may not be there at the project's conclusion.
Statement of work is reasonably well defined.	Statement of work is ill defined and subject to numerous scope changes.
Target is stationary.	Target may be moving.
There are few stakeholders.	There are multiple stakeholders.
There are few metrics and KPIs.	There can be numerous metrics and KPIs.

With projects that last 18 months or less, it is assumed that technology is known with some degree of assurance and technology may undergo little change over the life of the project. The same holds true for the assumptions. Project managers tend to believe that the assumptions made at the beginning of the project will remain intact for the duration of the project unless a crisis occurs.

People who are assigned to the project will most likely stay on board the project from beginning to end. The people may be full time or part time. This includes the project sponsor as well as the team members.

Because the project lasts 18 months or less, the statement of work is usually reasonably well defined, and the project plan is based on reasonably well-understood and proven estimates. Cost overruns and schedule slippages can occur, but not to the degree that they will happen on complex projects. The objectives of the project, as well as critical milestone or deliverable dates, are reasonably stationary and not expected to change unless a crisis occurs.

In the past, the complexities of nontraditional projects seem to have been driven by time and cost. Some people believe that these are the only two metrics that need to be tracked on a continuous basis. Complex projects may run as long as 10 years or even longer. Because of the long duration, the assumptions made at the initiation of the project will most likely not be valid at the end of the project. The assumptions will have to be revalidated throughout the project. There can be numerous metrics, and the metrics can change over the duration of the project. Likewise, technology can be expected to change throughout the project. Changes in technology can create significant and costly scope changes to the point where the final deliverable does not resemble the initially planned deliverable.

People on the governance committee and in decision-making roles most likely are senior people and may be close to retirement. Based on the actual length of the project, the governance structure can be expected to change throughout the project if the project's duration is 10 years or longer.

Because of scope changes, the statement of work may undergo several revisions over the life cycle of the project. New governance groups and new stakeholders can have their own hidden agendas and demand that the scope be changed; they might even cancel their financial support for the project. Finally, whenever there is a long-term complex project where continuous scope changes are expected, the final target may move. In other words, the project plan must be constructed to hit a moving target.

SITUATION: A project manager was brought on board a project and provided with a project charter that included all of the assumptions made in the selection and authorization of the project. Partway through the project, some of the business assumptions changed. The project manager assumed that the project sponsor would be monitoring the enterprise environmental factors for changes in the business assumptions. That did not happen. The project was eventually completed, but there was no real market for the product.

Given the premise that project managers are now more actively involved in the business side of projects, the business assumptions must be tracked the same way that budgets and schedules are tracked. If the assumptions are wrong or no longer valid, then either the statement of work may need to be changed or the project may need to be canceled. The expected value at the end of the project also must be tracked because unacceptable changes in the final value may be another reason for project cancellation.

Examples of assumptions that are likely to change over the duration of a project, especially on a long-term project, include these:

- The cost of borrowing money and financing the project will remain fixed.
- Procurement costs will not increase.
- Breakthroughs in technology will take place as scheduled.
- The resources with the necessary skills will be available when needed.
- The marketplace will readily accept the product.
- The customer base is loyal to the company.
- Competitors will not catch up to the company.
- The risks are low and can be easily mitigated.
- The political environment in the host country will not change.

The problem with having faulty assumptions is that they can lead to bad results and unhappy customers. The best defense against poor assumptions is good preparation at project initiation, including the development of risk mitigation strategies and tracking metrics for critical assumptions. However, it may not be possible to establish metrics for the tracking of all assumptions.

Most companies either have or are in the process of developing an enterprise project management (EPM) methodology. EPM systems usually are rigid processes designed around policies and procedures, and they work efficiently when the statement of work is well defined. With the new type of projects currently being used when techniques such as Agile Project Management are applicable, these rigid and inflexible processes may be more of a hindrance and costly to use on small projects.

EPM systems must become more flexible in order to satisfy business needs. The criteria for good systems will lean toward forms, guidelines, templates, and checklists rather than policies and procedures. Project managers will be given more flexibility in order to make the decisions necessary to satisfy the project's business needs. The situation is further complicated because all active stakeholders may wish to use their own methodology, and having multiple methodologies on the same project is never a good idea. Some host countries may be quite knowledgeable in project management, whereas other may have just cursory knowledge.

TIP Metrics and KPIs must be established for those critical activities that can have a direct impact on project success or failure. This includes the tracking of assumptions and the creation of business value.

Over the next decade, having a fervent belief that the original plan is correct may be a poor assumption. As the project's business needs change, the need to change the plan will be evident. Also, decision making based entirely on the triple constraints, with little regard for the project's final value, may result in a poor decision. Simply stated, today's view of project management is quite different from the views in the past, and this is partially because the benefits of project management have been recognized more over the past two decades.

TIP The more flexibility the methodology contains, the greater the need for additional metrics and KPIs.

Some of the differences between managing traditional and complex projects are summarized in Table 1.3. Perhaps the primary difference is whom the project manager must interface with on a daily basis. With traditional projects, the project manager interfaces with the sponsor and the client, both of whom may provide the only governance on the project. With complex projects, governance is by committee and there can be multiple stakeholders whose concerns need to be addressed.

Defining Complexity

Complex projects can differ from traditional projects for a multitude of reasons, including:

- Size
- Dollar value
- Uncertain requirements

TABLE 1.3   Summarized Differences between Traditional and Nontraditional Projects

MANAGING TRADITIONAL PROJECTS	MANAGING NONTRADITIONAL PROJECTS
Single-person sponsorship	Governance by committee
Possibly a single stakeholder	Multiple stakeholders
Project decision making	Both project and business decision making
An inflexible project management methodology management methodology	Flexible or "fluid" project
Periodic status reporting	Real-time reporting
Success defined by the triple constraints	Success defined by competing constraints, value, and other factors
Metrics and KPIs derived from the earned value measurement system	Metrics and KPIs may be unique to the particular project and even to a particular stakeholder

- Uncertain scope
- Uncertain deliverables
- Complex interactions
- Uncertain credentials of the labor pool
- Geographical separation across multiple time zones
- Use of large virtual teams
- Other differences

There are numerous definitions of a "complex" project, based on the interactions of two or more of the preceding elements. Even a small, two-month infrastructure project can be considered complex according to the definition. Project complexity can create havoc when selecting and using metrics. The projects that project managers manage within their own companies can be regarded as complex projects if the scope is large and the statement of work is only partially complete. Some people believe that research and development (R&D) projects are always complex because, if a plan for R&D can be laid out, then there probably is not R&D. R&D is when the project manager is not 100% sure where the company is heading, does not know what it will cost, and does not know if and when the company will get there.

Complexity can be defined according to the number of interactions that must take place for the work to be executed. The greater the number of functional units that must interact, the harder it is to perform the integration. The situation becomes more difficult if the functional units are dispersed across the globe and if cultural differences makes integration difficult. Complexity can also be defined according to size and length. The larger the project is in scope and cost and the greater the time frame, the more likely it is that scope changes will occur, significantly affecting the budget and schedule. Large, complex projects tend to have large cost overruns and schedule slippages. Good examples of this are Denver International Airport, the Channel Tunnel between England and France, and the "Big Dig" in Boston.

Trade-Offs

Project management is an attempt to improve efficiency and effectiveness in the use of resources by getting work to flow multidirectionally through an organization, whether traditional or complex projects. Initially, this flow might seem easy to accomplish, but typically a number of constraints are imposed on projects. The most common constraints are time, cost, and performance (also referred to as scope or quality), which are known as the *triple constraints*.

Historically, from an executive-level perspective, the goal of project management was to meet the triple constraints of time, cost, and performance while maintaining good customer relations. Unfortunately, because most projects have some unique characteristics, highly accurate time and cost estimates were not always possible, and trade-offs between the triple constraints were necessary. As will be discussed later, today we focus on competing constraints and there may be significantly more than three constraints on a project, and metrics may have to be established to track each constraint. There may be as many as 10 or more competing constraints. Metrics provide the basis for informed trade-off decision making. Executive management, functional management, and key stakeholders must be involved in almost all trade-off discussions to ensure that the final decision is made in the best interests of the project, the company, and the stakeholders. If multiple stakeholders are involved, as occurs on complex projects, then agreement from all of the stakeholders may be necessary. Project managers may possess sufficient knowledge for some technical decision making but may not have sufficient business or technical knowledge to adequately determine the best course of action to address the interests of the parent company as well as the individual project stakeholders.

TIP Because of the complex interactions of the elements of work, a few simple metrics may not provide a clear picture of project status. The combination of several metrics may be necessary in order to make informed decisions based on evidence and facts.

Skill Set

All project managers have skills, but not all project managers may have the right skills for the given job. For projects internal to a company, it may be possible to develop a company-specific skill set or company-specific body of knowledge. Specific training courses can be established to support company-based knowledge requirements.

For complex projects with a multitude of stakeholders, all from different countries with different cultures, finding the perfect project manager may be an impossible task. Today the understanding of complex projects and the accompanying metrics is in its infancy, and it is still difficult to determine the ideal skill set for managing complex projects. Remember that project management existed for several decades before the first Project Management Body of Knowledge

(*PMBOK® Guide**) was created, and even now with the seventh edition, it is still referred to as a "guide."

We can, however, conclude that there are certain skills required to manage complex projects. Some of those skills are:

■ Knowing how to manage virtual teams
■ Understanding cultural differences
■ The ability to manage multiple stakeholders, each of whom may have a different agenda
■ Understanding the impact of politics on project management
■ How to select and measure project metrics

Governance

Cradle-to-grave user involvement in complex projects is essential. Unfortunately, user involvement can change because of politics and project length. It is not always possible to have the same user community attached to the project from beginning to end. Promotions, changes in power and authority positions because of elections, and retirements can cause shifts in user involvement.

Governance is the process of decision making. On large complex projects, governance will be in the hands of the many rather than the few. Each stakeholder may either expect or demand to be part of all critical decisions on the project. Governance must be supported by proper metrics that provide meaningful information. The channels for governance must be clearly defined at the beginning of the project, possibly before the project manager is assigned. Changes in governance, which are increasingly expected the longer the project takes, can have a serious impact on the way the project is managed as well as on the metrics used.

Decision Making

Complex projects have complex problems. All problems generally have solutions, but not all solutions may be good or even practical. Good metrics can make decision making easier. Also, some solutions to problems can be more costly than other solutions. Identifying a problem is usually easy. Identifying alternative solutions may require the involvement of many stakeholders, and each stakeholder may have a different view of the actual problem and the possible alternatives. To complicate matters, some host countries have very long decision-making cycles for problem identification and for the selection of the best alternative. Each stakeholder may select an alternative that is in the best interests of that particular stakeholder rather than in the best interests of the project.

Obtaining approval also can take a long time, especially if the solution requires that additional capital be raised and if politics play

---

* PMBOK is a registered mark of the Project Management Institute, Inc.

an active role. In some emerging countries, every complex project may require the signature of a majority of the ministers and senior government leaders. Decisions may be based on politics and religion as well.

Fluid Methodologies

With complex projects, the project manager needs a fluid or flexible project management methodology capable of interfacing with multiple stakeholders. The methodology may need to be aligned more with business processes than with project management processes, since the project manager may need to make business decisions as well as project decisions. Complex projects seem to be dictated more by business decisions than by pure project decisions.

Complex projects are driven more by the project's end business value than by the triple or competing constraints. Complex projects tend to take longer than anticipated and cost more than originally budgeted because of the need to guarantee that the final result will have the business value desired by customers and stakeholders. Simply stated, complex projects tend to be value-driven rather than driven by the triple or competing constraints.

TIP Completing a project within the triple constraints is not necessarily success if perceived stakeholder value is not there at project completion.

TIP The more complex the project, the more time is needed to select metrics, perform measurements, and report on the proper mix of metrics.

TIP The longer the project, the greater the flexibility needed to allow for different metrics to be used over the life of the project.

Given the fact that project complexity will continue to increase in the future, we can now identify the impact on project metrics:

- Significantly more metrics will be required on complex projects compared to traditional projects
- Business, strategic, and intangible metrics will be essential
- The metrics can change in each life cycle phase or from changes in the enterprise environmental factors
- Dashboards must be customized possibly for each viewer

1.3 GLOBAL PROJECT MANAGEMENT

Every company in the world has complex projects that it would have liked to undertake but was unable to because of limitations, such as:

- No project portfolio management function to evaluate projects
- A poor understanding of capacity planning
- A poor understanding of project prioritization
- A lack of tools for determining the project's business value
- A lack of project management tools and software

TABLE 1.4 Nonglobal versus Global Company Competencies

FACTOR	NONGLOBAL	GLOBAL
Core business	Sell products and services	Sell business solutions
Project management satisfaction level	Must be good at project management	Must excel at project management
P management methodology	Rigid	Flexible and fluid
Metrics/KPIs	Minimal	Extensive
Supporting tools	Minimal	Extensive
Continuous improvement	Follow the leader	Capture best practices and lessons learned
Business knowledge	Know your company's business	Understand the client's business model as well as your company's business model
Type of team	Colocated	Virtual

- A lack of sufficient resources
- A lack of qualified resources
- A lack of support for project management education
- A lack of a project management methodology
- A lack of knowledge in dealing with complexity
- A fear of failure
- A lack of understanding of metrics needed to track the project

Because not every company has the capability to manage complex projects, companies must look outside for suppliers of project management services. Companies that provide these services on a global basis consider themselves to be business solution providers and differentiate themselves from localized companies according to the elements in Table 1.4.

Those companies that have taken the time and effort to develop flexible project management methodologies and become solution providers are companies that are competing in the global marketplace. Although these companies may have as part of their core business the providing of products and services, they may view their future as being a global solution provider for the management of complex projects.

TIP Competing globally requires a different mind-set from competing locally. An effective project management information system based on possibly project-specific metrics may be essential.

For these companies, being good at project management is not enough; they must *excel* at project management. They must be innovative in their processes to the point that all processes and methodologies are highly fluid and easily adaptable to a particular client. They have an extensive library of tools to support the project management processes. Most of the tools were created internally with ideas discovered through captured lessons learned and best practices.

1.4 PROJECT MANAGEMENT METHODOLOGIES AND FRAMEWORKS

Most companies today seem to recognize the need for one or more project management methodologies but either create the wrong methodologies or misuse the methodologies that have been created. Many times companies rush into the development or purchasing of a methodology without any understanding of the need for one other than the fact that their competitors have a methodology. As Jason Charvat states:

> Using project management methodologies is a business strategy allowing companies to maximize the project's value to the organization. The methodologies must evolve and be "tweaked" to accommodate a company's changing focus or direction. It is almost a mind-set, a way that reshapes entire organizational processes: sales and marketing, product design, planning, deployment, recruitment, finance, and operations support. It presents a radical cultural shift for many organizations. As industries and companies change, so must their methodologies. If not, they're losing the point.[1]

There are significant advantages to the design and implementation of a good, flexible methodology:

- Shorter project schedules
- Better control of costs
- Fewer or no unwanted scope changes
- Can plan for better execution
- Results can be predicted more accurately
- Improves customer relations during project execution
- The project can be adjusted during execution to fit changing customer requirements
- Better visibility of status for senior management
- Execution is standardized
- Best practices can be captured

Rather than using policies and procedures, some methodologies are constructed as a set of forms, guidelines, templates, and checklists that can and must be applied to a specific project or situation. It may not be possible to create a single enterprise-wide methodology that can be applied to each and every project. Some companies have been successful doing this, but many companies successfully maintain more than one methodology. Unless project managers are capable of tailoring the EPM methodology to their needs, more than one methodology may be necessary.

---

1 Jason Charvat, *Project Management Methodologies* (Hoboken, NJ: John Wiley & Sons, 2003), p. 2.

There are several reasons why good intentions often go astray. At the executive levels, methodologies can fail if the executives have a poor understanding of what a methodology is and believe that a methodology is:

- A quick fix
- A silver bullet
- A temporary solution
- A cookbook approach for project success[2]

At the working levels, methodologies can also fail if they:

- Are abstract and high level
- Contain insufficient narratives to support these methodologies
- Are not functional or do not address crucial areas…
- Ignore the industry standards and best practices
- Look impressive but lack real integration into the business
- Use nonstandard project conventions and terminology
- Compete for similar resources without addressing this problem
- Do not have any performance metrics
- Take too long to complete because of bureaucracy and administration[3]

Methodologies also can fail because the methodology:

- Must be followed exactly even if the assumptions and environmental input factors have changed
- Focuses on linear thinking
- Does not allow for out-of-the-box thinking
- Does not allow for value-added changes that are not part of the original requirements
- Does not fit the type of project
- Is too abstract (rushing to design it)
- Development team neglects to consider bottlenecks and the concerns of the user community
- Is too detailed
- Takes too long to use
- Is too complex for the market, clients, and stakeholders to understand
- Does not have sufficient or correct metrics

Deciding on what type of methodology is not an easy task. There are many factors to consider, such as:[4]

- The overall company strategy—how competitive are we as a company?
- The size of the project team and/or scope to be managed

2 Ibid., p. 4.
3 Ibid., p. 5.
4 Ibid., p. 66.

- The priority of the project
- How critical the project is to the company
- How flexible the methodology and its components are

Numerous other factors can influence the design of a methodology. Some of these factors include:

- Corporate strategy
- Complexity and size of the projects in the portfolio
- Management's faith in project management
- Development budget
- Number of life cycle phases
- Technology requirements
- Customer requirements
- Training requirements and costs
- Supporting tools and software costs

Project management methodologies are created around the project management maturity level of the company and the corporate culture. If the company is reasonably mature in project management and has a culture that fosters cooperation, effective communication, teamwork, and trust, then a highly flexible methodology can be created based on guidelines, forms, checklists, and templates. As stated previously, the more flexibility that is added into the methodology, the greater the need for a family of metrics and KPIs. Project managers can pick and choose the parts of the methodology and metrics that are appropriate for a particular client. Organizations that do not possess either of these two characteristics rely heavily on methodologies constructed with rigid policies and procedures, thus creating significant paperwork requirements with accompanying cost increases and removing the flexibility that the project manager needs to adapt the methodology to the needs of a specific client. These rigid methodologies usually rely on time and cost as the only metrics and can make it nearly impossible to determine the real status of the project.

Charvat describes these two types as light methodologies and heavy methodologies.[5]

Light Methodologies

Ever-increasing technological complexities, project delays, and changing client requirements brought about a small revolution in the world of development methodologies. A totally new breed of methodology—which is agile, is adaptive, and involves the client every part of the way—is starting to emerge. Many of the heavyweight

---

5 The next two subsections are taken from Charvat, *Project Management Methodologies,* pp. 102–104.

methodologists were resistant to the introduction of these "light-weight" or "agile" methodologies.[6] These methodologies use an informal communication style. Unlike heavyweight methodologies, lightweight projects have only a few rules, practices, and documents. Projects are designed and built on face-to-face discussions, meetings, and the flow of information to the clients. The immediate difference of using light methodologies is that they are much less documentation-oriented, usually emphasizing a smaller amount of documentation for the project.

Heavy Methodologies

The traditional project management methodologies (i.e., the systems development life cycle [SDLC] approach) are considered bureaucratic or "predictive" in nature and have resulted in many unsuccessful projects. These heavy methodologies are becoming less popular. These methodologies are so laborious that the whole pace of design, development, and deployment slows down—and nothing gets done. Project managers tend to predict every milestone because they want to foresee every technical detail (i.e., software code or engineering detail). This leads managers to start demanding many types of specifications, plans, reports, checkpoints, and schedules. Heavy methodologies attempt to plan a large part of a project in great detail over a long span of time. This works well until things start changing, and the project managers inherently try to resist change.

Frameworks

More and more companies today, especially those that wish to compete in the global marketplace as business solution providers, are using frameworks rather than methodologies.

- **Framework**: The individual segments, principles, pieces, or components of the processes needed to complete a project. This can include forms, guidelines, checklists, and templates.
- **Methodology**: The orderly structuring or grouping of the segments or framework elements. This can appear as policies, procedures, or guidelines.

Frameworks focus on a series of processes that must be done on all projects. Each process is supported by a series of forms, guidelines, templates, checklists, and metrics that can be applied to a particular client's business needs. The metrics will be determined jointly by the project manager, the client, and the various stakeholders.

---

6 Martin Fowler, *The New Methodology, Thought Works*, 2001. Available at www.martinfowler.com/articles.

As stated previously, a methodology is a series of processes, activities, and tools that are part of a specific discipline, such as project management, and are designed to accomplish a specific objective. When the products, services, or customers have similar requirements and do not require significant customization, companies develop methodologies to provide some degree of consistency in the way that projects are managed. With these methodologies, the metrics, once established, usually remain the same for every project.

As companies become reasonably mature in project management, the policies and procedures are replaced by forms, guidelines, templates, and checklists. These tools provide more flexibility for the project manager in how to apply the methodology to satisfy a specific customer's requirements. This flexibility leads to a more informal application of the project management methodology, and significantly more metrics are now required.

Today, this informal project management approach has been somewhat modified and is referred to as a framework. A framework is a basic conceptual structure that is used to address an issue, such as a project. It includes a set of assumptions, project-specific metrics, concepts, values, and processes that provide the project manager with a means for viewing what is needed to satisfy a customer's requirements. A framework is a skeletal support structure for building the project's deliverables. Agile and Scrum are heavy users of frameworks.

Frameworks work well as long as the project's requirements do not impose severe pressure on the project manager. Unfortunately, in today's chaotic environment, this pressure appears to be increasing because:

- Customers are demanding low-volume, high-quality products with some degree of customization.
- Project life cycles and new product development times are being compressed.
- Enterprise environmental factors are having a greater impact on project execution.
- Customers and stakeholders want to be more actively involved in the execution of projects.
- Companies are developing strategic partnerships with suppliers, and each supplier can be at a different level of project management maturity.
- Global competition has forced companies to accept projects from customers that are all at a different level of project management maturity.

These pressures tend to slow down the decision-making processes at a time when stakeholders want the processes to be accelerated. This slowdown is the result of:

- Project managers being expected to make decisions in areas where they have limited knowledge.
- Project managers hesitating to accept full accountability and ownership for the projects.

- Excessive layers of management being superimposed on the project management organization.
- Risk management being pushed up to higher levels in the organizational hierarchy.
- Project managers demonstrating questionable leadership ability.

Both methodologies and frameworks are mechanisms by which we can obtain best practices and lessons learned in the use of metrics and KPIs. Figure 1.1 illustrates the generic use of a methodology or framework. Once the clients and stakeholders are identified, then the requirements, business case, and accompanying assumptions can be input. The methodology serves as a guide through the *PMBOK® Guide* process groups of initiation (I), planning (P), execution (E), monitoring and controlling (M), and closure (C). The methodology also provides us with guidance in the identification of metrics, KPIs, and dashboard reporting techniques for a particular client.

Some people believe that, once the deliverables are provided to the client and project closure takes place, the project is completed. This is not the case. More companies today are adding, at the end of the life cycle phases of the methodology, another life cycle phase, entitled "Customer Satisfaction Management." The purpose of this phase is to meet with the client and the stakeholders and discuss what was learned on the project regarding best practices, lessons learned, metrics, and KPIs. The intent is to see what can be done better for that client on future projects. Today, companies maintain metric and KPI libraries the same way that they maintain libraries for best practices and lessons learned.

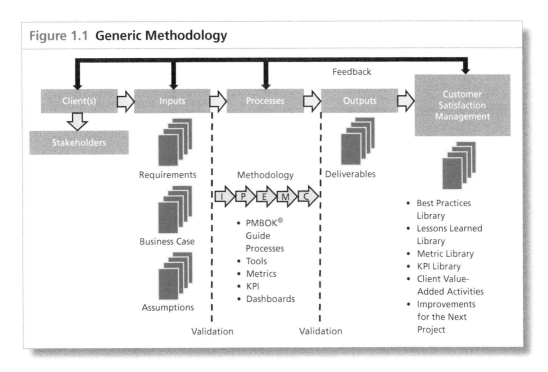

Figure 1.1 Generic Methodology

1.5 THE NEED FOR EFFECTIVE GOVERNANCE

The problems just described can be resolved by using effective project governance. Project governance is actually a framework by which decisions are made. Governance relates to decisions that define expectations, accountability, responsibility, the granting of power, or the verifying of performance. Governance also relates to consistent management, cohesive policies, processes, and decision-making rights for a given area of responsibility, and enables efficient and effective decision making.

Every project can have different governance, even if each project uses the same EPM methodology. The governance function can operate as a separate process or as part of project management leadership. Governance is not designed to replace project decision making but to prevent undesirable decisions from being made. Effective governance requires information and data. For decades, we focused on the use of a project management information system (PMIS) that contained almost exclusively information from the earned value measurement system (EVMS). Unfortunately, the EVMS did not contain all the necessary information for effective problem analysis and decision making. Today, governance personnel rely upon data visualization and business intelligence systems that can be customized for each member of a governance committee.

Historically, governance was provided by a single person acting as the project sponsor. Today, governance is provided by a committee. Committee membership can change from project to project and industry to industry. Membership may also vary according to the number of stakeholders and whether the project is for an internal or an external client.

1.6 ENGAGEMENT PROJECT MANAGEMENT

With project management now viewed as a strategic competency, it is natural for companies that wish to compete in a global marketplace to be strong believers in engagement project management or engagement selling. Years ago, the sales force would sell a product or services to a client and then move on to find another client. Today, the emphasis is on staying with clients and looking for additional work from the same clients.

In a marital context, an engagement can be viewed as the beginning of a lifelong partnership. The same holds true with engagement project management. Companies like IBM and HP no longer view themselves as selling products or services. Instead, they see themselves as business solution providers for their clients, and a business solution provider cannot remain in business without having superior project management capability.

As part of engagement project management, companies must convince clients that they have the project management capability to provide solutions to their business needs on a repetitive basis. In exchange for this, companies want clients to treat them as strategic partners rather than as just another contractor. This is shown in Figure 1.2.

Previously, it was stated that those companies that wish to compete in a global environment must have superior project management capability. This capability must appear in the contractor's response to a request for proposal issued by the client. Clients today are demanding that companies provide the following in proposals:

- The number of PMP® credential holders in the company and which ones will manage the contract if a company wins through competitive bidding.
- An EPM methodology or framework with a history of providing repeated successes.
- A willingness to customize the framework or methodology to fit the client's environment.
- The maturity level of project management in the company and which project management maturity model was used to perform the assessment.
- A best practices library for project management and a willingness to share this knowledge with the client, as well as the best practices discovered during the project.

Decades ago, the sales force (and marketing) had very little knowledge about project management. The role of the sales force was to win contracts, regardless of the concessions that had to be made. The project manager then "inherited" a project with an underfunded budget and an impossible schedule. Today, sales and marketing must understand project management and be able to sell it to clients as part of engagement selling. The sales force must sell the company's project management methodology or framework and the accompanying best practices. Sales and marketing are now involved in project management.

Engagement project management benefits both the buyer and the seller, as shown in Table 1.5.

Figure 1.2 **"Engagement" Project Management** *Source: Moodboard/Adobe Stock*

Customer's Expectations
Business Solutions

Contractor's Expectations
Long-Term Strategic Partnerships

TABLE 1.5 Before and After Engagement Project Management	
BEFORE ENGAGEMENT PROJECT MANAGEMENT	**AFTER ENGAGEMENT PROJECT MANAGEMENT**
Continuous competitive bidding	Sole-source or single-source contracting (fewer suppliers to deal with)
Focus on the near-term value of the deliverable	Focus on the lifetime value of the deliverable
Contractor provides minimal lifetime support for client's customers	Contractor provides lifetime support for customer value analyses and customer value measurement
Utilize one inflexible system	Access to contractor's many systems
Limited metrics	Use of the contractor's metrics library

The benefits of engagement project management are clear:

- Both the buyer and the seller save on significant procurement costs by dealing with single-source or sole-source contracts without having to go through a formal bidding process for each project.
- Because of the potential long-term strategic partnership, the seller is interested in the lifetime value of the business solution rather than just the value at the end of the project.
- Companies can provide lifelong support to clients as the latter try to develop value-driven relationships with their own clients.
- The buyer will get access to many of the project management tools used by the seller. The corollary is also true.

There is a risk in hiring consultants to manage projects if they bring their own methodology and accompanying metrics that are not compatible to the needs of the business or the person who hires them. Business solution providers must demonstrate that:

- Their approach is designed for the client's business model and strategy.
- The metrics they bring with them fit the client's business model and strategy.
- The client understands the metrics they are proposing.
- If necessary, they are willing to create additional metrics that fit the client's needs.

1.7 CUSTOMER RELATIONS MANAGEMENT

Engagement project management is forcing project managers to become active participants in customer relations management (CRM) activities. CRM activities focus on:

- Identifying the right customers
- Developing the right relationship with the customers
- Maintaining customer retention

TABLE 1.6 Engagement Manager versus Project Manager		
CUSTOMER VALUE MANAGEMENT	ENGAGEMENT MANAGER	PROJECT MANAGER
Phase 1: Identifying the right customers	• Strategic marketing • Proposal preparation • Engagement selling	• Assist in proposal preparation • May report to engagement manager
Phase 2: Developing the right relationship	• Defining acceptance criteria (metrics/KPIs) • Risk mitigation planning • Client briefings • Client invoicing • Soliciting satisfaction feedback and CRM	• Supporting CRM • Establishing performance metrics • Measuring customer value and satisfaction • Improving customer satisfaction management
Phase 3: Maintaining retention	• Conducting customer satisfaction management meeting • Updating client metrics and KPIs	• Attending customer satisfaction management meetings • Looking for future areas of improvement

CRM activities cannot be done entirely by the project manager. Some companies have both engagement managers and project managers. These two individuals must work together to maintain customer satisfaction. Table 1.6 shows the partial responsibilities of each.

1.8 OTHER DEVELOPMENTS IN PROJECT MANAGEMENT

For companies to be successful at managing complex projects on a repetitive basis and to function as solution providers, the project management methodology and accompanying tools must be fluid or adaptive. This means that companies may need to develop a different project management approach when interfacing with each stakeholder, given the fact that each stakeholder may have different requirements and expectations and the fact that most complex projects have long time spans. Figure 1.3 illustrates some of the new developments in project management, which apply to both traditional and nontraditional projects.

The five items in the figure fit together when done properly.

1. **New success criteria**: At the initiation of the project, the project manager will meet with the client and the stakeholders to come to stakeholder agreements on what constitutes success on the project. Initially, many of the stakeholders may have their own definition of success, but the project manager must forge an agreement, if possible.

Figure 1.3 **New Developments in Project Management**

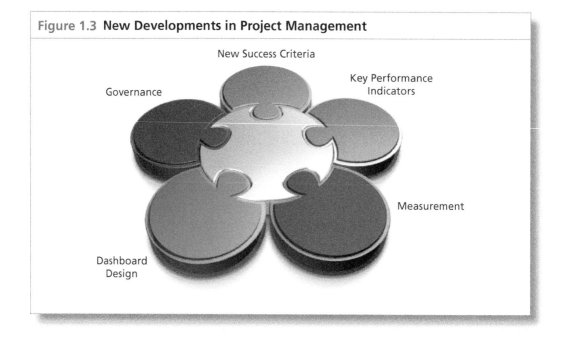

2. **Key performance indicators**: Once the success criteria are agreed upon, the project manager and the project team will work with the stakeholders to define the metrics and KPIs that each stakeholder wishes to track. It is possible that each stakeholder will have different KPI requirements.

3. **Measurement**: Before the metrics and KPIs are agreed to and placed on the dashboards, the project manager must be sure all team members know how to perform the measurements. This is the hardest part because not all team members or strategic partners may have the capability or skills to measure all of the KPIs.

4. **Dashboard design**: Once the KPIs are identified and measurement techniques are identified, the project manager, along with the appropriate project team members, will design a dashboard for each stakeholder. Some of the KPIs in the dashboards will be updated periodically, whereas others may be updated on a real-time basis.

5. **Governance**: Once the measurements are made, critical decisions may have to be supervised by the governance board. The governance board can include key stakeholders as well as stakeholders who are functioning just as observers.

1.9 A NEW LOOK AT DEFINING PROJECT SUCCESS

The ultimate purpose of project management is to create a continuous stream of project successes. This can happen provided that a good definition of "success" is available on each project.

SITUATION: Many years ago, as a young project manager, I asked a vice president in my company, "What is the definition of success on my project?" He responded, "The only definition in this company is meeting the target profit margin in the contract." I then asked him, "Does our customer have the same definition of success?" That ended our conversation.

For years, customers and contractors each worked toward different definitions for success. The contractor focused on profits as the only success factor, whereas the customer was more concerned with the quality of the deliverables. As project management evolved, all of that began to change.

Success Is Measured by the Triple Constraints

The triple constraints can be defined as a triangle with the three sides representing time, cost, and performance (which may include quality, scope, and technical performance). This was the basis for defining success during the birth of project management. This definition was provided by the customer, where cost was intended to mean "within the contracted cost." The contractor's interpretation of cost was profit.

Historically, only the triple constraints were used to define project success. Unfortunately, even if all of the deliverables are completed on time and within cost, the project may still be a failure if:

- There is no market demand for the product or services created.
- The products and services did not satisfy the customer's needs.
- The product and services appeared to satisfy the customer's needs but the customer was unhappy with the performance of the deliverables.
- The benefits defined in the business case were not achieved.
- The resulting financial value expected from the benefits was significantly less than anticipated.

It became apparent that metrics other than those used to track the triple constraints were needed to define project success.

Customer Satisfaction Must Be Considered as Well

Managing a project within the triple constraints is always a good idea, but the customer must be satisfied with the end result. A contractor can complete a project within the triple constraints and still find that the customer is unhappy with the end result. So, we have now placed a circle around the triple constraints, entitled "customer satisfaction." The president of an aerospace company stated, "The only definition of success in our business is customer satisfaction." That brought the customer and the contractor a little closer together. In the early years of using project management techniques, aerospace and defense contractors were incurring large cost overruns, and it was almost impossible to define success according to the triple constraints. Numerous scope changes were initiated by both customers and contractors. Because of the numerous scope changes, the only two metrics used on projects were related to time and cost. Success, however, was measured by follow-on business, which was an output of customer satisfaction.

Other (or Secondary) Factors Must Be Considered as Well

SITUATION: Several years ago, I met a contractor that had underbid a job for a client by almost 40 percent. When I asked why the company was willing to lose money on the contract, the person responded, "Our definition of success on this project is being able to use the client's name as a reference in our sales brochures."

There can be secondary success factors that, based on the project, are more important than the primary factors. These secondary factors include using the customer's name as a reference, corporate reputation, and image, compliance with government regulations, strategic alignment, technical superiority, ethical conduct, and other such factors. The secondary factors may end up being more important than the primary factors of the triple constraints.

Success Must Include a Business Component

By the turn of the twenty-first century, companies were establishing PMOs. One of the PMO's primary activities was to make sure that each project was aligned to strategic business objectives. The definition of success, thus, included a business component as well as a technical component. As an example, consider the following components included in the definition of success provided by a spokesperson from Orange Switzerland:

- The delivery of the product within the scope of time, cost, and quality characteristics
- The successful management of changes during the project life cycle
- The management of the project team
- The success of the product against criteria and target during the project initiation phase (e.g., adoption rates and ROI[7])

As another example, consider the following provided by Colin Spence, project manager/partner at Convergent Computing (CCO). General guidelines for a successful project are as follows:

- Meeting the technology and business goals of the client on time, on budget, and on scope.

7 Quoted in H. Kerzner, *Project Management Best Practices: Achieving Global Excellence* (Hoboken, NJ: John Wiley & Sons, 2006), pp. 22–23. 8 Quoted in ibid., p. 23.

- Setting the resource or team up for success, so that all participants have the best chance to succeed and have positive experiences in the process.
- Exceeding the client's expectations in terms of abilities, teamwork, and professionalism and generating the highest level of customer satisfaction.
- Winning additional business from the client, and being able to use them as a reference account and/or agree to a case study.
- Creating or fine-tuning processes, documentation, and deliverables that can be shared with the organization and leveraged in other engagements.[8]

The definition of the role of the project manager also changed. Project managers were managing part of a business rather than merely a project, and they were expected to make sound business decisions as well as project decisions. There must be a business purpose for each project. Each project is expected to make a contribution of business value to the company when the project is completed.

Prioritization of Success Constraints May Be Necessary

Not all project constraints are equal. The prioritization of constraints is performed on a project-by-project basis. Sponsors' involvement in this process is essential. Secondary factors are also considered to be constraints and may be more important than the primary constraints. For example, years ago, at Disneyland and Disney World, the project managers designing and building the attractions at the theme parks had six constraints:

1. Time
2. Cost
3. Scope
4. Safety
5. Aesthetic value
6. Quality

At Disney, the last three constraints, those of safety, aesthetic value, and quality, were considered locked-in constraints that could not be altered during trade-offs. All trade-offs were made on time, cost, and scope.

The importance of the components of success can change over the life of the project. For example, in the initiation phase of a project, scope may be the critical factor for success, and all trade-offs are made on the basis of time and cost. During the execution phase of the project, time and cost may become more important, and then trade-offs will be made on the basis of scope.

---

8 Quoted in ibid., p. 23.

SITUATION: The importance of the components of success at a point in time can also determine how decisions are made. As an example, a project sponsor asked a project manager when the project's baseline schedules would be prepared. The project manager responded, "As soon as you tell me what is most important to you, time, cost, or risk, I'll prepare the schedules. I can create a schedule based on least time, least cost, or least risk. I can give you only one of those three in the preparation of the schedule." The project sponsor was somewhat irate because he wanted all three. The project manager knew better, however, and held his ground. He told the sponsor that he would prepare one and only one schedule, not three schedules. The project sponsor finally said, rather reluctantly, "Lay out the schedule based on least time."

As previously stated, the definition of project success has a business component. That is true for both the customer and contractor's definition of success. Also, each project can have a different definition of success. There must be up-front agreement between the customer and the contractor at project initiation or even at the first meeting between them on what constitutes success at the end of or during the project. In other words, there must be a common agreement on the definition of success, especially the business reason for working on the project.

The Definition of Success Must Include a "Value" Component

Previously it was stated that there must be a business purpose for working on a project. Now, however, it is understood that, for real success to occur, there must be value achieved at the completion of the project. Completing a project within the constraints of time and cost does not guarantee that business value will be there at the end of the project. In the words of Warren Buffett, one of the world's most successful investors and chairman and chief executive of Berkshire Hathaway, "Price is what you pay. Value is what you get."

One of the reasons why it has taken so long to include a value component in the definition of success is that it is only in the last several years we have been able to develop models for measuring the metrics to determine the value on a project. These same models are now being used by PMOs in selecting a project portfolio that maximizes the value the company will receive. Also, as part of performance reporting, we are now reporting metrics on time at completion, cost at completion, value at completion, and time to achieve value.

TIP The definition of success must be agreed upon between the customer and the contractor.

Determining the value component of success at the completion of the project can be difficult, especially if the true value of the project cannot be determined until well after the project is completed. Some criteria on how long to wait to assess the true value may need to be established.

Multiple Components for Success

Today, project managers have come to the realization that there are multiple constraints on a project. More complex projects, where the traditional triple constraints success factors are constantly changing, are being worked on. For example, in Figure 1.4, for traditional projects, time, cost, and scope may be a higher priority than the constraints within the triangle. However, for more complex projects, the constraints within the triangle may be more important.

Beginning with the fourth edition of the *PMBOK® Guide* the term "triple constraints" was no longer used. Because there can be more than three constraints, the term "competing constraints" is now used, in recognition of the fact that the exact number of success constraints and their relative importance can change from project to project. What is important is that metrics must be established for each constraint on a project. However, not all of the metrics on the constraints will be treated as KPIs.

The Future

So, what does the future look like? The following list is representative of some of the changes that are now taking place:

- The project manager will meet with the client at the very beginning of the project, and they will come to an agreement on what constitutes project success.
- The project manager will meet with other project stakeholders and get their definition of success. There can and will be multiple definitions of success for each project.
- The project manager, the client, and the stakeholders will come to an agreement on what metrics they wish to track to verify that success will be achieved. Some metrics will be treated as KPIs.

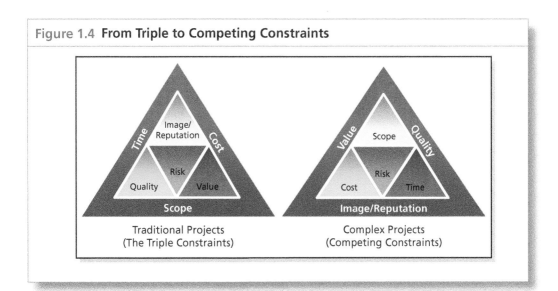

Figure 1.4 **From Triple to Competing Constraints**

Traditional Projects
(The Triple Constraints)

Complex Projects
(Competing Constraints)

- The project manager, assisted by the PMO, will prepare dashboards for each stakeholder. The dashboards will track each of the requested success metrics in real time rather than relying on periodic reporting.
- At project completion, the PMO will maintain a library of project success metrics that can be used on future projects.

In the future, the PMO can be expected to become the guardian of all project management intellectual property. The PMO will create templates to assist project managers in defining success and establishing success metrics.

1.10 THE GROWTH OF PAPERLESS PROJECT MANAGEMENT

Making informed decisions requires information. In its early years, project management relied heavily on legacy systems for the information needed. Over the past several decades, other information systems have emerged, as seen in Figure 1.5. PMIS evolved to provide information solely for the project at hand. Later, enterprise resource planning (ERP) systems and CRM systems appeared that provided project management with sufficient information such that they could now make business- as well as project-based decisions. Today, the amount of information that a company can generate is overwhelming, and all of this information will be stored in data or information warehouses. With pure legacy systems that tracked business metrics, the

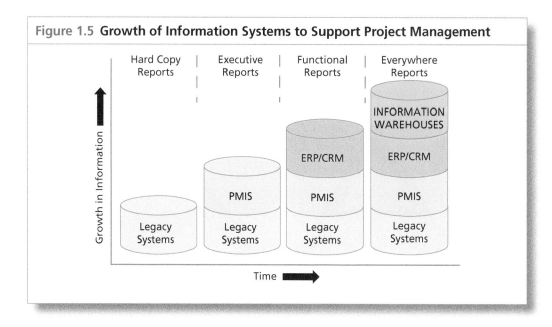

Figure 1.5 **Growth of Information Systems to Support Project Management**

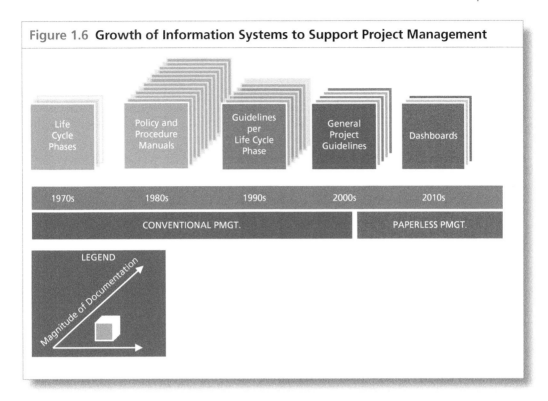

Figure 1.6 **Growth of Information Systems to Support Project Management**

information was reported mainly vertically up the organizational hierarchy. Today, project-based information can be reported everywhere including to organizations external to the company.

Having more information comes with a price: more costly reporting and larger and more frequent reports. This is shown in Figure 1.6. As the cost of paperwork grew, companies began looking at the possibility of paperless project management. This would necessitate identification of just the critical information and presenting the information using dashboards.

Initially, reporting was done at the end of each life cycle phase. Unfortunately, this meant that some customers would not see project status until the end-of-phase gate review meetings. To solve this problem, policy and procedure manuals were created that dictated how and when reporting should take place. Unfortunately, this system placed restriction on the project managers, and eventually the policies and procedures were replaced with guidelines. Today, the focus is on dashboards.

1.11 PROJECT MANAGEMENT MATURITY AND METRICS

All companies desire maturity and excellence in project management. Unfortunately, not all companies recognize that the time frame can be shortened by performing strategic planning for project management maturity and excellence. The simple use of project management, even for an extended period of time, does not lead to excellence. Instead, it can result in repeated mistakes and, what's worse, learning from your own mistakes rather than the mistakes of others.

Strategic planning for project management is unlike other forms of strategic planning in that it is most often performed at the middle and lower levels of management. Executive management is still involved, mostly in a supporting role, and provides funding together with employee release time for the effort.

There are models that can be used to assist in achieving excellence. One such model is the Project Management Maturity Model, shown in Figure 1.7. Each of the five levels represents a different degree of maturity in project management.

Level 1—Common Language: In this level, the organization recognizes the importance of project management and the need for a good understanding of the basic knowledge on project management, along with the accompanying language and terminology.

Level 2—Common Process: In this level, the organization recognizes that common processes need to be defined and developed such that the successes on one project can be repeated on other projects. Also included in this level is the recognition that project management can be applied to and support other methodologies employed by the company.

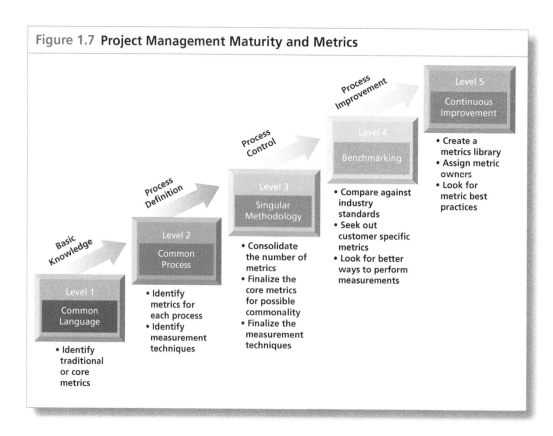

Figure 1.7 **Project Management Maturity and Metrics**

Level 3—**Singular Methodology**: In this level, the organization recognizes the synergistic effect of combining all corporate methodologies and processes into a singular methodology, the center of which is project management. The synergistic effects also make process control easier with a single methodology than with multiple methodologies. However, as companies evolve into the advanced stages of growth and maturity, the singular methodology is replaced by more than one flexible methodology that can be customized to the needs of each project.

Level 4—**Benchmarking**: This level contains the recognition that process improvement is necessary to maintain a competitive advantage. Benchmarking should be performed on a continuous basis. The company must decide whom to benchmark against and what to benchmark.

Level 5—**Continuous Improvement**: In this level, the organization evaluates the information obtained through benchmarking and must then decide whether this information will enhance the singular methodology or not.

Although these five levels normally are accomplished with forms, guidelines, templates, and checklists, the growth in metrics management has allowed further enhancement of the model by including in each level the necessity for metrics, as shown in Figure 1.7. Metrics can serve as a sign of organizational maturity. The need for paperless project management will require that more emphasis be placed on metrics management as part of the project management maturity process.

Maturity in project management allows companies to recognize that project management is a strategic competency, as shown in Figure 1.8. For companies that promote their project management capabilities to external clients, competency in project management is viewed as a sustained competitive advantage. However, ineffective metrics management can increase the risks in maintaining a sustained competitive advantage, as shown in Figure 1.9. These risks are covered in detail in later chapters.

Figure 1.8 shows that excellence in project management is achieved when project management is seen as a strategic competency and the company recognizes that its project management capability has become a competitive advantage. Unfortunately, competitive advantages are not always sustainable, as can be seen in Figure 1.10. As a company exploits its competitive advantage, competitors counterattack to reduce or eliminate that advantage. Therefore, as illustrated in Figure 1.11, a company must have continuous improvement for the competitive advantage to grow into a sustained competitive advantage.

Having a sustained competitive advantage in project management does not come just from being on time and on budget at the end of each project. Rather, offering clients something that competitors cannot do may help. But in project management, a true competitive advantage occurs when efforts are directly linked to the customers' perception of

Figure 1.8 **Project Management Competitiveness**

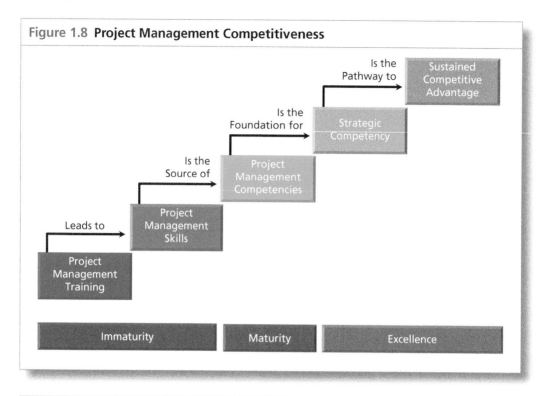

Figure 1.9 **Metric risks to Maintain a Sustained Competitive Advantage**

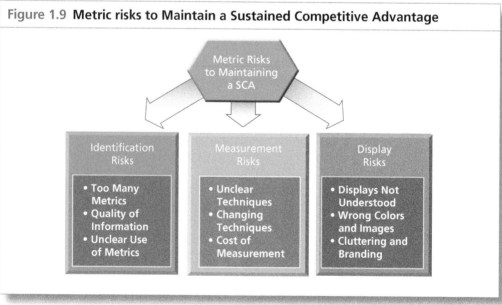

value. Whatever means the company uses to show this, such as through the use of value-reflective metrics, gives it a sustainable competitive advantage. Value-reflective metrics, which are discussed in Chapter 5, show how to create value. If these metrics undergo continuous improvement, then users may be adding value for the customers.

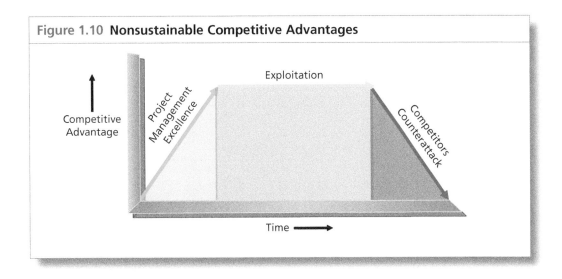

Figure 1.10 **Nonsustainable Competitive Advantages**

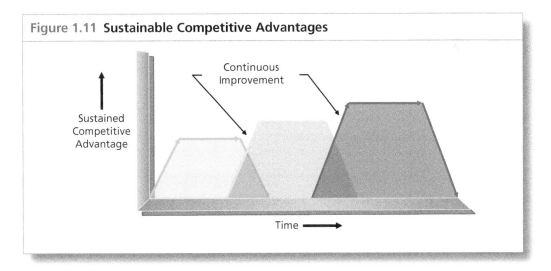

Figure 1.11 **Sustainable Competitive Advantages**

There is no point in wasting resources on value metrics unless the client understands the metrics and perceives the value that is being created. Therefore, client input into the selection of the attributes for the value metrics is essential. Table 1.7 shows some typical value-reflective metrics and the accompanying strategic competitive advantage.

1.12 PROJECT MANAGEMENT BENCHMARKING AND METRICS

One of the fastest ways to reach maturity and excellence in project management is through the use of benchmarking. A benchmark is a measurement or standard against which comparisons can be made.

TABLE 1.7   Competitive Advantages from Value-Reflective Metrics

METRICS WITH VALUE ATTRIBUTES	POSSIBLE COMPETITIVE ADVANTAGE
Deliverables produced	Efficiency
Product functionality	Innovation
Product functionality	Product differentiation
Support response time	Service differentiation
Staffing and employee pay grades	People differentiation
Quality	Quality differentiation
Action items in the system and how long	Speed of problem resolution and decisions
Cycle time	Speed to market
Failure rates	Quality differentiation and innovation

Benchmarking is the process of comparing business processes and performance metrics to industry bests or best practices from other industries. Dimensions typically measured are quality, time, and cost. In the process of benchmarking, management identifies the best firms in its industry, or in another industry where similar processes exist, and compares the results and processes of those studied (the "targets") to its own company's results and processes. In this way, management learns how well the targets perform and, more important, the business processes that explain why these firms are successful.

Best Practice versus Proven Practice

In project management, the terms "best practice benchmarking" or "process benchmarking" are used, referring to how organizations evaluate various aspects of their processes in relation to the practices of best practice companies, usually within a peer group defined for the purposes of comparison. This evaluation process allows organizations to develop plans on how to make improvements or adapt specific best practices, usually with the aim of increasing some aspect of project management performance. Benchmarking often is treated as a continuous process in which organizations continually seek to improve their practices.

For more than a decade, companies have been fascinated with the term "best practices." Best practices are generally those practices that have been proven to produce superior results. But now, after a decade or more of use, the term is being scrutinized and it is recognized that perhaps better expressions exist. When a company says that it has a best practice, it really means that there is a technique, process, metric, method, or activity that can be more effective at delivering an outcome than any other approach and that it provides the company with the desired outcome with fewer problems and unforeseen complications. As a result, the company ends up with the most efficient and effective way of accomplishing a task based on a repeatable process that has been proven over time for a large number of people and/or projects.

There are several arguments why the words "best practice" should not be used. First is the argument that the identification of a best practice may lead some to believe that they were performing some activities incorrectly in the past, and that may not be the case. What was known as a best practice may simply be a more efficient and effective way of achieving a deliverable. Another argument is that some people believe that best practices imply that there is one and only one way of accomplishing a task. This also may be a faulty interpretation. Third, and perhaps most important, is the argument that a best practice is the "best" way of performing an activity and, since it is the best, no further opportunities for improvement are possible.

Once a best practice has been identified and proven to be effective, normally it is integrated into project management processes so that it becomes a standard way of doing business. Therefore, after acceptance and proven use of the idea, the better expression possibly should be "proven practice" rather than best practice. This leaves the door open for further improvements.

These are just some arguments why "best practices" may be just a buzzword and should be replaced. Perhaps this will happen in the future. However, for the remainder of this text, the term "best practices" is used, with the caveat that other terms may be more appropriate.

Benchmarking Methodologies

No single benchmarking process has been universally adopted. The wide appeal and acceptance of benchmarking has led to the emergence of a variety of benchmarking methodologies. However, with regard to project management, benchmarking activities usually are easier to implement and accept because of the existence of the *PMBOK® Guide* and a PMO. The *PMBOK® Guide* helps to identify areas where benchmarking would be beneficial, and people understand that the PMO is responsible for continuous improvements in project management.

The following is an example of a typical benchmarking methodology.

- **Identify problem areas**: Because benchmarking can be applied to any business process or function, a range of research techniques may be required. They include informal conversations with customers, employees, or suppliers; exploratory research techniques such as focus groups; and in-depth marketing research, quantitative research, surveys, questionnaires, reengineering analysis, process mapping, quality control variance reports, financial ratio analysis, or simply reviewing cycle times or other performance indicators.

- **Identify others that have similar processes**: Because project management exists in virtually every industry, benchmarking personnel should not make the mistake of looking only at their own industry.
- **Identify organizations that are leaders in these areas**: Look for the very best in any industry and in any country. Consult customers, suppliers, financial analysts, trade associations, and magazines to determine which companies are worthy of study. Symposiums and conferences sponsored by the Project Management Institute provide excellent opportunities to hear presentations from companies that are doing things exceptionally well. Even companies that are in financial distress may be outstanding in some areas of project management.
- **Visit the "best practice" companies to identify leading-edge practices**: Companies typically agree to mutually exchange information beneficial to all parties in a benchmarking group and share the results within the group.
- **Implement new and improved business practices**: Take the leading-edge practices and develop implementation plans that include identification of specific opportunities, funding the project, and selling the ideas to the organization for the purpose of gaining demonstrated value from the improvements.

Benchmarking Costs

The three main types of costs in benchmarking are:

1. **Visitation costs**: This includes hotel rooms, travel costs, meals, token gifts, and lost labor time.
2. **Time costs**: Members of the benchmarking team will be investing time in researching problems, finding exceptional companies to study, visits, and implementation. This will take them away from their regular tasks for part of each day so additional staff might be required.
3. **Benchmarking database costs**: Organizations that institutionalize benchmarking into their daily procedures find it is useful to create and maintain a database or library of best practices.

The cost of benchmarking can be reduced substantially by utilizing internet resources. These resources aim to capture benchmarks and best practices from organizations, business sectors, and countries to make the benchmarking process much quicker and cheaper.

Types of Benchmarking

There are several types of benchmarking studies:

- **Process benchmarking**: The initiating firm focuses its observation and investigation of project management and business processes with a goal of identifying and observing the best practices from one or more benchmark firms. Activity analysis will be required where the objective is to benchmark cost and efficiency in executing the processes that are part of a project management methodology. This is the most common form of benchmarking in project management. Process benchmarking cannot be successful if users do not fully understand their own processes.

- **Metric benchmarking**: The process of comparing the different metrics that organizations are using for continuous improvements. Time, cost, and quality are just three of the metrics that are being used. Additional metrics are being created to measure what is needed, not what is the easiest to measure. The intent is to identify the core metrics needed for project management. One of the biggest challenges for metric benchmarking is the variety of metric definitions used among companies or divisions. Definitions may change over time within the same organization due to changes in leadership and priorities. The most useful comparisons can be made when metrics definitions are common between compared units and do not change so improvements can be verified.
- **Financial benchmarking**: Performing a financial analysis and comparing the results in an effort to assess your overall competitiveness and productivity.
- **Benchmarking from an investor perspective**: Extending the benchmarking universe to also compare to peer companies that can be considered alternative investment opportunities from the perspective of an investor.
- **Performance benchmarking**: Allows the initiator firm to assess its competitive position by comparing products and services with those of target firms.
- **Product benchmarking**: The process of designing new products or upgrades to current ones. This process sometimes can involve reverse engineering, which involves taking apart competitors' products to find strengths and weaknesses.
- **Strategic benchmarking**: This involves observing how others compete. This type of benchmarking usually is not industry specific, meaning it is best to look at other industries.
- **Functional benchmarking**: A company focuses its benchmarking on a single function to improve the operation of that particular function. Complex functions such as human resources, finance and accounting, and information and communication technology are unlikely to be directly comparable in cost and efficiency terms and may need to be disaggregated into processes to make valid comparison.
- **Best-in-class benchmarking:** This involves studying the leading competitor or the company that carries out a specific function best.
- **Internal benchmarking:** A comparison of a business process to a similar process inside the organization. This is a quest for internal best practices.

- **Competitive benchmarking:** This is a direct competitor-to-competitor comparison of a product, service, process, or method.
- **Generic benchmarking:** This approach broadly conceptualizes unrelated business processes or functions that can be practiced in the same or similar ways regardless of industry.

Benchmarking Code of Conduct

Numerous problems can occur during benchmarking. Some problems result from misunderstandings, whereas other problems could involve legal issues. The Code of Conduct of the International Benchmarking Clearinghouse is an excellent starting point.

- **Legality:** Avoid any discussions that could be interpreted as illegal for you or your benchmarking partners.
- **Exchange:** Be prepared to answer the same questions you are asking. Letting partners review the questions in advance is helpful.
- **Confidentiality:** All information should be treated as proprietary information. You may wish to consider having everyone sign a nondisclosure agreement.
- **Use of Information:** There must be an agreement, preferably in writing, on how the information will be used.
- **Contact:** Follow your partners' protocols and customs on whom you are allowed to interface with.
- **Preparation:** Be fully prepared for partner interfacing and exchanges of information.
- **Completion:** Avoid making promises or commitments that cannot be kept.

Benchmarking Mistakes

Benchmarking mistakes can lead to benchmarking failures. Some of these mistakes include:

- Limiting benchmarking activities to just the company's own industry.
- Benchmarking only industry leaders; industry followers can provide just as much information as industry leaders.
- Failing to recognize that not all results are applicable, especially where organizational and cultural differences exist.
- Failing to have a benchmarking plan and not knowing what to look for.

Points to Remember

Some critical points must be remembered when performing benchmarking:

- It is necessary to understand the culture and circumstances behind the numbers to fully understand their meaning and use. The "how" is just as important as the "how much?"

- In project management, changes can occur quickly. It is important to set frequencies for the benchmarking studies, and each process studied may require different frequencies.
- The more rigorous the benchmarking process, the better the results.
- Regardless of how good a company thinks its project management systems are, there is always room for improvement.
- Those who do not believe in continuous improvement soon become industry followers rather than leaders.
- Executives who are not familiar with or supportive of benchmarking will always adopt the "not invented here" argument or "this is the way we have always done it."
- Successful benchmarking is "doing," not "knowing."
- Benchmarking allows users to learn from the mistakes of others rather than from their own mistakes.
- Because of the rate of change that takes place in project management, it is highly unlikely that the targets that are benchmarked with will be leaders in all areas of project management.
- Benchmarking can prevent surprises.
- People must recognize the need for change. This must be accomplished with benchmarking evidence rather than just claims or opinions.
- Change occurs quickly when the people who are needed to change or to make the change are involved in the benchmarking studies.
- Implementing change requires a champion. Having a PMO is almost always the right idea.

1.13 **CONCLUSIONS**

The future of project management may very well rest in the hands of the solution providers. These providers will custom-design project management frameworks and methodologies for each client and possibly for each stakeholder. They must be able to develop metrics that go well beyond the current *PMBOK® Guide* and demonstrate a willingness to make business decisions as well as project decisions. The future of project management looks quite good, but it will be a challenge.

2

THE DRIVING FORCES FOR BETTER METRICS

<table>
<tr><td>CHAPTER OVERVIEW</td><td>Today, more than ever before, a great percentage of projects are becoming distressed and possibly failing. Techniques such as project audits and health checks encourage the use of a more formalized metrics management system. Effective project decisions cannot be made without meaningful metrics. Stakeholder relations management thrives on meaningful metrics.</td></tr>
<tr><td>CHAPTER OBJECTIVES</td><td><ul><li>To understand the importance of metrics in dealing with stakeholders</li><li>To understand the importance of metrics when conducting project audits</li><li>To understand the importance of metrics when performing health checks</li><li>To understand the metrics can and will change when trying to recover a distressed project</li></ul></td></tr>
<tr><td>KEY WORDS</td><td><ul><li>Boundaries</li><li>Distressed projects</li><li>Project audits</li><li>Project health checks</li><li>Scope creep</li><li>Stakeholder relations management (SRM)</li></ul></td></tr>
</table>

2.0 INTRODUCTION

Companies do not simply add more metrics or key performance indicators (KPIs) by choice. Usually driving forces make it evident that such changes are needed. Complacency works when things are going well or as planned. When project managers start accepting more complex projects, however, as was discussed in Chapter 1, things have a tendency to go poorly, and the need for additional metrics is recognized.

By performing audits and health checks, project managers can prevent a project from becoming distressed, provided that the cause of the problem is detected early enough and options exist for corrective action. Unfortunately, the metrics used might not act as an early warning

Project Management Metrics, KPIs, and Dashboards: A Guide to Measuring and Monitoring Project Performance, Fourth Edition. Harold Kerzner.
© 2023 John Wiley & Sons, Inc. Published 2023 by John Wiley & Sons, Inc.
Companion Website: www.wiley.com/go/kerznermetrics4e

system. By the time new metrics for analysis of a potentially failing project are established, the damage may have been done and recovery may no longer be possible. The final result can be devastating if stakeholder relations management (SRM) fails and future business is not forthcoming. Situations like this may be attributed to improper identification, selection, implementation, and measurement of the right metrics and KPIs.

2.1 STAKEHOLDER RELATIONS MANAGEMENT

Stakeholders are, in one way or another, individuals, companies, or organizations that may be affected by the outcome of the project or the way in which the project is managed. Stakeholders may be either directly or indirectly involved throughout the project, or may function simply as observers. A stakeholder can shift from a passive role to being an active member of the team and participate in making critical decisions.

SITUATION: In order to impress the stakeholders, the project manager agrees to establish more metrics than needed. Once the project begins and the stakeholders begin examining the measurements of the metrics, the project manager realizes that some stakeholders are actively involved in the project to the point where they are trying to micromanage him or her.

SITUATION: As the project progresses, several stakeholders begin asking for additional metrics that were not part of the original plan. The project management methodology does not provide data for these metrics, and the cost of changing the methodology at this point is prohibitive.

SITUATION: The project manager tries to impress stakeholders by giving them more information than they actually need. The result is that they may not be able to identify what information they actually need and mistrust the data to the point where they do not use it for decision-making purposes.

On small or traditional projects, both internal and external ones, project managers generally interface with just the project sponsor as the primary stakeholder, and the sponsor usually is assigned from the organization that funds the project. However, the larger the project, the greater the number of stakeholders that project managers must interface with. Project managers may now have to deal with governance by committee. The situation becomes even more potentially problematic if there is a large number of stakeholders, geographically dispersed, all at different levels of management in their respective hierarchy, each with a different level of authority, and language and cultural differences. Trying to interface with all of these people on a regular basis and make decisions, especially on a large, complex project is very time-consuming.

This section is adapted from Harold Kerzner and Carl Belack, *Managing Complex Projects* (Hoboken, NJ: John Wiley & Sons, 2010), Chapter 10.

TIP Because of the potentially large number of stakeholders, project managers should not attempt to establish metrics that can satisfy all of the stakeholders all of the time.

TIP Passive stakeholders can become active stakeholders when the situation merits it. The project manager must consider metrics for passive stakeholders as well, but perhaps not the same number of metrics that would be provided for the active stakeholders.

TIP Not all of the stakeholders will be in agreement on the interpretation of the metrics and have the same conclusions on what action, if any, is necessary.

One of the complexities of SRM is figuring out how to satisfy everyone without sacrificing the company's long-term mission or vision. Also, the company may have long-term objectives in mind for this project, and those objectives may not necessarily be aligned to the project's objectives or each stakeholder's objectives. Lining up all of the stakeholders in a row and getting them to agree to all decisions is more wishful thinking than reality. Project managers may discover that it is impossible to get all of the stakeholders to agree, and can only hope to placate as many as possible at any given time.

SRM cannot work effectively without commitments from all of the stakeholders. Obtaining these commitments can be difficult if the stakeholders cannot see what is in it for them at the completion of the project, namely the value that they expect or other personal interest. The problem is that what one stakeholder perceives as value another stakeholder may have a completely different perception of or he or she may desire a different form of value. For example, one stakeholder could view the success of the project as a symbol of prestige, authority, power, or an opportunity for promotion. Another stakeholder could perceive the value as simply keeping people employed. A third stakeholder could see value in the final deliverables of the project and the inherent quality in it. A fourth stakeholder could see the project as an opportunity for future work with particular partners.

Another form of agreement involves developing a consensus on how stakeholders will interact with each other. It may be necessary for certain stakeholders to interact with each other and support one another by sharing resources, providing financial support in a timely manner, and sharing intellectual property. This support may necessitate a commonality among some metrics. Although all stakeholders recognize the necessity for these agreements, they can be affected by politics, economic conditions, and other enterprise environmental factors that may be beyond the control of the project manager. Certain countries may not be willing to work with other countries because of culture, religion, views on human rights, and other such factors.

For the project manager, obtaining these agreements right at the beginning of the project is essential. Some project managers are fortunate to be able to do this while others are not. Leadership changes in certain governments, or any organizational hierarchy, may make it difficult to enforce these agreements on complex projects. Changes in leadership can cause changes in the project's direction.

TIP Metrics systems, no matter how good, may not generate interaction among stakeholders. Metrics serve as a support tool and should not be viewed as a replacement for effective project communications.

TIP Changes in stakeholders may necessitate the creation of new metrics regardless how far the project has progressed.

It is important that everyone who has a stake in the project be willing to communicate what they believe are the factors critical for success. Success is easier to define and reach when there is broad agreement. However, there is no guarantee that there will be an agreement among the project manager, client(s), and stakeholders as to a core set of metrics and KPIs. Project managers must be prepared to construct a different set of dashboards possibly for each viewer. But there is a risk. If too many metrics are established, then too much data may be necessary, and the project team may find it difficult to capture all of the information that is needed. The metrics selected must be scaled to fit the needs of the project as well as the needs of individual stakeholders. Also, the metrics selected must be used to ensure that the time and cost involved with selecting and measuring the metrics has been worthwhile.

Some simple steps can be followed:

- The project team must brainstorm what metrics are appropriate for this project. This is easy to do if a metrics library exists for metric selection and recommendation purposes. This brainstorming may be accomplished prior to stakeholder involvement.
- The team must then assess the metrics to make sure they are needed and look for ways of combining metrics if appropriate.
- The team must make sure that the metrics are presented in terms that every viewer can understand (i.e., What is the meaning of a partial or incomplete deliverable? What is the meaning of an unacceptable deliverable?).
- The team must then decide what metrics to recommend to the stakeholders.

Sometimes the number of metrics is selected based on the personal whims of the stakeholders and the project team, and then it is discovered how costly measuring and reporting all of the metrics is. The number of metrics selected should be based on the size of the project, its complexity, the decisions that the governance committee must make, and the accompanying risks. There must be an informational value for each metric. Otherwise, the result can be information overload and an abundance of useless information.

On small projects, there is not sufficient time or funding to work with a large number of metrics. Success on a small project may be as simple as customer acceptance of the deliverables or no complaints by the customer. But for larger and more complex projects, success may not be able to be determined from a single metric. Core metrics may be necessary.

It is important for the project manager to fully understand the issues and challenges facing each of the stakeholders, especially their information needs. Although it may seem unrealistic, some stakeholders can have different views on the time requirements of the project. In some developing nations, the construction of a new hospital in a highly populated area may drive the commitment for the project, even though it could be late by a year or longer. People just want to know that the hospital will be built eventually.

In some cultures, workers cannot be fired. Because workers believe they have job security, it may be impossible to get them to work faster or better. In some countries, there may be as many as 50 paid holidays for workers, and this can have an impact on the project manager's schedule. Not all workers in each country have the same skill level even though they have the same title. For example, a senior engineer in an emerging nation may have the same skills as a lower-grade engineer in another country. In some locations where labor shortages may exist, workers are assigned to tasks based on availability rather than ability. Having sufficient headcount is not a guarantee that the work will get done in a timely manner and that the level of quality will be there.

In some countries, power and authority, as well as belonging to the right political party, are symbols of prestige. People in these positions may not view the project manager as their equal and may direct all of their communications to the project sponsor. In this case, it is possible that salary is less important than relative power and authority.

It is important to realize that not all of the stakeholders may want the project to be successful. This will happen if stakeholders believe that, at the completion of the project, they may lose power, authority, hierarchical positions in their company, or, in a worst case, even lose their job. Sometimes these stakeholders will either remain silent or even be supporters of the project until the end date approaches. If the project is regarded as unsuccessful, these stakeholders may respond by saying "I told you so." If it appears that the project may be a success, these stakeholders may suddenly be transformed from supporters or the silent majority to adversaries and encourage failure.

It is very difficult to identify stakeholders with hidden agendas. These people may hide their true feelings and be reluctant to share information. Often there are no telltale or early warnings signs that indicate their true belief in the project. However, if stakeholders are reluctant to approve scope changes, provide additional investment, or assign highly qualified resources, this could be an indication that they may have lost confidence in the project.

Not all stakeholders understand project management. Not all stakeholders understand the role of a project sponsor or their role as part of project governance. Not all stakeholders

TIP The project manager may find it necessary to establish country-specific metrics for his or her personal use.

understand how to interface with a project or the project manager, even though they may readily accept and support the project and its mission. Simply stated, the majority of the stakeholders quite often are never trained in how to properly function as a stakeholder or a member of a project governance committee. Unfortunately, this lack of training cannot be detected early on but may become apparent as the project progresses.

Some stakeholders may be under the impression that they are merely observers and do not need to participate in decision making or authorization of scope changes. Their new role could constitute a rude awakening. Some will accept it, whereas others will not. Those who do not accept the new role usually are fearful that participating in a decision that turns out to be wrong can be the end of their career.

Some stakeholders may believe they should be micromanagers, often usurping the authority of the project manager by making decisions that they may not necessarily be authorized to make, at least not alone. Stakeholders who attempt to micromanage can do significantly more harm to the project than stakeholders who remain observers.

It may be a good idea for the project manager to prepare a list of expectations that he or she has of stakeholders. This is essential even though the stakeholders visibly support the existence of the project. Role clarification for stakeholders should be accomplished early, in the same way that the project manager provides role clarification for team members at the initial project kickoff meeting.

The current view of stakeholder management in Table 2.1 results from the implementation of "engagement project management" practices. In the past, whenever a sale was made to the client, the salesperson would then move on to find a new client. Salespeople viewed themselves as providers of products and/or services.

Today, salespeople view themselves as providers of business solutions. In other words, salespeople now tell clients, "We can provide

TABLE 2.1   Changing Views in Stakeholder Relations Management

PAST VIEW	PRESENT VIEW
Manage existing relationships	Build relationships for the future; that is, engagement project management
Align the project to short-term business goals	Align the project to long-term strategic business goals
Provide ethical leadership when suits	Provide ethical leadership throughout the project
Project is aligned to profits	Project is aligned stakeholders' expectation of value
Identify profitable scope changes	Identify value-added scope changes
Provide stakeholders with the least number of metrics and KPIs	Provide stakeholders with sufficient metrics and KPIs such that they can make informed decisions

TIP Providing too many metrics and KPIs may be an invitation for stakeholders to micromanage the project. Another possibility is that the stakeholders may not trust the data and therefore not use the data for decision making.

you with a solution to all of your business needs and what we want in exchange is to be treated as a strategic business partner." This benefits both the buyer and the seller, as discussed previously.

Therefore, as a solution provider, the project manager focuses heavily on the future and on establishing a long-term partnership agreement with the client and the stakeholders. This focus is heavily oriented toward value rather than near-term profitability.

On the micro-level, SRM can be defined using the six processes shown in Figure 2.1.

1. **Identify the stakeholders:** This step may require support from the project sponsor, sales, and the executive management team. Even then, there is no guarantee that all of the stakeholders will be identified.
2. **Stakeholder analysis:** This analysis requires an understanding of which stakeholders are key stakeholders—those who have influence, the ability, and the authority to make decisions, and can make or break the project. This analysis also includes developing SRM strategies based on the results of the analysis.
3. **Perform stakeholder engagements:** During this step, the project manager and the project team get to know the stakeholders.
4. **Stakeholder information flow:** This step is the identification of the information flow network and the preparation of the necessary reports for each stakeholder.

Figure 2.1 Stakeholder Relations Management

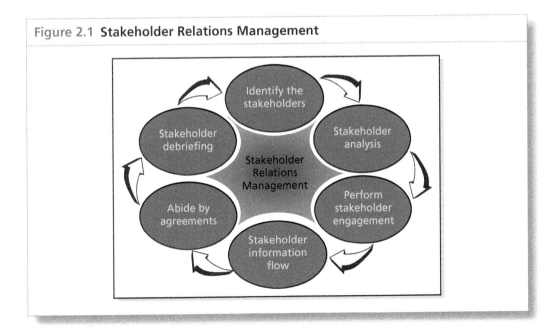

5. **Abide by agreements:** This step enforces stakeholder agreements made during the initiation and planning stages of the project.
6. **Stakeholder debriefings:** This step occurs after contract or life cycle phase closure and is used to capture lessons learned and best practices for improvements on the next project involving these stakeholders or the next life cycle phase.

Stakeholder management begins with stakeholder identification. This is easier said than done, especially if the project is multinational. Stakeholders can exist at any level of management. Corporate stakeholders are often easier to identify than political or government stakeholders. Each stakeholder is an essential piece of the project puzzle.

Stakeholders must work together and usually interact with the project through the governance process. Therefore, it is essential to know which stakeholders will participate in governance and which will not.

As part of stakeholder identification, the project manager must know whether he or she has the authority or perceived status to interface with the stakeholders. Some stakeholders perceive themselves as higher status than the project manager. In this case, the project sponsor may be the person to maintain interactions.

Stakeholders can be identified in several ways. More than one way can be used on projects.

- **Groups:** This could include financial institutions, creditors, regulatory agencies, and the like.
- **Individuals:** These could be identified by name or title, such as the CIO, COO, CEO, or just the name of the contact person in the stakeholder's organization.
- **Contribution:** This could include financial contributor, resource contributor, or technology contributor.
- **Other factors:** This could include the authority to make decisions or other such factors.

It is important to understand that not all stakeholders have the same expectations of a project. Some stakeholders may want the project to succeed at any cost, whereas other stakeholders may prefer to see the project fail even though they openly seem to support it. Some stakeholders view success as the completion of the project regardless of the cost overruns, whereas others may define success in financial terms only. Some stakeholders are heavily oriented toward the value they expect to see in the project, and this is the only definition of success for them. The true value may not be seen until months after the project has been completed. Some stakeholders may view the project as their opportunity for public notice and increased status and, therefore, want to be actively involved. Others may prefer a more passive involvement.

On large, complex projects with a multitude of stakeholders, it may be impossible for the project manager to cater to all of the stakeholders. Therefore, the project manager must know who the most influential stakeholders are and who can provide the greatest support on the project. Typical questions to ask include:

■ Who are powerful and who are not?
■ Who will have or require direct, or indirect, involvement?
■ Who has the power to kill the project?
■ What is the urgency of the deliverables?
■ Who may require more or less information than others?

Not all stakeholders are equal in influence, power, or the authority to make decisions in a timely manner. It is imperative for the project manager to know who sits on the top of the list as having these capabilities.

Finally, it is important for project managers to remember that stakeholders can change over the life of a project, especially if it is a long-term project. Also, the importance of certain stakeholders can change over the life of a project and in each life cycle phase. The stakeholder list is, therefore, an organic document subject to change.

Stakeholder mapping is most frequently displayed on a grid, comparing stakeholders' power and their level of interest. This is shown in Figure 2.2. The four cells can be defined as:

1. **Manage closely:** These are high-powered, interested people who can make or break a project. The project manager must put forth

Figure 2.2 Stakeholder Mapping

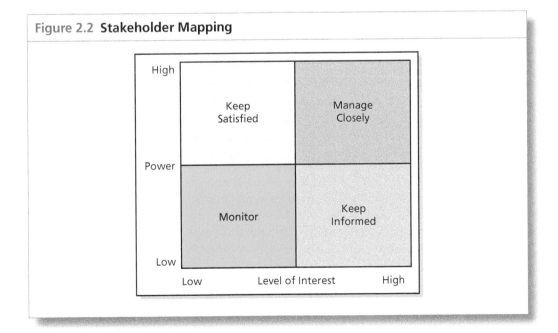

the greatest effort to satisfy them and be aware that there are factors that can cause them to change quadrants rapidly.

2. **Keep satisfied:** These are high-powered, less interested people who can also make or break a project. The project manager must put forth some effort to satisfy them but not with excessive detail that can lead to boredom and total disinterest. They may not get involved until the end of the project approaches.

3. **Keep informed:** These are people with limited power but keen interest in the project. They can function as an early warning system of approaching problems and may be technically astute and able to assist with some technical issues. These are the stakeholders who often provide hidden opportunities.

4. **Monitor only:** These are people with limited power and who may not be interested in the project unless a disaster occurs. The project manager should provide them with some information but not with so much detail that they will become disinterested or bored.

TIP Where the stakeholders are positioned on the grid could determine the number of metrics that they want or will be provided. As stakeholders change positions in the grid, so might the number of metrics they receive.

The larger the project, the more important it becomes to know who is and is not an influential or key stakeholder. Although the project manager must win the support of all stakeholders, or at least try to do so, the key stakeholders come first. Key stakeholders may be able to assist the project manager in identifying enterprise environmental factors that can have an impact on the project. These factors could include forecasts on the host country's political and economic conditions, the identification of potential sources for additional funding, and other such issues. In some cases, the stakeholders may have software tools that can supplement the project manager's available organizational process assets.

Thus far, we have discussed the importance of winning over the key or influential stakeholders. There is also a valid argument for winning over the stakeholders who are considered to be unimportant. Although some stakeholders may appear to be unimportant, this can change rapidly. For example, an unimportant stakeholder may suddenly discover that a scope change is about to be approved and that scope change can seriously affect him or her, perhaps politically. In such a case, the formerly unimportant stakeholder (originally deemed so for apparent lack of concern about the project) becomes a key stakeholder.

Another example occurs on longer-term projects, where stakeholders may change over time, perhaps because of politics, promotions, retirements, or reassignments. A new stakeholder may suddenly want to be an important stakeholder, whereas his or her predecessor was more of an observer. Finally, stakeholders may be relatively quiet in one life cycle phase because of limited involvement but become more active in other life cycle phase, when they must participate. The same may hold true for people who are key stakeholders in early life cycle phases and just observers in later phases. The project team must know who the stakeholders are. The team must also be able to determine which stakeholders are critical stakeholders at specific points in time. The criticality of the stakeholder determines what metrics will appear on his or her dashboard.

Stakeholder engagement is the point when the project manager physically meets with the stakeholders and determines their needs and expectations. As part of this, the project manager must:

- Understand them and their expectations
- Understand their needs
- Value their opinions
- Find ways to win their support on a continuous basis
- Identify early on any stakeholder problems that can influence the project

Even though stakeholder engagement follows stakeholder identification, it is often through stakeholder engagement that it is determined which stakeholders are supporters, advocates, neutral, or opponents. This identification may also be viewed as the first step in building a trusting relationship between the project manager and the stakeholders.

As part of stakeholder engagement, it is necessary for the project manager to understand each stakeholder's interests. One of the ways to accomplish this is to ask the stakeholders (usually the key stakeholders) what information they would like to see in performance reports. This information will help identify the KPIs needed to service that stakeholder. Each stakeholder may have a different set of KPI interests. It is a costly endeavor for the project manager to maintain multiple KPI tracking and reporting flows, but it is a necessity for successful SRM. Getting all of the stakeholders to agree on a uniform set of KPI reports and dashboards may be almost impossible.

There must be an agreement on what information is needed for each stakeholder, when the information is needed, and in what format the information will be presented. Some stakeholders may want a daily or weekly information flow, whereas others may be happy with monthly data. For the most part, the information will be provided via the internet.

TIP There is a high likelihood that stakeholders will change during the execution of the project, especially on long-term projects. This does not mean that changes in the metrics should also take place. Communication with these stakeholders is essential to see if their information needs have changed.

Project managers should use a communications matrix to carefully lay out planned stakeholder communications. Information in this matrix might include the following: the definition or title of the communication (e.g., status report, risk register), the originator, the intended recipients, the medium to be used, the rules for the access of information, and the frequency of publication or updates.

Previously, we discussed the complexities of determining the metrics for each stakeholder. Some issues that need to be addressed include:

- The potential difficulty in getting customer and stakeholder agreements on the metrics
- Determining if the metric data is in the system or needs to be collected
- Determining the cost, complexity, and timing for obtaining the data
- Considering the risks of information system changes and/or obsolescence that can affect metric data collection over the life of the project

Metrics have to be measurable, but some metric information may be difficult to quantify. For example, customer satisfaction, goodwill, and reputation may be important to some stakeholders, but they may be difficult to quantify. Some metric data may need to be measured in qualitative terms rather than quantitative terms.

The need for effective stakeholder communications is clear:

- Communicating with stakeholders on a regular basis is a necessity.
- Knowledge of the stakeholders may allow the project manager to anticipate their actions.
- Effective stakeholder communication builds trust.
- Virtual teams thrive on effective stakeholder communications.
- Although stakeholders are classified by groups or organizations, communications still occur between people.
- Ineffective stakeholder communications can cause a supporter to become a blocker.

Part of the process of stakeholder engagement involves the establishment of agreements between the individual stakeholders and the project manager, and among other stakeholders as well. These agreements must be enforced throughout the project. The project manager must identify:

- Any and all agreements among stakeholders (i.e., funding limitations, sharing of information, approval cycle for changes, etc.)
- How politics may change stakeholder agreements
- Which stakeholders may be replaced during the project (i.e., retirement, promotion, change of assignment, politics, etc.)

The project manager must be prepared for the fact that not all agreements will be honored.

Three additional critical factors must be considered for successful SRM:

1. Effective SRM takes time. It may be necessary to share this responsibility with sponsor, executives, and members of the project team.
2. Based on the number of stakeholders, it may not be possible to address their concerns face to face. The project manager must maximize his or her ability to communicate via the internet. This is also important when managing virtual teams.
3. Regardless of the number of stakeholders, documentation on the working relationships with the stakeholders must be archived. This is critical for success on future projects.

Effective stakeholder management can be the difference between an outstanding success and a terrible failure. Successful stakeholder management can result in binding agreements. The resulting benefits may be:

- Better decision making and in a more timely manner
- Better control of scope changes; prevention of unnecessary changes
- Follow-on work from stakeholders
- End user satisfaction and loyalty
- Minimization of the impact that politics can have on a project

Sometimes, regardless of how hard project managers try, they will fail at SRM. Typical reasons include:

- Inviting stakeholders to participate too early, thus encouraging scope changes and costly delays
- Inviting stakeholders to participate too late so that their views cannot be considered without costly delays
- Inviting the wrong stakeholders to participate in critical decisions, thus leading to unnecessary changes and criticism by key stakeholders
- Key stakeholders becoming disinterested in the project
- Key stakeholders who are impatient with the lack of progress
- Allowing key stakeholders to believe that their contributions are meaningless
- Managing the project with an unethical leadership style or interfacing with the stakeholders in an unethical manner

2.2 PROJECT AUDITS AND THE PMO

The necessity for a structured independent review of various parts of a business, including projects, has taken on a more important role. Part of this can be attributed to the Sarbanes-Oxley Act compliance requirements. These audits are now part of the responsibility of the project management office (PMO).

These independent reviews are audits that focus on either discovery or decision making. They also can focus on determining the health of a project. The audits can be scheduled or random and can be performed by in-house personnel or external examiners.

SITUATION: The project manager has been notified by the PMO that part of her new responsibility is to audit projects on a regular basis. In order to do this, the PMO has requested that all projects track certain metrics that are of interest to the PMO. The tracking of some of these will be costly and was not included when determining the original cost baseline.

There are several types of audits. Some common types include:

- **Performance audits:** These audits are used to appraise the progress and performance of a given project. The project manager, project sponsor, or an executive steering committee can conduct this audit.
- **Compliance audits:** These audits are usually performed by the PMO to validate that the project is using the project management methodology properly. Usually the PMO has the authority to perform the audit but may not have the authority to enforce compliance.
- **Quality audits:** These audits ensure that the planned project quality is being met and that all laws and regulations are being followed. The quality assurance group performs this audit.
- **Exit audits:** These audits are usually for projects that are in trouble and may need to be terminated. Personnel external to the project, such as an exit champion or an executive steering committee, conduct the audits.
- **Best practices audits:** These audits can be conducted at the end of each life cycle phase or at the end of the project. Some companies have found that project managers may not be the best individuals to perform such audits. In these situations, the company may have professional facilitators trained in conducting best practices reviews.
- **Metric and KPI audits:** These audits are similar to best practices audits and are used to establish a library for metrics. They are also used to validate and/or update that the metric is still of value for reporting purposes, the metric measurement technique is correct, and the metric display on a dashboard is done with the proper images.

2.3 INTRODUCTION TO SCOPE CREEP

There are three things that most project managers know will happen with almost certainty: death, taxes, and scope creep. Scope creep is the continuous enhancement of the project's requirements as the project's deliverables are being developed. Scope creep is viewed as the growth in the project's scope.

Although scope creep can occur in any project in any industry, it is most frequently associated with information systems development projects. Scope changes can occur during any project life cycle phase. Scope changes occur because it is the nature of humans not to be able to completely describe the project or the plan to execute the project at the start. This is particularly true on large, complex projects. As a result, we gain more knowledge as the project progresses, and this leads to creeping scope and scope changes.

Scope creep is a natural occurrence for project managers. We must accept the fact that this will happen. Some people believe that there are magic charms, potions, and rituals that can prevent scope creep. This is certainly not true. Perhaps the best we can do is to establish processes, such as configuration management systems or change control boards, to get some control over scope creep. However, these processes are not designed to prevent scope creep but to prevent unwanted scope changes from taking place.

Therefore, it can be argued that scope creep is not just allowing the scope to change but an indication of how well we manage changes to the scope. If all of the parties agree that a scope change is needed, then perhaps it can be argued that the scope simply changed rather than crept. Some people view scope creep as scope changes not approved by the sponsor or the change control board.

Scope creep often is viewed as detrimental to project success because it increases the cost and lengthens the schedule. Although this is true, scope creep can also produce favorable results, such as add-ons that give products a competitive advantage. Scope creep can also please the customer if the scope changes are seen as providing additional value for the final deliverable.

Defining Scope Creep

Perhaps the most critical step in the initiation phase of a project is defining the project scope. The first attempt at scope definition may occur as early as the proposal or competitive bidding stage. At this point, sufficient time and effort may not be devoted to an accurate determination or understanding of the scope and customer requirements. To make matters worse, all of this may be done well before the project manager is brought on board.

TIP Metrics can be established for the tracking of scope creep. However, the usefulness of these metrics is questionable because of the many causes of scope creep.

Once the project manager is brought on board, he or she must either become familiar with and validate the scope requirements if they have already been prepared or interview the various stakeholders and gather the necessary information for a clear understanding of the scope.

In doing so, the project manager prepares a list of what is included and excluded from his or her understanding of the requirements. Yet no matter how meticulously the project manager attempts to do this, the scope is never known with 100% clarity. This is one of the primary reasons why metrics may need to change over the life of the project.

The project manager's goal is to establish the boundaries of the scope. To do this, the project manager's vision of the project and each stakeholder's vision of the project must be aligned. There must also be an alignment with corporate business objectives because there must be a valid business reason for undertaking the project. If the alignments do not occur, then the boundary for the project will become dynamic or constantly changing rather than remaining stationary. The same can be said to hold true when selecting metrics.

Figure 2.3 shows the boundaries of the project. The project's overall boundary is designed to satisfy both business objectives established by the project manager's company and technical/scope objectives established by the customer, assuming it is an external client. The project manager and the various stakeholders, including the customer, can have different interpretations of the scope boundary and the business boundary. Also, the project manager may focus heavily on the technology that the customer needs rather than business value that the project manager's company desires. Simply stated, the project manager may seek to exceed the specifications, whereas the stakeholders and the project manager's company want to meet the minimum specification levels in the shortest amount of time.

When scope creep occurs and scope changes are necessary, the scope boundary can move. However, the scope boundary may not be

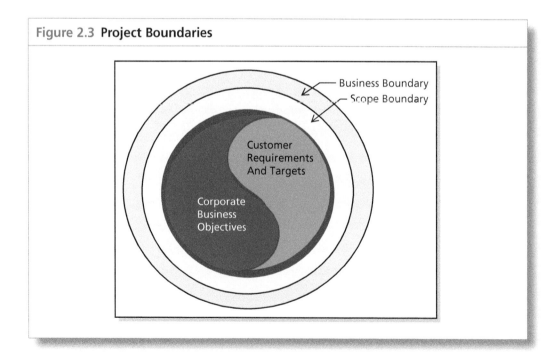

Figure 2.3 **Project Boundaries**

able to move if it alters the business boundary and corporate expectations. As an example, a scope change to add value to a product might not be approved if it extends the launch date of the product.

It is important to understand that the scope of the project is not what the customer asked for but what we agree to deliver. What we agree to can include and exclude things that the customer asked for.

There are certain facts that we now know:

- The scope boundary is what the project manager commits to delivering.
- The boundary is usually never clearly defined at the start of the project.
- Sometimes the boundary may not be clearly defined until we are well into the project.
- Progressive or rolling-wave planning (i.e. planning that becomes more detailed as the project progresses) may have to be used to clearly articulate the scope.
- Sometimes the scope is not fully known until the deliverables are completed and tested.
- Finally, even after stakeholders' acceptance of the deliverables, the interpretation of the scope boundary still can be up for debate.

The scope boundary can drift during the implementation of the project because, as we get further into the project and gain more knowledge, we identify unplanned additions to the scope. This scope creep phenomenon is then accompanied by cost increases and schedule extensions. But is scope creep really evil? Perhaps not; it is something we must live with as project managers. Some projects may be fortunate enough to avoid scope creep. In general, the larger the project, the greater the likelihood that scope creep will occur.

The length of the project also has an impact on scope creep. If the business environment is highly dynamic and continuously changing, products and services must be developed to satisfy market needs. Therefore, on long-term projects, scope creep may be seen as a necessity for keeping up with customer demands, and project add-ons may be required to obtain customer acceptance.

TIP Metrics must be established to track alignment to both the business boundary and the scope boundary. However, not all of these metrics are KPIs that are reported to the stakeholders. Most of these metrics are for the project manager's use only.

Some of the typical metrics that are used from measuring scope creep focus on looking for trends such as:

- Erratic cycle time such as on projects using an agile approach
- Lack of progress
- Chronic scope creep
- Scope creep occurring faster than the team can absorb the added work

Scope Creep Dependencies

Often, scope changes are approved without evaluating the downstream impact that the scope change can have on work packages that have not started yet. As an example, making a scope change early in the project to change the design of a component may result in a significant cost overrun if long-lead raw materials that were ordered and paid for are no longer needed. Also, there could be other contractors who have begun working on their projects, assuming that the original design was finalized. Now a small scope change by one contractor could have a serious impact on other downstream contractors. Dependencies must be considered when approving a scope change because the cost of reversing a previous decision can have a severe financial impact on the project.

Causes of Scope Creep

In order to prevent scope creep from occurring, the project manager must begin by understanding the causes of scope creep. The causes are numerous, and it is wishful thinking to believe that all of these causes can be prevented. Many of the causes are well beyond the control of the project manager, even though for some of these we can establish metrics that function as early warning signs. Some causes are related to business scope creep, and others are part of technical scope creep.

- **Poor understanding of requirements:** This occurs when a project is accepted or rushed into without fully understanding what must be done.
- **Poorly defined requirements:** Sometimes the requirements are so poorly defined that numerous assumptions must be made, and as project managers get into the later stages of the project, it is discovered that some of the assumptions are no longer valid.
- **Complexity:** The more complex the project, the greater the impact of scope creep. Being too ambitious and believing that project managers can deliver more than they can offer on a complex project can be disastrous.
- **Failing to "drill down."** When a project is initiated using only highlevel requirements, scope creep can be expected when project managers get involved in the detailed activities in the work breakdown structure.
- **Poor communications:** Poor communication between the project manager and the stakeholders can lead to ill-defined requirements and misinterpretation of the scope.
- **Misunderstanding expectations:** Regardless of how the scope is defined, stakeholders and customers have expectations of the outcome of the project. Failure to understand these expectations up front can lead to costly downstream changes.

- **Featuritis:** Also called gold-plating a project, this occurs when the project team adds in its own, often unnecessary features and functionality in the form of "bells and whistles."
- **Perfectionism:** This occurs when the project team initiates scope changes in order to exceed the specifications and requirements rather than just meet them. Project teams may see this as a chance for glory.
- **Career advancement:** Scope creep may require additional resources, perhaps making the project manager more powerful in the eyes of senior management. Scope creep also lengthens projects and provides team members with a much longer tenure in a temporary home if they are unsure about their next assignment.
- **Time-to-market pressure:** Many projects start out with highly optimistic expectations. If the business exerts pressure on the project manager to commit to an unrealistic product launch date, then he or she may need to reduce functionality. This could be less costly or even more costly based on where the descoping takes place.
- **Government regulations:** Compliance with legislation and regulatory changes can cause costly scope creep.
- **Deception:** Sometimes project managers know well in advance that the customer's statement of work has "holes" in it. Rather than inform the customer about the additional work that will be required, project managers underbid the job based on the original scope and, after contract award, push through profitable scope changes.
- **Penalty clauses:** Some contracts have penalty clauses for late delivery. By pushing through (perhaps unnecessary) scope changes that will lengthen the schedule, the project manager may be able to avoid penalty clauses.
- **Placating the customer:** Some customers will request nice-to-have-but-not-necessary scope changes after the contract begins. Although it may seem nice to placate the customer, always saying yes does not guarantee follow-on work.
- **Poor change control:** The purpose of a change control process is to prevent unnecessary changes. If the change control process is merely a rubber stamp that approves all of the project manager's requests, then continuous scope creep will occur.

Need for Business Knowledge

Scope changes must be properly targeted prior to approval and implementation, and this is the weakest link because it requires business knowledge as well as technical knowledge. As an example, scope changes should not be implemented at the expense of risking exposure to product liability lawsuits or safety issues. Likewise, making scope changes exclusively for the sake of enhancing one's image or reputation should be avoided if it could result in an unhappy client. Also, scope changes should not be implemented if the payback period for the product is drastically extended in order to capture the recovery costs of the scope change. Scope changes should be based on a solid business foundation. For example, developing a very high-quality product may seem nice at the time, but there must be customers willing to pay the higher price. The result might be a product that nobody wants or can afford.

There must exist a valid business purpose for a scope change. This purpose includes assessment of the following factors, at a minimum:

- Customer needs and the added value that the scope change will provide
- Market needs, including the time required to make the scope change, the payback period, return on investment, and whether the final product's selling price will be overpriced for the market
- Impact on the length of the project and product life cycle
- Competition's ability to imitate the scope change
- Product liability associated with the scope change and the impact on the company's image

Business Side of Scope Creep

In the eyes of the customer, scope creep is viewed as a detriment to success unless it provides add-ons or added value. For contractors, however, scope creep has long been viewed as a source of added profitability on projects. Years ago, it was common practice on some Department of Defense contracts to underbid the original contract during competitive bidding to ensure the award of the contract and then push through large quantities of lucrative scope changes. Scope creep was planned for.

Customers were rarely informed of gaps in their statements of work that could lead to scope creep. Even if the statement of work was clearly written, often it was intentionally or unintentionally misinterpreted for the benefit of seeking out profitable scope changes, whether those changes actually were needed or not. For some companies, scope changes were the prime source of corporate profitability, more so than the initial contract. During competitive bidding, executives would ask the bidding team two critical questions before submitting a bid: (1) What is our cost of doing the work we are promising? and (2) How much money can we expect from scope changes once the contract is awarded to us? Often the answer to the second question determined the size of the initial bid. In other words, contractors may plan for significant scope creep before the project even begins.

Ways to Minimize Scope Creep

Some people believe that scope creep should be prevented at all costs. However, not allowing necessary scope creep to occur can be dangerous and possibly detrimental to business objectives. Furthermore, it may be impossible to prevent scope creep. Perhaps the best that can be done is to control scope creep by minimizing its amount and extent. Some of the activities that may be helpful include these:

- **Realize that scope creep will happen:** Scope creep is almost impossible to prevent. Rather, attempts should be made to control scope creep.
- **Know the requirements:** The project manager must fully understand the requirements of the project and must communicate with the stakeholders to make sure both partners both have the same understanding.
- **Know the client's expectations:** The client and the stakeholders can have expectations that may not be in alignment with the project manager's interpretation of the requirements on scope. The project manager must understand the expectations, and continuous communication is essential.
- **Eliminate the notion that the customer is always right:** Constantly saying yes to placate the customer can cause sufficient scope creep that a good project becomes a distressed project. Some changes probably could be clustered together and accomplished later as an enhancement project.
- **Act as the devil's advocate:** The project manager should not take for granted that all change requests are necessary, even if they are internally generated by the project team. Question the necessity for the change. Make sure that there is sufficient justification for the change.
- **Determine the effect of the change:** Scope creep will affect the schedule, cost, scope/requirements, and resources. See whether some of the milestone dates can or cannot be moved. Some dates are hard to move, whereas others are easy. See if additional resources are needed to perform the scope change and if the resources will be available.
- **Get user involvement early:** Early user involvement may prevent some scope creep or at least identify the scope changes early enough such that the effects of the changes are minimal.
- **Add in flexibility:** It may be possible to add some flexibility into the budget and schedule if a large amount of scope creep is expected. This could be in the form of a management/contingency monetary reserve for cost issues and a reserve activity built into the project schedule for timing issues.
- **Know who has signature authority:** Not all members of the scope change control board possess signature authority to approve a scope change. The project manager must know who possesses this authority.

In general, people who request scope changes do not do so in an attempt to make the project manager's life miserable. They do so

TIP It is very difficult, if not impossible, to determine the real health of a project with only the traditional triple constraint metrics that are in common use today.

because of a desire to please, through a need for perfection, to add functionality, or to increase the value in the eyes of the client. Some scope changes are necessary for business reasons, such as add-ons for increased competitiveness. Scope creep is a necessity and cannot be eliminated, but it can be controlled.

2.4 PROJECT HEALTH CHECKS

Projects seem to progress quickly until they are about 60–70% complete. During that time, everyone applauds that work is progressing as planned. Then, perhaps without warning, the truth comes out, and project managers discover that the project is in trouble. This occurs because of:

- Disbelief in the value of using project's metrics
- Selecting the wrong metrics
- Fear of what project health checks may reveal

Some project managers have an incredible fixation with project metrics and numbers, believing that metrics are the holy grail in determining status. Most projects seem to focus on only two metrics: time and cost. These are the primary metrics in all earned value measurement systems. Although these two metrics *may* give the project manager a reasonable representation of where the project is today, using these two metrics to provide forecasts into the future produces gray areas and may not indicate future problem areas that could prevent a successful and timely completion of the project. At the other end of the spectrum, there are managers who have no faith in the metrics and who therefore focus on vision, strategy, leadership, and prayers.

Rather than relying on metrics alone, the simplest solution might be to perform periodic health checks on the project. In doing this, three critical questions must be addressed:

1. Who will perform the health check?
2. Will the interviewees be honest in their responses?
3. Will management and stakeholders overreact to the truth?

The surfacing of previously unknown or hidden issues could lead to loss of employment, demotions, or project cancellation. Yet project health checks offer the greatest opportunity for early corrective action to save a potentially failing project. Health checks can also discover future opportunities. It is essential to use the right metrics.

Understanding Project Health Checks

People tend to use the terms "audits" and "health checks" synonymously. Both are designed to ensure successful repeatable project outcomes, and both must be performed on projects that appear to be heading for a successful outcome as well as those that seem destined to fail. There are lessons learned and best practices that can be discovered from both successes and failures. Also, detailed analysis of a project that appears to be successful at the moment might bring to the surface issues that show that the project is really in trouble.

Table 2.2 shows some of the differences between audits and health checks. Although some of the differences may be subtle, we focus our attention on health checks.

SITUATION: During a team meeting, the project manager asks the team, "How's the work progressing?" The response is: "We're doing reasonably well. We're just a little bit over budget and a little behind schedule, but we think we've solved both issues by using lower-salaried resources for the next month and having them work overtime. According to our enterprise project management methodology, our unfavorable cost and schedule variances are still within the threshold limits, and the generation of an exception report for management is not necessary. The customer should be happy with our results thus far."

These comments are representative of a project team that has failed to acknowledge the true status of the project because they are too involved in its daily activities. Project managers, sponsors, and

TABLE 2.2 Audits versus Health Checks

VARIABLE	AUDIT	HEALTH CHECKS
Focus	On the present	On the future
Intent	Compliance	Execution effectiveness and deliverables
Timing	Generally scheduled and infrequent	Generally unscheduled and done when needed
Items to be searched	Best practices	Hidden, possible destructive issues and possible cures
Interviewer	Usually someone internal	External consultant
How interview is led	With entire team	One-on-one sessions
Time frame	Short term	Long term
Depth of analysis	Summary	Forensic review
Metrics	Use of existing or standard project metrics	Special health check metrics may be necessary

executives who are caught up in their own daily activities readily accept comments like these with blind faith, thus failing to see the big picture. If an audit had been conducted, the conclusion might have been the same—namely that the project is successfully following the enterprise project management methodology (EPMM) and that the time and cost metrics are within the acceptable limits. A forensic project health check, however, might disclose the seriousness of the issues.

Just because a project is on time and/or within the allotted budget does not guarantee success. The end result could be that the deliverable has poor quality so that it is unacceptable to the customer. In addition to time and cost, project health checks focus on quality, resources, benefits, and requirements to name just a few factors. The need for more metrics than are now used should be apparent. The true measure of the project's future success is the value that the customers see at the completion of the project. Health checks must therefore be value focused. Audits, in contrast, usually do not focus on value.

Health checks can function as an ongoing tool, being performed randomly when needed or periodically throughout various life cycle stages. However, specific circumstances indicate that a health check should be accomplished quickly. These circumstances include:

- Significant scope creep
- Escalating costs accompanied by a deterioration in value and benefits
- Schedule slippages that cannot be corrected
- Missed deadlines
- Poor morale accompanied by changes in key project personnel
- Metric measurements that fall below the threshold levels

Periodic health checks, if done correctly and using good metrics, eliminate ambiguity so that the project's true status can be determined. The benefits of health checks include these:

- Determining the current status of the project
- Identifying problems early enough that sufficient time exists to take corrective action
- Identifying the critical success factors that will support a successful outcome or the critical issues that can prevent successful delivery
- Identifying lessons learned, best practices, and critical success factors that can be used on future projects
- Evaluating compliance with and improvements to the EPMM
- Validating that the project's metrics are correct and provide meaningful data
- Identifying which activities may require or benefit from additional resources

- Identifying current and future risks as well as possible risk mitigation strategies
- Determining if the benefits and value will be there at completion
- Determining if euthanasia is required to put the project out of its misery
- Developing or recommending a fix-it plan

Some misconceptions about project health checks are described next.

- The person doing the health check does not understand the project or the corporate culture and is thus wasting time.
- The health check is too costly for the value received by performing it.
- The health check ties up critical resources in interviews.
- By the time we get the results from the health check, either it will be too late to make changes or the nature of the project may have changed.

Who Performs the Health Check?

One of the challenges facing companies is whether the health check should be conducted by internal personnel or by external consultants. The risk with using internal personnel is that they may have loyalties or relationships with people on the project team and therefore may not be totally honest in determining the true status of the project or in deciding who was at fault.

Using external consultants or facilitators is often the better choice. External facilitators can bring to the table:

- A multitude of forms, guidelines, templates, and checklists used in other companies and similar projects
- A promise of impartiality and confidentiality
- A focus on only the facts, hopefully free of politics
- An environment where people can speak freely and vent their personal feelings
- An environment that is relatively free from other day-to-day issues
- New ideas for project metrics

Life Cycle Phases

The three life cycle phases for project health checks include:

1. Review of the business case and the project's history
2. Research and discovery of the facts
3. Preparation of the health check report

Reviewing the business case and project's history may require the health check leader to have access to proprietary knowledge and financial information. This person may have to sign nondisclosure agreements and noncompete clauses before being allowed to perform the health check.

In the research and discovery phase, the health check leader prepares a list of questions that need to be answered. The list can be prepared from the *PMBOK® Guide's*[1] domain areas or areas of knowledge. The questions can also come from the knowledge repository in the consultant's company and from business case analysis templates, guidelines, checklists, or forms. The questions can change from project to project and industry to industry.

Some of the critical areas that must be investigated are listed next.

- Performance against baselines
- Ability to meet forecasts
- Benefits and value analyses
- Governance
- Stakeholder involvement
- Risk mitigation
- Contingency planning

If the health check requires one-on-one interviews, the health check leader must be able to extract the truth from interviewees who have different interpretations or conclusions about the status of the project. Some people will be truthful, whereas others either will say what they believe the interviewer wants to hear or will distort the truth as a means of self-protection.

The final phase is the preparation of the report. The report should include:

- A listing of the issues
- Root cause analyses, possibly including identification of the individuals who created the problems
- Gap analysis
- Opportunities for corrective action
- A get-well or fix-it plan

Project health checks are not "Big Brother Is Watching You" activities. Rather, they are part of project oversight. Without these health checks, the chances for project failure increase significantly. Project health checks also provide us with insight on how to keep risks under control. Performing health checks and taking corrective action early is certainly better than having to manage a distressed project.

---

1 PMBOK is a registered mark of the Project Management Institute, Inc.

2.5 **MANAGING DISTRESSED PROJECTS**

Professional sports teams treat each new season as a project. For some teams, the only definition of success is winning the championship, whereas for others, success is viewed as just a winning season. Not all teams can win the championship, but having a winning season is certainly within reach.

At the end of the season, perhaps half of the teams will have won more games than they lost. However, for the other half of the teams, those that had losing records, the season (i.e., project) was a failure. When a project failure occurs in professional sports, managers and coaches are fired, there is a shakeup in executive leadership, some players are traded or sold to other teams, and new players are brought on board. These same tactics can be used to recover failing or distressed projects in an industrial setting.

Some general facts about troubled projects are listed next.

- Some projects are doomed to fail regardless of recovery attempts.
- The chances of failure on any given project may be greater than the chances of success.
- Failure can occur in any life cycle phase; success occurs at the end of the project.
- Troubled projects do not go from "green" to "red" overnight.
- There are early warning signs, but often they are overlooked or misunderstood.
- Most companies have a poor understanding of how to manage troubled projects.
- Not all project managers possess the skills to manage a troubled project.

Not all projects will be successful. Companies that have a very high degree of project success probably are not working on enough projects and certainly are not taking on very much risk. These types of companies eventually become followers rather than leaders. For companies that desire to be leaders, knowledge on how to turn around a failing or troubled project is essential.

Projects do not get into trouble overnight. There are early warning signs, but most companies seem to overlook or misunderstand them. Some companies simply ignore the telltale signs and plow ahead, hoping for a miracle. Failure to recognize these signs early can make the downstream corrections a very costly endeavor. Also, the longer a company waits to make corrections, the more costly the changes become.

Some companies perform periodic project health checks. These health checks, when applied to healthy-seeming projects, can lead to the discovery that the project may be in trouble even though on the surface it looks healthy. Outside consultants often are hired to perform the health checks in order to get impartial assessments. The consultant rarely takes over the project once the health check is completed but may have made recommendations for recovery.

When a project gets way off track, the cost of recovery is huge, and vast or even new resources may be required for corrections. The ultimate goal for recovery is no longer to finish on time but to finish with reasonable benefits and value for the customer and the stakeholders. The project's requirements may change during recovery to meet new goals, if they have changed. Regardless of what is done, however, not all troubled projects can be recovered.

Root Causes of Failure

There are numerous causes of project failure. Some causes are quite common in specific industries, such as information technology, whereas others can appear across all industries. Here is a generic list of common causes of failure.

- End user stakeholders not involved throughout the project
- Minimal or no stakeholder backing; lack of ownership
- Weak business case
- Corporate goals not understood at lower organizational levels
- Plan asks for too much in too little time
- Poor estimates and projections, especially financial
- Unclear stakeholder requirements
- Passive user stakeholder involvement after handoff
- Unclear expectations
- Unrealistic assumptions, if they exist at all
- Plans based on insufficient data
- No systemization of the planning process
- Planning performed by a planning group
- Inadequate or incomplete requirements
- Lack of resources
- Assigned resources lack experience.
- Staffing requirements not fully known
- Constantly changing resources
- Poor overall project planning
- Changed enterprise environmental factors, causing outdated scope
- Missed deadlines and no recovery plan
- Exceeded and out-of-control budgets
- Lack of replanning on a regular basis
- Lack of attention to the human and organizational aspects of the project
- Best-guess project estimates not based on history or standards
- Not enough time provided for proper estimating
- Ignorance of the exact major milestone dates or due dates for reporting
- Team members working with conflicting requirements
- People shuffled in and out of the project with little regard for its schedule

- Poor or fragmented cost control
- Stakeholders use different organizational process assets, which may be incompatible with the assets of project partners
- Weak project and stakeholder communications
- Poor assessment of risks if done at all
- Wrong type of contract
- Poor project management; team members, especially virtual ones, possess a poor understanding of project management
- Technical objectives more important than business objectives

These causes of project failure can be sorted into three broad categories:

1. **Management mistakes:** These are the result of a failure in stakeholder management, perhaps by allowing too many unnecessary scope changes, failing to provide proper governance, refusing to make decisions in a timely manner, and ignoring the project manager's requests for help. These mistakes also can be the result of wanting to gold-plate the project, which is the result of not performing project health checks.
2. **Planning mistakes:** These are the result of poor project management, perhaps not following the principles stated in the *PMBOK® Guide*, not having a timely kill switch in the plan, not planning for project audits or health checks, and not selecting the proper tracking metrics.
3. **External influences:** These are normally failures in assessing the environmental input factors correctly. Environmental input factors include the timing for getting approvals and authorization from third parties and a poor understanding of the host country's culture and politics.

Definition of Failure

Historically, the definition of success on a project was viewed as accomplishing the work within the triple constraints and obtaining customer acceptance. Today the triple constraints are still important, but they have taken a backseat to the business and value components of success. In today's definition, success is when the planned business value is achieved within the imposed constraints and assumptions, and the customer receives the desired value.

Although most companies seem to have a reasonably good understanding of project success, the same companies seem to have a poor understanding of project failure. The project manager and the stakeholders can be working with different definitions of project failure. The project manager's definition might be just not meeting the competing constraints criteria. Stakeholders, in contrast, might seem more interested in business value than the competing constraints once the project actually begins. Here are some stakeholders' perceptions of failure:

- The project has become too costly for the expected benefits or value.
- The project will be completed too late.
- The project will not achieve its targeted benefits or value.
- The project no longer satisfies the stakeholders' needs.

Early Warning Signs of Trouble

Projects do not become distressed overnight. They normally go from "green" to "yellow" to "red," and along the way are early warning signs or metrics indicating that failure may be imminent or that immediate changes may be necessary.

Typical early warning signs include these:

- Business case deterioration
- Different opinions on project's purpose and objectives
- Unhappy/disinterested stakeholders and steering committee members
- Continuous criticism by stakeholders
- Changes in stakeholders without any warning
- No longer a demand for the deliverables or the product
- Invisible sponsorship
- Delayed decisions resulting in missed deadlines
- High-tension meetings with team and stakeholders
- Finger-pointing and poor acceptance of responsibility
- Lack of organizational process assets
- Failing to close life cycle phases properly
- High turnover of personnel, especially critical workers
- Unrealistic expectations
- Failure in progress reporting
- Technical failure
- Having to work excessive hours and with heavy workloads
- Unclear milestones and other requirements
- Poor morale
- Everything is a crisis
- Poor attendance at team meetings
- Surprises, slow identification of problems, and constant rework
- Poor change control process

The earlier the warning signs are discovered the more opportunities exist for recovery. When warning signs appear, a project health check should be conducted. Successful identification and evaluation of the early warning signs can indicate that the distressed project:

- Can succeed according to the original requirements, but some minor changes are needed
- Can be repaired, but major changes may be necessary
- Cannot succeed and should be killed

There are three possible outcomes when managing a troubled project:

1. The project must be completed; that is, it is required by law.
2. The project can be completed, but with major costly changes to the requirements.
3. The project should be canceled when:
   - Costs and benefits are no longer aligned.
   - What was once a good idea no longer has merit.

Some projects cannot be canceled because they are required by law. These include projects necessary for compliance with government laws on environmental issues, health, safety, pollution, and the like. For these projects, failure is not an option. The hardest decision to make is obviously to hit the kill switch and cancel the project. Companies that have a good grasp on project management establish processes to make it easy to kill a project that cannot be saved. There is often a great deal of political and cultural resistance to killing a project. Stakeholder management and project governance are important factors in the ease with which a project can be terminated.

Selecting the Recovery Project Manager

Companies often hire outside consultants to perform project health checks. If the health check report indicates that an attempt should be made to recover the troubled project, then perhaps a new project manager should be brought on board with skills in project recovery. Outside consultants normally do not take over troubled projects because they may not have a good grasp of the company's culture, business and project management processes, politics, and employee working relationships. Not all project managers possess the skills to be effective recovery project managers (RPMs). In addition to possessing project management knowledge, typical skills needed include:

- Strong political courage and political savvy
- A willingness to be totally honest when attacking and reporting the critical issues
- Tenacity to succeed even if it requires a change in resources
- Understanding that effective recovery is based on information, not emotions
- Ability to deal with stress, personally and with the team

Recovering a failing project is like winning the World Series of Poker. In addition to having the right poker skills, some degree of luck is also required.

Taking over a troubled project is not the same as starting up a new project. RPMs must have a good understanding of what they are about to inherit, including high levels of stress. What they inherit includes:

- A burned-out team
- An emotionally drained team
- Poor morale
- An exodus of the talented team members, who are always in high demand elsewhere
- A team that may have a lack of faith in the recovery process
- Furious customers
- Nervous management
- Invisible sponsorship and governance
- Either invisible or highly active stakeholders

Project managers who do not understand what is involved in the recovery of a troubled project can make matters worse by hoping for a miracle and allowing the death spiral to continue to a point where recovery is no longer possible. The death spiral continues if project managers:

- Force employees to work excessive hours unnecessarily
- Create unnecessary additional work
- Replace team members at inappropriate times
- Increase team stress and pressure without understanding the ramifications
- Search for new "miracle" tools to resolve some of the issues
- Hire consultants who cannot help or make matters worse by taking too long to understand the issues

Recovery Life Cycle Phases

A company's existing EPMM may not help it recover a failing project. After all, the company's standard EPMM, which may not have been appropriate for this project, may have been a contributing factor to the project's decline. It is a mistake to believe that any methodology is the miracle cure. Projects are "management by people," not tools or methodologies. A different approach may be necessary for the recovery project to succeed.

Figure 2.4 shows the typical life cycle phases for a recovery project. These phases can differ significantly from the company's standard methodology life cycle phases. The first four phases in the figure are used for problem assessment and to evaluate and, it is hoped, verify that the project may be able to be saved. The last two phases are where the actual recovery takes place.

The Understanding Phase

The purpose of the understanding phase is for the newly assigned RPM to review the project and its history. To do this, the RPM will need some form of mandate or a project charter that may be different from that of the RPM's predecessor. This mandate must be provided as quickly as possible because time is a constraint rather than a luxury. Typical questions that may be addressed in the mandate include:

Figure 2.4 Recovery Life Cycle Phases

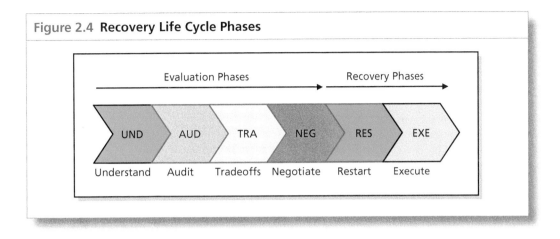

- What authority will the RPM have to access proprietary or confidential information? This includes information that may not have been available to the RPM's predecessor, such as contractual agreements and actual salaries.
- What support will the RPM be given from the sponsor and the stakeholders? Are there any indications that they will accept less-than optimal performance and a descoping of the original requirements?
- Will the RPM be allowed to interview the team members in confidence?
- Will the stakeholders overreact to brutally honest findings, even if the problems were caused by the stakeholders and governance groups?

Included in this phase are the following:

- Understanding of the project's history
- Reviewing the business case, expected benefits, and targeted value
- Reviewing the project's objectives
- Reviewing the project's assumptions
- Familiarizing oneself with the stakeholders and their needs and sensitivities
- Seeing if the enterprise environmental factors and organizational process assets are still valid

The Audit Phase

Now that there is an understanding of the project's history, the audit phase begins, which is a critical assessment of the project's existing status. The following is part of the audit phase:

- Assessing the actual performance to date
- Identifying the flaws

- Performing a root cause analysis
- Looking for surface (or easy-to-identify) failure points
  - Looking for hidden failure points
  - Determining what are the "must have," "nice to have," "can wait," and "not needed" activities or deliverables
- Looking at the issues log and seeing if the issues are people issues. If there are people issues, can people be removed or replaced?

The audit phase also includes validation that:

- The objectives are still correct.
- The benefits and value can be met but perhaps to a lesser degree.
- The assigned resources possess the proper skills.
- Roles and responsibilities are assigned to the correct team members.
- The project's priority is correct and will support the recovery efforts.
- Executive support is in place.

The recovery of a failing project cannot be done in isolation. It requires a recovery team and strong support/sponsorship.

The timing and quality of the executive support needed for recovery most often is based on the perception of the value of the project. Five important questions that need to be considered as part of value determination are listed next.

1. Is the project still of value to the client?
2. Is the project still aligned to the company's corporate objectives and strategy?
3. Is the company still committed to the project?
4. Are the stakeholders still committed?
5. Is there overall motivation for rescue?

Since recovery cannot be accomplished in isolation, it is important to interview the team members as part of the audit phase. This may very well be accomplished at the beginning of the audit phase to answer the previous questions. The team members may have strong opinions on what went wrong as well as good ideas for a quick and successful recovery. The RPM must obtain support from the team if recovery is to be successful. This support includes:

- Analyzing the culture
- Data gathering and assessment involving the full team
- Making it easy for the team to discuss problems without finger-pointing or the laying of blame
- Interviewing team members, perhaps on a one-on-one basis

- Reestablishing work-life balance
- Reestablishing incentives, if possible

It can be difficult to interview people and get their opinions on where we are, what went wrong, and how to correct it. This is especially true if the people have hidden agendas. Say the RPM has a close friend associated with the project; how will the RPM react if that friend is found guilty of being part of the problem? This is referred to as an emotional cost.

Another problem is that people may want to hide critical information if something went wrong and they could be identified with it. They might view the truth as affecting their chances for career advancement. The RPM may need a comprehensive list of questions to ask to extract the right information.

When a project gets into trouble, people tend to play the blame game, trying to make it appear that someone else is at fault. This may be an attempt to muddy the waters and detract the interviewer from the real issues. It is done out of a sense of self-preservation. It may be difficult to decide who is telling the truth and who is fabricating information.

The RPM may conclude that certain people must be removed from the project if it is to have a chance for recovery. Regardless of what the people did, they should be allowed to leave the project with dignity. The RPM might say, "Annie is being reassigned to another project that needs her skills. We thank her for the valuable contribution she has made to this project."

Perhaps the worst situation is when the RPM discovers that the real problems were with the project's governance. Stakeholders and governance groups may not receive the news that they were part of the problem well. The author's preference is always to be honest in defining the problems, even if it hurts. This response must be handled with tact and diplomacy.

The RPM also must assess the team's morale. This includes:

- Looking at the good things first to build morale
- Determining if the original plan was overly ambitious
- Determining if there were political problems that led to active or passive resistance by the team
- Determining if the work hours and workloads were demoralizing

The Trade-Off Phase

It is hoped that, by this point, the RPM has the necessary information for decision making as well as the team's support for the recovery. It may be highly unlikely that the original requirements can be met without some serious trade-offs. The RPM now must work with the team and determine the trade-off options that the RPM will present to the stakeholders.

When the project first began, the triple constraints most likely looked like what was shown previously in Figures 2.1–2.3. Time,

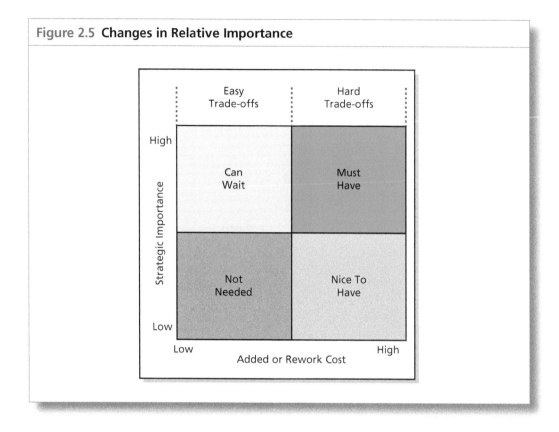

Figure 2.5 **Changes in Relative Importance**

cost, and scope were the primary constraints and trade-offs would have been made on the secondary constraints of quality, risk, value, and image/reputation. When a project becomes distressed, stakeholders know that the original budget and schedule may no longer be valid. The project may take longer and may cost significantly more money than originally thought. Therefore, the primary concerns for the stakeholders regarding whether to support the project further may change to value, quality, and image/reputation, as shown in right side of Figures 2.1–2.3. The trade-offs that the team will present to the customer and stakeholders will then be trade-offs on time, cost, scope, and possibly risk.

One way of looking at trade-offs is to review the detailed work breakdown structure and identify all activities remaining to be accomplished. The activities are then placed on the grid in Figures 2.2–2.5. The "must have" and "nice to have" work packages or deliverables are often the most costly and the hardest to use for trade-offs. If vendors are required to provide work package support, then we must perform vendor trade-offs as well, which include:

- Assessing vendor contractual agreements
- Determining if the vendor can fix the problems

- Determining if vendor concessions and trade-offs are possible
- Establishing new vendor schedules and pricing

Once all of the elements are placed on the grid in Figures 2.2–2.5, the team will assist the RPM with trade-offs by answering the following questions:

- Where are the trade-offs?
- What are the expected casualties?
- What can and cannot be done?
- What must be fixed first?
- Can we stop the bleeding?
- Have the priorities of the competing constraints changed?
- Have the features changed?
- What are the risks?

Once the trade-offs have been discovered, the RPM and the team must prepare a presentation for the stakeholders. The RPM will need to discuss two primary questions with the stakeholders:

1. Is the project worth saving? If the project is not worth saving, then the RPM must have the courage to say so. Unless a valid business reason exists for continuation, the RPM must recommend cancellation.
2. If the project is worth saving, can the stakeholders expect a full or partial recovery, and by when?

Other factors, which most likely are concerns of the stakeholders, must be addressed. These factors include changes in the:

- Political environment
- Enterprise environmental factors
- Organizational process assets
- Business case
- Assumptions
- Expected benefits and final value

TIP When things go bad and the project manager is trying to recover a potentially failing project, concessions may have to be made by allowing additional metrics and KPIs to be introduced into project. This may be the only way the project can be saved.

Existing or potential lawsuits also must be considered.

The Negotiation Phase
At this point, the RPM is ready for stakeholder negotiations. Items that must be addressed as part of stakeholder negotiations include:

- The items that are important to the stakeholders (i.e., time, cost, value, etc.)
- Prioritization of the trade-offs
- Honesty in the RPM's beliefs for recovery
- Not giving stakeholders unrealistic expectations
- Getting stakeholder buy-in
- Negotiating for the needed sponsorship and stakeholder support

SITUATION: A project manager has been placed in charge of a distressed project. She did her homework correctly and is ready for negotiations with the stakeholders and the client. They inform the project manager that they now wish to be more actively involved in the project and want additional metrics to be included, especially metrics directly related to project success and/or failure. Inserting these metrics may be costly and were not priced out as part of the original cost baseline.

Restart Phase

Assuming the stakeholders have agreed to a recovery process, the RPM is now ready to restart the project. This includes:

- Briefing the team on stakeholder negotiations
- Making sure the team learns from past mistakes
- Introducing the team to the stakeholders' agreed-upon recovery plan, including the agreed-upon milestones
- Identifying any changes to the way the project will be managed
- Fully engaging the project sponsor as well as the key stakeholders for their support
- Identifying any changes to the roles and responsibilities of the team members
- Getting the team to support the use of any new metrics that were proposed during the negotiation phase

Three restarting options include:

1. **Full anesthetic:** Bring all work to a standstill until the recovery plan is finalized.
2. **Partial anesthetic:** Bring some work to a standstill until the scope is stabilized.
3. **Scope modification:** Continue work but with modifications as necessary.

Albert Einstein once said: "We cannot solve our problems with the same thinking we used when we created them." It may be necessary to bring on board new people with new ideas. However, there are risks. The RPM may want these people full time on the project, but retaining highly qualified workers who may be in high demand elsewhere could be difficult. Since this project most likely will slip, some of the team members may be committed to other projects about to begin. However, the RPM may be lucky enough to have strong executive-level sponsorship and retain these people. This could allow the RPM to use a colocated team organization.

The Execution Phase

During the execution phase, the recovery project manager must focus on certain back-to-work implementation factors. These include:

- Learning from past mistakes
- Stabilizing scope
- Rigidly enforcing the scope change control process
- Performing periodic critical health checks and using earned value measurement reporting
- Providing effective and essential communications
- Maintaining positive morale
- Adopting proactive SRM
- Not relying on or expecting the company's EPMM to save the recovery project manager
- Not allowing unwanted stakeholder intervention, which increases pressure
- Carefully managing stakeholder expectations
- Insulating the team from politics

Recovery project management is not easy, and there is no guarantee that a project manager can or will succeed. The recovery project manager will be under close supervision and be scrutinized by superiors and stakeholders. He or she may even be required to explain all actions, but saving a potentially troubled project from disaster is certain worth the added effort.

3

METRICS

CHAPTER OVERVIEW

This chapter describes the characteristics of a metric. Not all metrics are equal. The value of a metric must be well understood in order for it to be used correctly and for it to provide the necessary information for informed decision making.

CHAPTER OBJECTIVES

- To understand the complexities in determining project status
- To understand the meaning and use of a metric
- To understand the benefits of using metrics
- To understand the components and types of metrics
- To understand the resistance to using metrics

KEY WORDS

- Information systems
- Measurement
- Metrics

3.0 INTRODUCTION

Metrics keep stakeholders informed as to the status of the project. Stakeholders must be confident that the correct metrics are used and that the measurement portrays a clear and truthful representation of the status. Metrics may determine if it is feasible to take on a certain project or if a certain course of action should be taken. Metrics can be developed to track organizational maturity in project management as well as innovation progress.

The project manager and the appropriate stakeholders must come to an agreement on which metrics to be used and how measurements will be made. There must also be agreement on which metrics will be part of the dashboard reporting system and how the metric measurement will be interpreted. Recently metrics management has taken on a much higher level of importance, so a metrics management expert may be part of the project management office (PMO).

Project Management Metrics, KPIs, and Dashboards: A Guide to Measuring and Monitoring Project Performance, Fourth Edition. Harold Kerzner.
© 2023 John Wiley & Sons, Inc. Published 2023 by John Wiley & Sons, Inc.
Companion Website: www.wiley.com/go/kerznermetrics4e

3.1 PROJECT MANAGEMENT METRICS: THE EARLY YEARS

In the early years of project management, the United States government discovered that project managers in contractors' firms were functioning more as project monitors than as project managers. Monitors simply recorded information and passed it along to higher levels of management for consideration. Project managers were reluctant to take any action as a result of negative information because they either lacked sufficient authority to implement change or did not know what actions to take. The result was often customer micromanagement of projects. Unfortunately, as projects became larger and more complex, government micromanagement became exceedingly difficult to deal with.

Determining the true project status became difficult, as shown humorously in Figure 3.1. On some larger projects, it became difficult to determine who was controlling costs. This is shown in Figure 3.2.

The solution was to get the contractors to learn and implement project management rather than project monitoring. Project managers were now expected to:

- Establish boundaries, baselines, and targets for performance.
- Measure the performance.
- Determine the variances from the baselines or targets.

Figure 3.1 **Determining Project Status**

Figure 3.2 **Who Controls Costs?**

- Develop contingency plans to reduce or eliminate unfavorable variances.
- Obtain approval of the contingency plans.
- Implement the contingency plans.
- Measure the new variances.
- Repeat the process when necessary.

For this to work, metrics would be needed. The government then created the Cost/Schedule Control Systems and later the Earned Value Measurement System (EVMS).[1] The metrics that were established as part of these systems allowed project managers to determine project status, at least what they thought was the status. As an example:

The Project

- A budget of $1.2 million.
- Time duration is 12 months.
- Production requirement is 10 deliverables.

Timeline

- Elapsed time is six months.
- Money spent to date is $700,000.
- Deliverables produced: 4 complete and 2 partial.

---

1 In 1967, DOD Instruction 7000.2 identified 35 Cost/Schedule Control System Criteria (C/SCSC). In 1997, DOD Regulation 5000.2-R identified 32 EVMS criteria.

With the EVMS metrics, project managers were able to reasonably determine status. The metrics helped them determine both current and an often questionable prediction of the future. The metrics provided an early warning system that allowed project managers sufficient time to make course corrections in small increments. The metrics emphasized prevention over cures by identifying and resolving problems early. Metrics can function as risk triggers such that the impact of downstream risks can be minimized.

The benefits of using these metrics now became abundantly clear:

- Accurate displaying of project status
- Early and accurate identification of trends
- Early and accurate identification of problems
- Reasonable determination of the project's health
- A source of critical information for controlling projects
- Basis for course corrections

The prolonged use of these metrics led project managers to identify several best practices that had to take place:

- Thorough planning of the work to be performed to complete the project
- Good estimating of time, labor, and costs
- Clear communication of the scope of the required tasks
- Disciplined budgeting and authorization of expenditures
- Timely accounting of physical progress and cost expenditures
- Frequent, periodic comparison of actual progress and expenditures to schedules and budgets, both at the time of comparison and at project completion
- Periodic reestimation of time and cost to complete the remaining work

For more than 40 years, these metrics have been treated as the gospel. Among others, here are the limitations to the use of the metrics in EVMS:

- Time and cost are basically the only two metrics. Most of the other metrics being reported are derivatives of time and cost.
- Measurements of time and cost can be inaccurate, thus leading to faulty status reports.
- The quality and value of the project cannot be calculated using time and cost metrics alone.
- Completing a project within time and cost does not imply that the project is a success.
- Time and cost information can be fudged.
- Unfavorable metrics do not necessarily mean that a project is in trouble.
- Unfavorable metrics do not provide information for corrective action.
- Customers and stakeholders do not always understand the meaning of these metrics, whether favorable or unfavorable.

The conclusion is clear. In today's project management environment where projects are becoming more complex, the metrics of EVMS, by themselves, may not be sufficient for managing projects. EVMS does not address some fundamental issues that can lead to project failure, such as unrealistic planning, poor governance, low quality of resources, and poor estimates. This is not to say that EVMS does not work but rather that additional metrics are needed. Project managers need to establish metrics that cover the big picture, namely business value to be delivered, benefits achieved, quality of the results, effort, productivity, and team performance, to name just a few. In another example, in information technology projects today, there is a push for the establishment of code-related metrics. A grouping of metrics rather than a single metric may be needed to determine performance accurately.

3.2 PROJECT MANAGEMENT METRICS: CURRENT VIEW

One of the reasons why the government created the EVMS was for standardization in status reporting. Companies then followed the government's lead, using time and cost as the primary metrics. If other metrics were considered, they were for internal use only and not often shared with customers.

Today, part of the project manager's role is to understand what critical metrics need to be identified and managed for the project to be viewed as a success by all of the stakeholders. Project managers have come to the realization that defining project-specific metrics and key performance indicators are joint ventures among the project manager, client, and stakeholders. Getting stakeholders to agree on the metrics is difficult, but it must be done as early as possible in the project.

Unlike financial metrics used for the Balanced Scorecard, project-based metrics can change during each life cycle phase as well as from project to project. This can be seen in Figure 3.3. Therefore, the establishment and measurement of metrics may be an expensive necessity to validate the critical success factors (CSFs) and maintain customer satisfaction. Many people believe that the future will be metric-driven project management.

There are several reasons for the growth in project management metrics:

- There is a need for "paperless" project management. The cost of paperwork has become quite expensive.
- New techniques, such as agile project management, are pressuring project managers to reduce costs and eliminate unnecessary waste.
- The growth in complex projects requires more frequent health checks, which, in turn, require a better understanding of metrics.

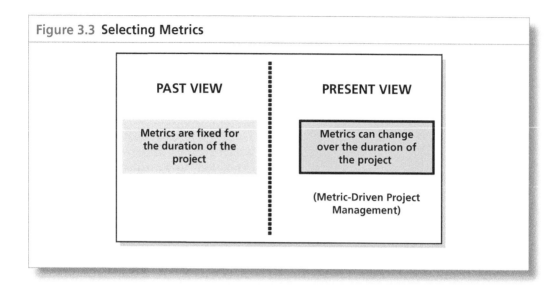

Figure 3.3 **Selecting Metrics**

- Project success is now defined as the delivery of both business value and project value. Therefore, we need metrics that can track and report business value as well as the traditional project value.
- Collecting metrics may be the only way that project managers can validate performance and success.

Metrics and Small Companies

The implementation of metrics management and the use of dashboards is relatively inexpensive. This makes it attractive to small companies. Many small companies are afraid that implementing metrics management will entail a large cost expenditure, but that certainly is not the case. A successful metrics management program builds customer loyalty, and this in turn generates new business. The only major expense might be the hiring of someone familiar with dashboard design.[2]

3.3 METRICS MANAGEMENT MYTHS

- There are several myths concerning the use of metrics for project management. Several of the myths are listed next and will be discussed in later sections.
- Metrics management is easy.
- Dashboard viewers always seem to understand the metrics they are viewing.
- Dashboards can be designed with minimal effort.

2 For an excellent paper on the application of metrics to small businesses, see "5 KPI Metrics Every Small Business Owner Should Monitor," posted by DashboardSpy around November 1, 2011 (www.DashboardSpy.com).

■ With a little effort, a perfect set of metrics for each project can be selected.
■ Once metrics are selected, anyone can perform the measurements.
■ There can never be too many metrics.
■ Project management metrics cannot change over the duration of the project.
■ Not all metrics can be measured.
■ Metrics tell users what steps to take to remedy an unfortunate situation.
■ Good project metrics should be tied to employee performance reviews.

3.4 SELLING EXECUTIVES ON A METRICS MANAGEMENT PROGRAM

Selling metrics management to executives, and receiving a timely response, requires showing both the benefits and the cost savings. There are several approaches:

■ If the company is already using financial metrics for business strategy, and having reasonable success, relate the implementation of project metrics to financial metrics.
■ Demonstrate quantitatively the cost savings from fewer meetings and less paperwork as a result of using metrics management.
■ Perform benchmarking studies to show how many other companies are using metrics management and obtaining favorable results.
■ Show the executives how it will work using a pilot project.

A Detroit-based company went to metrics management to reduce the amount of paperwork appearing on projects. Previously, executive project review meetings were preceded by the preparation of reports that included 14 steps for each page in the report:

1. Organizing
2. Writing
3. Typing
4. Proofing
5. Editing
6. Retyping
7. Graphic arts
8. Approvals
9. Reproduction
10. Classification
11. Distribution
12. Security
13. Storage
14. Disposal

The company eliminated all of these reports and instead conducted the meetings with dashboards and traffic lights for each major work package in the work breakdown structure. In the first year alone, the PMO estimated the cost savings to be more than $1 million resulting from fewer meetings and a reduction in paperwork.

While it is true that there will be a start-up cost for implementing a metrics management program, the downstream rewards and financial savings can be orders of magnitude greater than the start-up costs. Convincing an organization of this is often difficult, however, because senior managers are often resistant to challenging the status quo.

Executives must understand that successful metrics management programs require the establishment of a metrics culture. To do so:

- The organization must understand that effective metrics management is based on employee buy-in, not on purchasing software and hiring consultants.
- Executives must make sure that they do not inadvertently undermine metrics management implementation.
- Executives must be willing to see or change bad habits that can be corrected using metrics management.
- Metrics management implementation must focus on what needs to be improved the most.
- Metrics management implementation can fail if the data is not reliable.
- However, expecting 100% reliability may be unrealistic because of human error.
- The results of metrics management may require that people move out of their comfort zones.

Executives will not support a metrics management system that looks like pay for performance for executives and can affect their bonuses and chances for promotion. In such cases, executives may then select only those metrics that make them look good.

Establishing the culture may necessitate employee training that includes the following:

- Metrics management training programs must encourage participants to focus on the discovery of continuous improvement opportunities.
- Continuous improvements and best practices related to metrics and KPIs must be captured throughout the project rather than just at the end. By doing this, the benefits of the best practices can be seen rapidly.
- The participants must understand that the goal of "data mining" is to extract information about performance and transform it into an understandable display that can be used for informed decision making.
- Executive support must be visible during the training.

3.5 **UNDERSTANDING METRICS**

TIP Be prepared for new or changing metrics as the project progresses.

TIP Metrics may not provide any real value unless they can be measured.

Although most companies use some type of metrics for measurement, they seem to have a poor understanding of what constitutes a metric, at least for use in project management. There may also be a poor understanding of why we need good metrics. A project cannot be managed effectively without metrics and accompanying measurements capable of providing complete or almost complete information. Without effective metrics, project managers tend to wait until the project is far off track before considering relevant action. By that time, it may be too late to rescue it, and the only solution is cancellation. Therefore, the simplest definition of a metric is something that is measured. Consider the following:

- If it cannot be measured, then it cannot be managed.
- What gets measured gets done.
- No one ever really understands anything fully unless it can be measured.

Metrics can be measured and recorded as:

- Observations
    - Ordinal (i.e., four or five stars) and nominal (i.e., male or female) data tables
    - Ranges/sets of value
    - Simulation
    - Statistics
    - Calibration estimates and confidence limits
    - Decision models (earned value, expected value of perfect information, etc.)
    - Sampling techniques
    - Decomposition techniques
    - Human judgment

TIP When performance is measured, performance generally improves.

TIP You may never know with any reasonable degree of precision the exact performance of the project. However, good metrics can provide a close estimate.

If the project manager cannot offer a stakeholder something that can be measured, then how can he or she promise that the stakeholder's expectations will be met? What cannot be measured cannot be controlled. Good metrics lead to proactive project management rather than reactive project management, if the metrics are timely and informative. Likewise, some desired metrics simply may not work and should be abandoned.

Despite having a variety of measurement techniques, project managers must be careful when promising that they can measure efficiency, effectiveness, and productivity. Some goals are difficult to measure accurately.

For years, measurement itself was not well understood. Project managers avoided metrics management because it was not understood, but authors such as Douglas Hubbard have helped to resolve the problem. According to Hubbard:

- Your problem is not as unique as you think.
- You have more data than you think.
- You need less data than you think.
- There is useful measurement that is much simpler than you think.[3]

Not all metrics have the same time frames for measurements and life expectancies. This will impact the frequency with which metrics are measured. Also, performance does not improve immediately and can fluctuate throughout the life of the project. Just because we establish a target for measurement does not mean that it must be measured frequently. Some metrics may be measured and reported in real time, whereas others may be looked at weekly or monthly. For simplicity's sake, project-based metrics can be broken down according to the following time frames for measurement purposes:

Metrics with full project duration measurements: These are metrics, such as cost and schedule variances, that are used for the entire duration of the project and measured either weekly or monthly.

Metrics with life cycle phase measurements: These are metrics that exist only during a particular life cycle phase. As an example, metrics that track the amount or percentage of direct labor dollars used for project planning would probably be measured just in the project planning phase.

Metrics with limited life measurements: These are metrics that exist for the life of an element of work or work package. As an example, we could track the manpower staffing rate for specific work packages, or the number of deliverables produced in a specific month.

Metrics that use rolling-wave or moving-window measurements: These are metrics where the starting and finishing measurement dates can change as the project progresses. As an example, calculations for the cost performance index and schedule performance index are used to measure trends for forecasting. On long-term projects, a moving window of the most recent six data points

---

3 Douglas W. Hubbard, *How to Measure Anything; Finding the Value of Intangibles in Business* (Hoboken, NJ: John Wiley & Sons, 2007), p. 31.

(monthly measurements) may be used to obtain a linear curve fit for the trend line.

Alert metrics and measurements: These metrics are used to indicate that an out-of-tolerance condition exists. The metrics may exist just until the out-of-tolerance condition is corrected, but they may appear later on in the project if the situation appears again. Alert metrics could also be metrics that are used continuously but are highlighted differently when an out-of-tolerance condition exists.

Over the years, numerous benefits have surfaced from the use of metrics management. Some benefits are:

- Metrics tell project managers if they are hitting the targets/milestones, getting better or getting worse.
- Metrics allow project managers to catch mistakes before they lead to other mistakes; early identification of issues.
- Good metrics lead to informed decision making, whereas poor or inaccurate metrics lead to bad management decisions.
- Good metrics can assess performance accurately.
- Metrics allow for proactive management in a timely manner.
- Metrics improve future estimating.
- Metrics improve performance in the future.
- Metrics make it easier to validate baselines and maintain the baselines with minimal disruptions.
- Metrics can more accurately assess success and failure.
- Metrics can improve client satisfaction.
- Metrics are a means of assessing the project's health.
- Metrics track the ability to meet the project's CSFs.
- Good metrics allow the definition of project success to be made in terms of factors other than the traditional triple constraints.
- Metrics help in resolving crises.
- Metrics allow project managers to identify and mitigate risks.

It is important to remember that metrics are measurements and, therefore, provide project managers with opportunities for continuous improvements to the project management processes. Selecting metrics without considering a plan for future action is a waste of time and money. If a measurement indicates that the metric is significantly far away from the target, then the team must investigate the root cause of the deviation, determine what can be done to correct the deviation, get the plan to correct the deviation approved, and then implement the new plan. Metrics also allow project managers to create a database of historical information from which to analyze trends and improve future estimating.

However, there is a dark side to using metrics. Metrics management requires an understanding of human behavior. Care must be taken that the use of metrics and measurement techniques does not encourage unintended behavior. Therefore, it is important to understand the need for metrics and measurement. Some important considerations for metrics and their measurement include:

- Expecting to find meaningful measurement techniques quickly or easily is wishful thinking.
- Measurement is an observation designed to reduce uncertainty.
- Measurement supports informed decisions and informed decisions support better measurements.
- Metrics should not be measured for sake of making some measurements. Measurements should be made to identify problems quickly and then trying to fix the problem.
- Metrics can tell project managers if resources are being wasted.
- Metrics can bring forth factual evidence that a problem exists and must be solved. Metrics provide justification for tough choices.

Some projects will meet expectations while others will fail to meet expectations. Performance metrics are used to better understand the uncertainties that led to differences between planned and actual performance. Although metrics are most frequently used to validate the health of a project, they can also be used to discover best practices in the processes. It is necessary to capture best practices and lessons learned for long-term continuous improvement. Without effective use of metrics, companies could spend years trying to achieve sustained improvements. In this regard, metrics are a necessity because:

- Project approvals are often based on insufficient information and poor estimating.
- Project approvals are based on unrealistic return on investment (ROI), net present value, and payback period calculations.
- Project approvals are often based on a best-case scenario.
- The true time and cost requirements may be either hidden or not fully understood during the project approval process.

Metrics require a:

- Need or purpose
- Target, baseline, or reference point (that is meaningful rather than targets that are easy to achieve)
- Means of measurement
- Means of interpretation
- Reporting structure

Even with good metrics, metrics management can fail. The most common causes of failure are:

- Poor governance, especially by stakeholders
- Slow decision-making processes

- Overly optimistic project plans
- Trying to accomplish too much in too little time
- Poor project management practices and/or methodology
- Poor understanding of how the metrics will be used

Sometimes the failure of metrics management is the result of poor stakeholder relations management. Typical issues that can lead to failure include failing to:

- Resolve disagreements among the stakeholders
- Resolve mistrust among the stakeholders
- Define CSFs
- Get an agreement on the definition of project success
- Get an agreement on the metrics needed to support the CSFs and the definition of success
- See if the CSFs are being met
- Get an agreement on how to measure the metrics
- Understand the metrics
- Use the metrics correctly

Unless there is stakeholder agreement on how the metrics will be used to define or predict success and failure, all project managers can hope for are best guesses. Disagreements on the use of the metrics and their meaning can result in a loss of credibility for the use of metrics.

3.6 CAUSES FOR LACK OF SUPPORT FOR METRICS MANAGEMENT

During the past decade, one of the drivers for effective metrics management has been the growth in complex projects. The larger and more complex the project is, the greater the difficulty in measuring and determining success. Therefore, the larger and more complex the project is, the greater the need for metrics.

Determining the metrics requires answering certain critical questions, however:

- Measurements
  - What should be measured?
  - When should it be measured?
  - How should it be measured?
  - Who will perform the measurement?
- Collecting information and reporting
  - Who will collect the information?
  - When will the information be collected?
  - When and how will the information be reported?

For many companies, answering these questions, especially on complex projects, was a challenge. As a result, metrics were often ignored because they were hard to define and collect. There was an inherent fear in the level of detail that might be needed.

Other reasons for the lack of support included:

- Metrics management was viewed as extra work and a waste of productive time.
- There was no guarantee that the correct metrics would be selected.
- If the wrong metrics are selected, then the company is wasting time collecting the wrong data.
- Metrics management is costly, and the benefits do not justify the cost.
- Metrics are expensive and useless.
- Metrics require change, and people often dislike changing their work habits.
- Metrics encourage unintended and/or unwanted behavior.

Metrics management is often seen as an add-on to the existing work of the project team, but without these metrics, often reactive rather than proactive management is focused on. As a result, the focus is on the completion of individual work packages rather than on completion of the business solution for the client.

Everyone understands the value in using metrics, but there is still the inherent fear among team members that metrics will be seen as "Big Brother is watching you!" Employees will not support a metrics management effort that looks like a spying machine. Metrics should track the performance of a project rather than the performance of individuals. Metrics should never be used as justification for punishment.

Executives often will not support metrics management because of horror stories they heard about other companies or the fact that their pet projects will now be exposed.

3.7 USING METRICS IN EMPLOYEE PERFORMANCE REVIEWS

Job descriptions for project managers now include a requirement for some knowledge about metrics and metrics management. However, using the value of the project-based metrics, whether favorable or unfavorable, as part of employee performance reviews may be a bad idea. Some reasons for this include:

- Metrics are usually the results of more than one person's contribution. It may be impossible to isolate individual contributions.
- Unfavorable metrics may be the result of circumstances beyond the employee's control.

- The employee may fudge the numbers in the metrics to look good during performance reviews, and stakeholders may not get a true representation of the project's status.
- Real problems may be buried so that they do not appear in performance reviews.
- The person doing the performance review may not understand that the true value of the metric may not be known until sometime in the future.
- Employees working on the same project may end up competing with one another rather than collaborating, and the project's results could end up being suboptimal.

3.8 CHARACTERISTICS OF A METRIC

A metric should possess certain basic characteristics. For example, a metric should:

- Have a need or a purpose
- Provide useful information
- Focus toward a target
- Be able to be measured with reasonable accuracy
- Reflect the true status of the project
- Support proactive management
- Assist in assessing the likelihood of success or failure
- Be accepted by the stakeholders as a tool for informed decision making

There are several types of metrics. For simplicity's sake, they can be broadly identified as:

- **Results indicators (RIs):** These tell what has been accomplished.
- **Key performance indicators (KPIs):** These are the critical performance indicators that can drastically increase performance or accomplishment of the project's objectives.

Most companies use an inappropriate mix of these two types of metrics and call them all KPIs. However, there is a difference between a metric and a KPI.

- Metrics generally focus on the accomplishment of performance objectives, focusing on "Where are we today?"
- KPIs focus on future outcomes and address "Where will we end up?"

For simplicity's sake, we consider only metrics in this chapter. KPIs are discussed in more depth in Chapter 4.

Financial metrics have been used for more than a decade for analyzing business strategies. Financial metrics are business-related metrics and are based on measurements of how well business goals are being met as part of a corporate strategy. Even with the long-term use of business and financial metrics, limitations still exist:

- Metrics such as profitability tell if things look good or bad, but they do not necessarily provide meaningful information on what must be done to improve performance.
- Business-based or financial metrics are usually the result of many factors, and it is therefore difficult to isolate what must be done to implement change.
- Business-based or financial metrics are linked to long-term strategic objectives and usually do not change much.
- Words such as "customer satisfaction" and "reputation" have no real use as a metric unless they can be measured with some precision.
- Some business based metrics cannot be measured until well into the future because it is the beneficial use of the deliverable that determines success.

Over the years, most of these limitations have been overcome by using KPIs, which are specialized metrics. Although business metrics work well when focusing on a business strategy, there are significant differences between business metrics and project management metrics. This is shown in Table 3.1.

In contrast to business environments, which are long term, project environments are much shorter and, therefore, more susceptible to changing metrics. In a project environment, metrics can

TABLE 3.1   Business versus Project Management Metrics

VARIABLE	BUSINESS/FINANCIAL	PROJECT
Focus	Financial measurement	Project performance
Intent	Meeting strategic goals	Meeting project objectives, milestones, and deliverables
Reporting	Monthly or quarterly	Real-time data
Items to be looked at	Profitability, market share, repeat business, number of new customers, etc.	Adherence to competing constraints, validation and verification of performance
Length of use	Decades or longer	Life of the project
Use of the data	Information flow and changes to the strategy	Corrective action to maintain baselines
Target audience	Executive management	Stakeholders and working levels

change from project to project, during each life cycle phase, and at any time because of:

- The way the company defines value internally
- The way the customer and the contractor jointly define success and value at project's initiation
- The way the customer and contractor come to an agreement at project's initiation as to what metrics should be used on a given project
- New or updated versions of tracking software
- Improvements to the enterprise project management methodology and accompanying project management information system
- Changes in the enterprise environmental factors
- Changes in the project's business case assumptions

3.9 METRIC CATEGORIES AND TYPES

In the previous sections, metrics were defined as being either business metrics or project management metrics. This list can be expanded further to include four broad categories:

1. Business-based or financial metrics
2. Success-based metrics
3. Project-based metrics
4. Project management process metrics

Historically, metrics were most commonly used to evaluate a business strategy. Thus, typical business-based metrics included:

- ROI
- Net present value
- Payback period
- Cost reduction
- Improved efficiency
- Paperwork reduction
- Future opportunities
- Accuracy and timing of information
- Profitability
- Market share
- Sales growth rate
- Number of new customers
- Amount of repeat business

Another category includes those metrics directly related to the success of the project. Examples of these are:

- Benefits achieved
- Value achieved

- Goals/milestones achieved
- Stakeholder satisfaction
- User satisfaction

Project-based metrics can be large in number. They are discussed in more depth in Chapter 4. However, for simplicity's sake, they might include:

- Time
- Cost
- Scope and the number of scope changes
- Rate of change in the requirements (i.e., requirements growth over time)
- Quality
- Customer satisfaction with project performance
- Safety considerations
- Risk mitigation

Regardless of how metrics are selected, there must exist a tracking metric for each constraint on the project. Since the number and criticality of the constraints can change from project to project, pressure is placed on the project team to establish more sophisticated metrics for the tracking and reporting of each constraint. A constraint that cannot be monitored, measured, and reported cannot be controlled.

Project management process metrics are directly related to lessons learned and best practices. Included in this category might be metrics related to:

- Continuous improvements
- Benchmarking
- Accuracy of the estimates
- Accuracy of the measurements
- Accuracy of the targets for the metrics and the KPIs

Thus far, the four broad categories of metrics have been identified. Within each category there are and can be subcategories or types of metrics, based on how the metric will be used. As an example, these seven types of metrics or metric indicators could appear in each major category:

1. Quantitative metrics (planning dollars or hours as a percentage of total labor)
2. Practical metrics (improved efficiencies)
3. Directional metrics (risk ratings getting better or worse)
4. Actionable metrics (affect change as the number of unstaffed hours)
5. Financial metrics (profit margins, ROI, etc.)

6. Milestone metrics (number of work packages on time)
7. End result or success metrics (customer satisfaction)

As mentioned previously, and as is shown in later chapters, almost endless metrics can be established for project management applications. Only a few of these, however, can be classified as KPIs. Typical metrics that may be established as KPIs, depending on the use, include:

- Cost variance
- Schedule variance
- Cost performance index
- Schedule performance index
- Resource utilization
- Number of unstaffed hours
- Percentage of milestones missed
- Management support hours as a percentage of labor
- Planning cost as a percentage of labor
- Percentage of assumptions that have changed
- Customer loyalty
- Percentage of turnover of key workers
- Percentage of labor hours spent on overtime
- Cost per page for customer reporting

Care must be taken when setting up a classification system for metrics. Project managers should note these important factors:

- It does not matter which classification system of metrics is used as long as some system is used.
- It is important not to get stuck in a "metrics mania" mode, where significantly more metrics than needed must be created just for the sake of the metrics.
- The metrics identified must be used. If a metric cannot continue to be used, then the metric may be flawed.
- Consideration should be given to combining or removing metrics that provide little or no value.

3.10 SELECTING THE METRICS

Quite often, the wrong project metrics are selected because the selection is based on who is doing the asking. Selecting a commonly used metric is easy, but it may be inappropriate for the project at hand. The result will be useless data. Another reason why the wrong metrics are selected is the law of least resistance, whereby metrics are selected based on the ease and speed with which they can be measured.

According to Owen Head:

> When establishing the processes to be used in gathering metrics, it is important to prioritize the list in order of importance, and avoid processing any that aren't truly needed. No metric should be included, for example, unless the team will actually be able to take the time to react to it. If the team won't be able to take action in response to an undesirable metric, then it would be a waste of time to gather it. It's always possible to increase the number of metrics as the project moves forward, but attempting to take on too much at the beginning can steal time and attention from other critical project activities.[4]

Sometimes the people doing the selection do not understand the use of the metric. Metrics do not tell us what action to take or whether the success or recovery of a failing project is possible.

Figure 3.4 shows the metrics/KPI spectrum. On the left-hand side are people who want an abundance of metrics regardless of the value of the information the metrics provide. On the right hand side are the people who ignore the use of metrics because they are hard to define and collect. Finding a compromise is not easy, but project managers must determine how many metrics are needed.

With too many metrics:

- Metrics management steals time from other work.
- So much information is provided to stakeholders that they cannot determine what information is critical.
- Information that has limited value is provided.

With too few metrics:

- Not enough critical information is provided.
- Informed decision making becomes difficult.

Figure 3.4 **Metrics Value Spectrum**

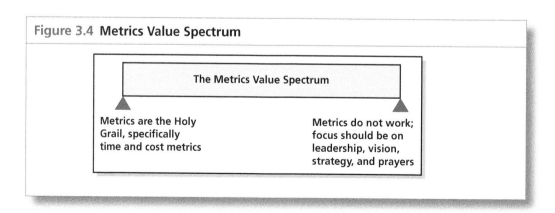

The Metrics Value Spectrum

Metrics are the Holy Grail, specifically time and cost metrics

Metrics do not work; focus should be on leadership, vision, strategy, and prayers

---

4 www.pmhut.com/a-minimalists-approach-to-project-metrics.

Project teams and stakeholders tend to select too many rather than too few metrics. According to Douglas Hubbard:

In business, only a few key variables merit deliberate measurement efforts. The rest of the variables have an "information value" at or near zero.[5]

Certain ground rules can be established as part of the metric selection process:

- Make sure the metrics selected are based on what is needed rather than what is wanted.
- If the metric does not make a difference in the actions necessary to improve the project or business, then it may be the wrong metric.
- Make sure that the metrics are worth collecting.
- Make sure that what is collected is used.
- Make sure that the metrics are informative.
- Train the team in the use and value of metrics.
- Metrics may have to be reevaluated as the project progresses to ensure that they are the correct metrics.

Selecting metrics is a lot easier when there exist competent baselines from which to make measurements. It is very difficult or even impossible to use metrics management effectively when the baselines undergo continuous transformation. For work that has not been planned yet, benchmarks and standards can be used instead of baselines.

Sometimes metric targets or goals are selected without a full understanding of how they will be measured. Fortunately, advances in measurement techniques enable just about anything to be measured, either qualitatively or quantitatively. Intangible goals may be tough to measure, but they are not unmeasurable. Tough things to measure include:

- Collaboration
- Commitment
- Creativity
- Culture
- Customer satisfaction
- Emotional maturity
- Employee morale
- Image/reputation
- Leadership effectiveness
- Motivation
- Quality of life
- Stress level
- Sustainability
- Teamwork

5 Hubbard, *How to Measure Anything*, p. 33.

TIP Effective selection of metrics cannot take place in a vacuum.

Metrics by themselves are just numbers or trends resulting from measurements. Metrics have no real value unless they can be properly interpreted by the stakeholders or subject matter experts and a corrective plan, if necessary, can be developed. It is important to know who will benefit from each metric. The level of importance can vary from stakeholder to stakeholder.

There is always the risk that the metrics information presented on the dashboard will be misunderstood and the wrong conclusions will be drawn. Some people argue that metrics should not exist without context. It is possible to use information drill-down buttons beside certain important metrics. There are two purposes for the drill-down buttons:

1. The information is optional and is used to reassure the project manager that the stakeholders understand the information that is being provided. This may be necessary in the early stages of using dashboards. The information can be in just cursory format.
2. The information is mandatory and must be provided to explain the meaning of the metric. As an example, assume a metric shows that there are action items that have been in the system for more than three months. This may not mean that the project team is derelict in their responsibility to resolve issues in a timely manner. The unresolved action items may be the result of special circumstances, such as waiting for the results of a certain test, the need for additional funding, and other such issues.

Although information drill-down buttons may be necessary, having too many or unnecessary drill-down buttons may end up creating rather than reducing paperwork. The intent of dashboards is to minimize or eliminate paperwork rather than increase it.

Several questions can be addressed during metrics selection:

- How knowledgeable are the stakeholders in project management?
- How knowledgeable are the stakeholders in metrics management?
- Do the necessary organizational process assets for metrics measurements exist?
- Will the baselines and standards undergo transformations during the project?

Two additional factors must be considered when selecting metrics. First, there is a cost involved in performing the measurements, and based on the frequency of the measurements, the costs can be quite large. Second, it must be recognized that metrics need to be updated. Metrics are like best practices; they age and may no longer provide the value or information that was expected. There are several reasons, therefore, for periodically reviewing the metrics:

- Customers may desire real-time reporting rather than periodic reporting, thus making some metrics inappropriate.
- The cost and complexity of the measurement may make a metric inappropriate for use.
- The metric does not fit well with the organizational process assets available for an accurate measurement.
- Project funding limits may restrict the number of metrics that can be used.

In reviewing the metrics, there are three possible outcomes:

1. Update a metric.
2. Leave a metric as is but possibly put it on hold.
3. Retire a metric from use.

Finally, metrics should be determined after the project is selected and approval is obtained. Selecting a project based on available or easy-to-use metrics often results in the selection of either the wrong project or metrics that provide useless data.

3.11 SELECTING A METRIC/KPI OWNER

Many of the companies that have recognized the importance of maintaining a best practices library have also created the position of a best practice owner. The concept of a KPI owner has been used for financial metrics, but only recently has it been adopted for project management metrics. Since companies are expected to maintain a metric/KPI library, they also should maintain the position of a metric/KPI owner. Each project team must accept ownership for the KPIs they use. Based on the number of KPIs, it may be advisable for each project team to assign a KPI owner to each KPI. The other choice, which is more common, is for one person in the company to be assigned as the KPI owner.

The metric/KPI owner:

- Must understand the company's culture
- Must have the respect of the labor force
- Must be able to foster support for the use of metrics
- May serve as a mentor for people using the metric
- Must perform continuous improvements on the metric, including improvements in metric measurement techniques
- Must support the PMO in determining whether the metric/KPI is still valid or has aged

3.12 METRICS AND INFORMATION SYSTEMS

It is possible on a given project to have several different project management information and reporting systems. As an example, on the same project, there can be an information system for the:

- Project manager's personal use
- Project manager's parent company
- Client and the stakeholders

There can be different metrics and KPIs in each of the information systems. The greatest number of metrics will appear in the project manager's information system. These metrics, which would be for the project manager's personal use, could include resource utilization, details related to work packages, risk-related activities, and cost-estimating accuracies. Executives in the parent company might focus on the project's profit margins, project headcount, customer satisfaction, and potential for future business. The information presented to the stakeholders is usually the KPIs, which are the critical metrics for informed decision making, and can include metrics on cost, schedule, value, and other such factors.

3.13 CRITICAL SUCCESS FACTORS

The ultimate purpose for working on projects, whether they are for internal or external clients, is to support some type of business strategy. This is shown in Figure 3.5. Once the project is selected as part of the portfolio selection of projects, a project strategy is

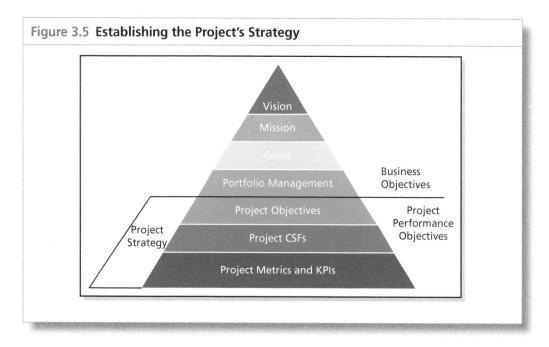

Figure 3.5 **Establishing the Project's Strategy**

developed around the project's objectives, criteria for success, and metrics/KPIs. The greater the agreement between the customer and the contractor on the project's strategy, the greater the chances for success.

The same holds true for agreements between the project manager and the stakeholders. Simply stated, it is important that everyone who has a stake in the project communicate what they believe to be success on the project.

As mentioned in Chapter 1, one of the first steps in the development of a customer–contractor relationship on a project, perhaps beginning even during the initial engagement with the client, is to come to an agreed-upon definition of success. Some companies first define success in terms of CSFs and then establish metrics and KPIs to determine whether these CSFs are being met. CSFs identify those activities necessary to meet the desired deliverables of the customer and maintain effective stakeholder relations management. Typical CSFs include:

- Adherence to schedules
- Adherence to budgets
- Adherence to quality
- Appropriateness and timing of signoffs
- Adherence to the change control process
- Add-ons to the contract
- Proper scoping of the existing environment
- Understanding the customer's requirements
- Early involvement by the customer and stakeholders
- Agreement upon and documenting of project objectives
- Provisioning all of the required resources
- Managing expectations
- Identifying all project risks and planned handling—with the customer
- Effective exception-handling process
- Control of requirements; preventing scope creep
- Explicit and specific communications with customer
- Defined processes and formalized gate reviews

Every company has its own definition of success because every company has different clients, different requirements with these clients, and different stakeholders. Some CSFs may be heavily oriented toward an internal definition of success rather than a customer's definition, although the ideal situation would be a compromise. Here are some of the CSFs at Convergent Computing,:

- Have experienced and well-rounded technical resources. These resources need to not only have outstanding technical skills, but also be good communicators, work well in challenging environments, and thrive in a team environment.

■ Make sure we understand the full range of the clients' needs, including both technical and business needs, and document a plan of action (the scope of work) for meeting these needs.

■ Have well-defined policies and processes for delivering technology services that leverage "best-practice" project management concepts and practices.

■ Have carefully crafted teams, with well-defined roles and responsibilities for the team members, designed to suit the specific needs of the client.

■ Enhance collaborations and communications both internally (within the team and from the team to chief compliance officer) and externally with our clients.

■ Leverage our experience and knowledge base as much as possible to enhance our efficiency and the quality of our deliverables.[6]

According to Bill Cattey:

> Stated simply and generically, it's important that everyone who has a stake in a project communicate about what is believed to be critical to the success of the project. Projects succeed better when there is broad agreement on what's needed for the project to be a success, and measurement is made to see if there is a gap between what is needed, and what's actually happening, and corrective action is taken to close the gap.[7]

CSFs measure the end result, usually as seen through the eyes of both the customer and the contractor. KPIs and metrics generally measure the quality of the processes used to achieve the end results and accomplish the CSFs. KPIs and metrics are internal measures and can be reviewed on a periodic basis throughout the life cycle of a project. Some people believe that CSFs are the same as metrics and KPIs and in confusion try to track them. CSFs are usually broad categories and difficult to track, whereas metrics and KPIs are more specific and are therefore more appropriate for measurement and then reporting through means such as dashboards. CSFs are often interim steps between the definition of success and the establishment of metrics.

Metrics measurements can be too costly. Even if all of the stakeholders are in agreement on the CSFs, the cost of measuring the metrics to support the CSFs can be prohibitive, and the benefits achieved may not support the cost. The measurements should pay for themselves in supporting the CSFs; overly expensive or useless measurements should not be made. Stakeholders must believe that the correct metrics were chosen and that the measurements accurately portray

---

6 H. Kerzner, *Project Management Best Practices: Achieving Global Excellence* (Hoboken, NJ: John Wiley & Sons, 2006), pp. 26–27.

7 Bill Cattey, "Project Management Metrics," http://web.mit.edu/wdc/www/project-metrics.html.

the true status. It is important to understand that some metrics for success cannot immediately determine the success of a project. True measurements may not be able to be made until well after the project is completed.

Projects often fail because the project manager and the stakeholders cannot agree on the CSFs and then may end up selecting useless metrics that cannot provide meaningful data. It is not uncommon for stakeholders to believe that having the fewest scope changes is a CSF, whereas the project manager believes that following the change control process rigidly is the true CSF, regardless of the number of scope changes. Some causes of failure include these:

- Improper understanding of the meaning of the CSFs. Failing to believe in the value of the CSFs.
- Each stakeholder is working toward his/her own definitions of the CSFs.
- Stakeholders refuse to come to an agreement on the correct CSFs for the project at hand. This can occur during the project as well as at the onset.
- Failure to understand the gap between the actual performance and the CSFs.
- Belief that the measurement costs for the metrics to support the CSFs are too great.
- Belief that measuring the metrics needed to support the CSFs is a waste of productive time.

3.14 METRICS AND THE PMO

Earlier we stated that a project cannot be managed effectively without having measurement and metrics capable of providing with complete information. Project managers do not always possess expertise in selecting the correct metrics, KPIs, and CSFs. This is where the PMO can be of assistance. With regard to metrics, the PMO can:

- Assist each project team with the development of project-based metrics
- Recognize that some metrics may be project or client specific
- Maintain a metrics library
- Develop a metric/KPI template for the library
- Recognize that metric and KPI improvements must evolve over time and that updating is necessary

As part of the PMO's responsibility as the guardian of all project management intellectual property, the PMO will coordinate the efforts for the updating of the metrics. Metric owners usually report dotted to the PMO. A periodic review of the effectiveness of each metric/KPI is essential because metrics have life cycles. As stated, the metrics can remain as is, be updated, or be retired from service. Metrics can become irrelevant without supporting data. Overseeing metrics and KPIs can be expensive and difficult. Creating unnecessary metrics should be avoided.

The PMO may have the responsibility of debriefing the project teams in order to capture lessons learned and best practices. A typical debriefing pyramid is shown in Figure 3.6. The PMO will look at metrics related to the project and use of organizational process assets, such as the enterprise project management systems, the business units, and perhaps even corporate strategic metrics.

At the same time, the PMO may evaluate all of the metrics used to see if the metrics should be part of the metrics library and if the cost of using a specific metric (i.e., measurement) was worth the effort. Figure 3.7 shows a typical way that the metrics may be evaluated. Below the risk boundary or the parity line, the value exceeds the cost, and the risks of using the metric are acceptable. Above the parity line, the measurement cost may exceed the value of using the metric, and careful risk analysis is necessary.

If the metrics are deemed to be of value, then they can be classified in the metrics library the same way that best practices often are classified. A typical example is shown in Figure 3.8. For companies that are new to project management, the focus is on promoting the use of metrics, but perhaps only a few. Companies that are reasonably mature in project management build and maintain metrics and KPI libraries.

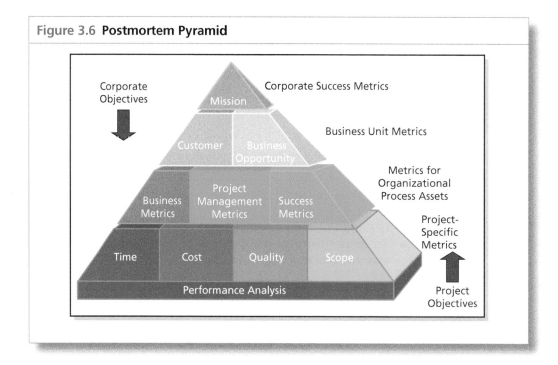

Figure 3.6 **Postmortem Pyramid**

Figure 3.7 Metric Cost versus Value

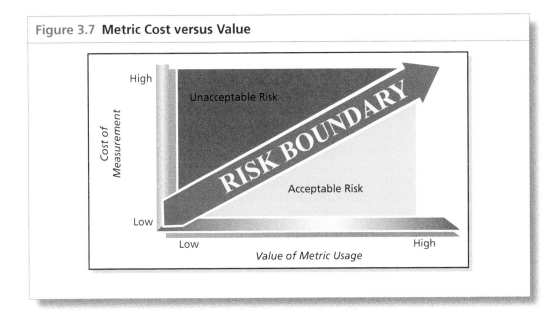

Figure 3.8 Best-Practices Classification

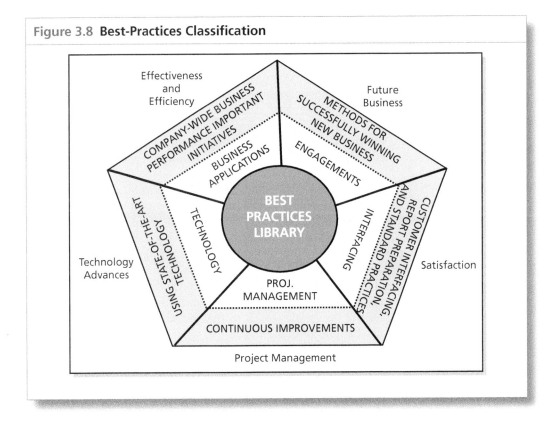

In determining the best possible metrics for a project, the PMO may find it necessary to perform metrics benchmarking. Two critical factors must be considered:

1. The project management maturity level of one's organization as well as the company against which one is benchmarking
2. The project management maturity level of the stakeholders

There are also misconceptions that must be avoided, such as:

- Metrics that work well for one company may not work well for another company, if the use of the metrics is based on in-house practices.
- Identifying the metrics is easy. Using them is difficult.
- Some metrics may be more a rough guide than a precise benchmark.

Since most PMOs are overhead rather than direct labor charges, it may be necessary for the PMO to establish its own metrics to show its contribution to the success of the company. Typical ROI metrics that the PMO use include:

- Percentage of projects using/following the enterprise project management system/framework
- Ratio of project manager to total project staff
- Customer satisfaction ratings
- Year-over-year throughput
- Percentage of projects at risk or in trouble
- Number of projects per headcount (staffing tolerance for projects)
- Ways to improve faster closure
- Percentage of scope changes per project
- Percentage of projects completed on time
- Percentage of projects completed within budget

It is important to understand that metrics management is an essential component of knowledge management and that involvement by the PMO is essential. It is very difficult to improve processes and workflow without gathering metrics and storing the results for traceability. Some companies maintain information warehouses that contain lots of data. Although much has been done to develop ways to input and store information into the warehouses, very little attention has been paid to how to extract the data that has value and is related to the metrics and KPIs that are selected. This is an area that will need attention in the future.

3.15 METRICS AND PROJECT OVERSIGHT/ GOVERNANCE

Every project governance group has its own distinguishing characteristics on its role, the types of decisions it is expected to make, and how it should interface with the project. In general, the governance group must be able to balance the risks of the project against the benefits or value of the final outcome. To do this, metrics are needed. It is important to understand what type of decision rights should be delegated to an executive oversight or governance group and what rights should go to the program team before selecting the metrics:

- **Program team:** Authority to maintain baselines
- **Oversight group:** Scope changes above a certain dollar level, additional funding, alignment to business objectives, health checks, and project/program termination (exit champion)

Although standard high-level metrics can be used for oversight groups, additional metrics may be necessary when a crisis occurs. The metrics needed to resolve a crisis are different from the traditional metrics that are used to monitor performance.

3.16 METRICS TRAPS

This chapter has discussed ways that project managers can get in trouble using the wrong metrics. These were examples of metrics traps. Fancy colors and charts displaying the metrics often lead to metrics traps. Some other common mistakes that people often make with metrics include:

- Believing that several fancy images are needed to display the same information that can be shown with one image
- Spending a great deal of time using trial-and-error solutions to determine which image is best
- Selecting metrics that cannot be measured effectively
- Promising stakeholders metrics before knowing how to perform the measurement
- Making stakeholders believe that metrics alone can predict the success or failure of a project

Metrics traps also can be created by others, such as stakeholders or members of the governance committee. Some of the more critical traps occur due to:

- Stakeholder infatuation with metrics and desire to see all of the metrics in the metrics library displayed on the dashboards
- Stakeholder requests for specific metrics that the project manager does not understand and does not have the organizational process assets to measure
- Stakeholder disagreement with what the metrics data show, causing conflicts to occur
- Stakeholders stating that they do not want to hear any bad news or see bad news displayed on the dashboards
- Stakeholders' desire to see the data before it appears on the dashboards and to filter the information such that they end up stretching the truth

3.17 **PROMOTING THE METRICS**

Most people who work on projects are motivated by seeing the results of their efforts. Promoting metrics creates awareness and support for the project, keeps people informed, and motivates workers. Historically, metrics were displayed only in the project's war room or command center. War rooms usually have one door and no windows. All of the walls are covered with charts and displays showing the health of the project. On large projects, it was common to have an assistant project manager whose job included updating all of the metrics and charts on a regular basis.

Some companies are creating a "wall of metrics" that everyone can see, including vendors and clients. Publishing these metrics outside of the traditional war room does not require additional effort, and the rewards can significantly outweigh the small cost.

3.18 **CHURCHILL DOWNS INCORPORATED'S PROJECT PERFORMANCE MEASUREMENT APPROACHES**

Since a PMO is the guardian of the company's project management intellectual property as well as the organization responsible for the project management methodology, the PMO can make it relatively painless to improve project management performance measurements. Centralizing these continuous improvement efforts in the PMO can accelerate the implementation and usage of metrics and KPIs. Such centralization can be accomplished in a simple manner rather than by using massive sophistication that could scare away possible users.

The remainder of this section has graciously been provided by Chuck Millhollan, formerly director of program management, Churchill Downs Incorporated.

The Churchill Downs Incorporated Project Management Office (PMO) has two basic premises that helped form our approach to project-related measurements. The first, and most important, is that the dashboard and report content is not as important as the discussion generated from the information. The second, with a genesis in lessons learned through our experiences, is that key performance indicators that are not defined, documented, and tracked have a much higher potential for being missed. Since most of the factors that influence what, when, and how we deliver are beyond the project manager's control, we focus on proactively managing expectations to ensure that those expectations match the delivered reality.

Since our stakeholder's define our success, we do not use project management "process" indicators to define project success. While schedule and budget targets are part of the criteria, sponsor acceptance, project completion, and ultimately project success, are based on meeting defined business objectives. To enhance the value of project performance measurement at Churchill Downs

Incorporated (CDI), we purposefully separate project management–related measurements and reporting from the benefit measurement associated with the delivered product or service. The remainder of this paper references to our project management–related metrics.

To understand our approach to defining critical-to-quality parameters and reporting processes, it is important to understand how CDI defines project success. When our PMO was chartered in April 2007, we developed a definition for project success with input from our executive leadership team. CDI considers a project as a success when the following are true:

1. Predefined business objectives and project goals were achieved or exceeded.
2. A high-quality product is fully implemented and utilized.
3. Project delivery meets or beats schedule and budget targets.
4. There are multiple winners:
    a. Project participants have pride of ownership and feel good about their work.
    b. The customer's (internal and/or external) expectations are met.
    c. Management has met its goals.
5. Project results helped build a good reputation for the project team and the product or service.
6. Methods are in place for continual monitoring and evaluation (benefit realization).

Reproduced by permission of CDI. For information on the CDI approach, contact Chuck Millhollan, PhD, MPM, PMP, PgMP, IIBA Certified Business Analysis Professional (CBAP), ASQ Certified Six Sigma Black Belt, ASQ Certified Manager of Quality/Organizational Excellence, and ASQ Certified Software Quality Engineer (Email: chuckfmill@insightbb.com).

Another key consideration when evaluating CDI's project performance reporting process is that we made a conscious decision not to invest in complex portfolio reporting tools that required enterprise acceptance and adoption. Instead, our modus operandi is to ensure that the heavy project management lifting is done behind the scenes and the information distributed leverages common desktop applications available and understood by a vast majority of our stakeholders. We are not opposed to using more sophisticated reporting tools that generate graphs, bar charts, etc.; however, we must first identify both the need and value. Finally, it is important to note that we are quick to evaluate and modify our reporting processes if we discover that senior/executive leadership is not leveraging the information to support their decision making. Our goal is to ensure that the PMO is not perceived as a score keeper, but instead that performance measurement and report is providing a defined benefit.

Toll Gates (Project Management–Related Progress and Performance Reporting)

Since PMO inception (April 2007), our project tracking process has evolved from a basic one-page document that displayed key milestones (such as Investment Council approval, completion, project health indicators, and projected completion dates) to a toll-gate process and associated graphic we refer to as the project quad. (See Figure 3.9.)

As our processes matured, so did the desire for and use of project-specific progress information. Since the initial Green/Yellow/Red project health indicators were subjective (however, valuable at the time), we had a need for more quantifiable performance and progress metrics. The project quad is divided into four intuitive sections that provide an executive summary level of information.

Quad Sections

I. **Project Overview:** This section provides project charter level information on the quantifiable business objectives, work authori-

Figure 3.9 Project Quad

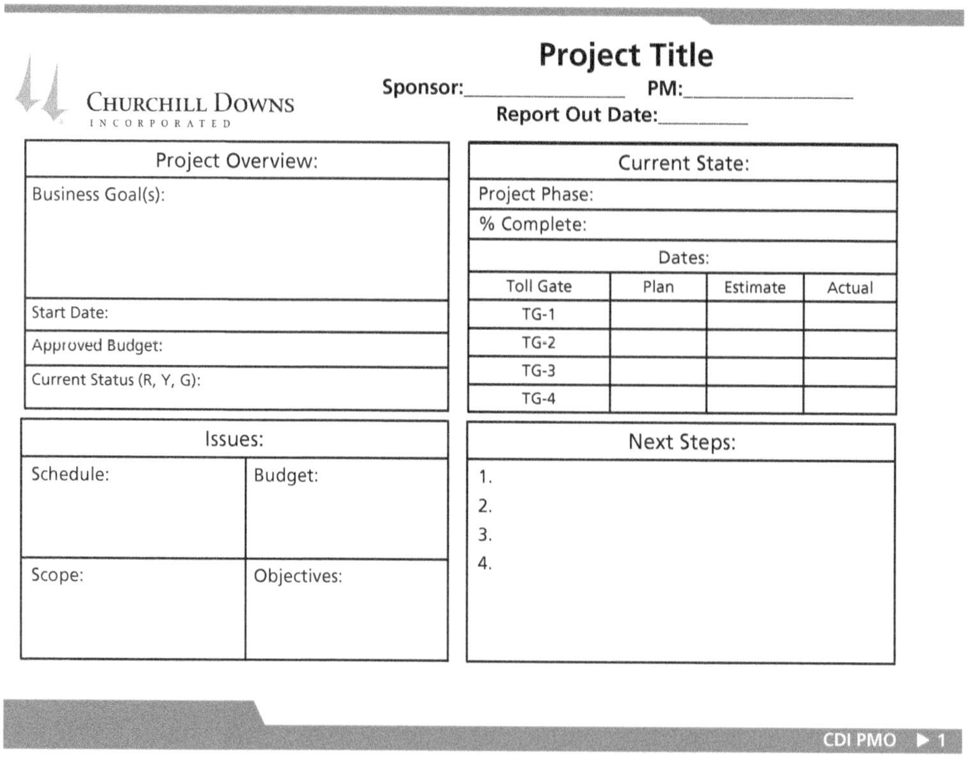
Churchill Downs Incorporated

Project Title

Sponsor:__________ PM:__________

Report Out Date:______

Project Overview:
Business Goal(s):
Start Date:
Approved Budget:
Current Status (R, Y, G):

Current State:
Project Phase:
% Complete:

Dates:			
Toll Gate	Plan	Estimate	Actual
TG-1			
TG-2			
TG-3			
TG-4			

Issues:
Schedule: Budget:
Scope: Objectives:

Next Steps:
1.
2.
3.
4.

CDI PMO ▶ 1

Figure 3.10 **Toll Gate Overview**

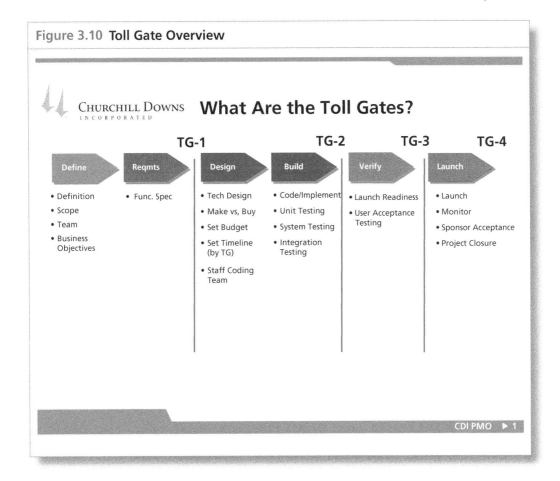

zation date, approved capital budget, and the standard project health indicator (now less subjective).

II. **Current State:** This section provides a quick reference for project progress through the defined Toll Gates and progress according to plan. The variances are immediately obvious and frequently initiate the discussion necessary to remove barriers to progress. Figure 3.10 is a training document we use to provide a high-level overview of the Toll Gates.

III. **Issues:** The project manager uses this section to provide bulletized statements about barriers to either progress or meeting the approved business objectives.

IV. **Next Steps:** This section is used to either identify the next milestones or for recommendations to bring performance back in line with the plan.

In order to support the Toll Gate process and facilitate the discussions necessary for accurate reporting and decision making, we use a standard list of questions that project managers must address as they move through the gates. Naturally, each question does not apply to each project and there is an iterative nature to many of the questions;

Figure 3.11 Toll Gate 2 Checklist

CHURCHILL DOWNS
I N C O R P O R A T E D

PROJECT NAME:

PROJECT NUMBER:

EXECUTIVE SPONSOR:

BUSINESS SPONSOR:

IT SPONSOR:

PM:

DESIGN	ANSWER	BUILD	ANSWER
Can we design & develop a solution to meet the requirements?		Are we ready for business user acceptance testing?	
What is the IT solution and it's long term supportability?		Do we have the resources to get the job done?	
How will we address any functionality gaps?		Do we have a well defined testing strategy/plans including	
Did we translate all the business requirements into detailed technical specifications?		Code Reviews & unit testing?	
Do we still have business buy in?		Performance and load testing?	
Do we have a firm project plan with commitment from all parties?		Integration testing?	
What about scope, budget, CTQs, and Benefits?		User acceptance testing?	
Does the design ensure we will meet out CTQs?		Pilot testing?	
		Did the testing show that we can meet our CTQs?	
		How will we commercialize?	

Approvals

_____ _____
Project Manager   Business Sponsor

_____ _____
Date   Date

_____
IT Sponsor

_____
Date

Figure 3.12 Project Toll Gate Dashboard

CHURCHILL DOWNS
INCORPORATED

PROJECT NAME	STATUS	TOLL GATE 1			TOLL GATE 2			TOLL GATE 3			TOLL GATE 4			NOTES / CRITICAL ISSUES
		Plan	Estimate	Actual	Plan	Estimate	Actual	Plan	Estimate	Actual	Plan	Estimate	Actual	
Enterprise Ticketing Solution	G													
Enterprise Vault	Hold													
Fair Grounds OTB Failover Testing	G													
Network Refresh	Hold													
Epiphany Service	R													
CDI Video Teleconferencing	G													
Youbet Merger Program	G													
Enterprise Printing Analysis	Y													
CDI Webex Services	G													
Twinspires Call Center Expansion	G													
Hullabalou & Ticketmaster Interface	G													
Peak-10 Clean-up	G													
Hullabalou Mobile App	G													
Digital Asset Management	G													
FGRC Oasis Upgrade	G													
SABO Derby Invitation Process Enhancements	G													
Avamar System Training	G													

however, the overall structure has proven effective for ushering projects from initiation to implementation. Figure 3.11 is the Toll Gate 2 checklist.

The nature of CDI's business makes delivery dates the primary constraint for the majority of our projects because of race meet openings dates for each track, marquee racing events (such as the Kentucky Derby), concerts, etc. During our biweekly project portfolio reviews, we use a one-page portfolio dashboard with summary-level data consisting primarily of Toll Gate delivery dates. This dashboard is used to drill down into the quads for specific projects that require additional discussion to facilitate decision making. Figure 3.12 illustrates the project dashboard used to facilitate the project reviews with the leadership team.

4 KEY PERFORMANCE INDICATORS

CHAPTER OVERVIEW

People tend to use the words "metrics" and "key performance indicators (KPIs)" interchangeably. Unfortunately, they mean different things. This chapter discusses the differences as well the role of KPIs in project management. At the end of the chapter is a white paper that provides a good summary of the information discussed in the chapter.

CHAPTER OBJECTIVES

- To understand the differences between metrics and KPIs
- To understand that KPIs are controllable factors
- To understand how to use KPIs correctly
- To understand the characteristics of a KPI
- To understand the components of a KPI
- To understand the categories of KPIs
- To understand the issues with KPI selection

KEY WORDS

- Actionable characteristics
- Critical success factors (CSFS)
- KPI owner
- Lagging indicators
- Leading indicators
- Relevant characteristics
- Smart rule
- Stakeholder classification

4.0 INTRODUCTION

As stated in previous chapters, part of the project manager's role is to understand what the critical metrics are that need to be identified, measured, reported, and managed so that the project will be viewed as a success by all of the stakeholders, if possible. The term "metric" is generic, whereas a key performance indicator (KPI) is specific. KPIs serve as early warning signs that, if an unfavorable condition exists and is not addressed, the results could be poor. KPIs and metrics can be displayed in dashboards, scorecards, and reports.

Project Management Metrics, KPIs, and Dashboards: A Guide to Measuring and Monitoring Project Performance, Fourth Edition. Harold Kerzner.
© 2023 John Wiley & Sons, Inc. Published 2023 by John Wiley & Sons, Inc.
Companion Website: www.wiley.com/go/kerznermetrics4e

Defining the correct metrics or KPIs is a joint venture of the project manager, client, and stakeholders and is a necessity in order to get stakeholder agreement. KPIs give everyone a clear picture of what is important on the project. One of the keys to a successful project is the effective and timely management of information, including the KPIs. KPIs give project managers information to make informed decisions and reduce uncertainty by managing risks.

Getting stakeholders' agreement on the KPIs is difficult. If stakeholders have 50 metrics to select from, they will somehow justify the need for all 50 of them. If they are shown 100 metrics, they will find a reason why all 100 should be reported. It is difficult to select from the metrics library those critical metrics that can function as KPIs.

For years, metrics and KPIs were used primarily as part of business intelligence (BI) systems. When applied to projects, KPIs answer this question: What is really important for different stakeholders to monitor on the project? In business, once a KPI is established, it becomes difficult to change as enterprise environmental factors change for fear that historical comparison data will be lost. Benchmarking industry KPIs, however, is still possible because the KPIs are long term. In project management, given of the uniqueness of projects, benchmarking is more complex because of the relatively short life span of the KPIs. In both business applications and projects, however, project managers assume that, if the KPI targets are being met or exceeded favorably, value is being added to the business or the project.

4.1 THE NEED FOR KPIS

Most often, items that appear in the dashboards are elements that both customers and project managers track. These items are referred to as KPIs. According to Wayne W. Eckerson:

> A KPI is a metric measuring how well the organization or an individual performs an operational, tactical or strategic activity that is critical for the current and future success of the organization.[1]

Although Eckerson's comment is more appropriate for business-oriented than for project-oriented metrics, the application to a project environment still exists. KPIs are high-level snapshots of how a project is progressing toward predefined targets. Some people confuse a KPI with a leading indicator. A leading indicator is actually a KPI that measures how the work done now will affect the future. KPIs can be treated as indicators but not necessarily leading indicators.

SITUATION: By the end of the second month of a 12-month project, the cost variance indicated that the project was over budget

---

1 Wayne W. Eckerson, *Performance Dashboards: Measuring, Monitoring and Managing Your Business* (Hoboken, NJ: John Wiley & Sons, 2006), p. 294.

by $40,000. The client then believed that if this continued until the end of the project, the final result would be a cost overrun of $240,000. The client became irate and called for a clear explanation as to why we were heading for a $240,000 cost overrun.

Although some metrics may appear to be leading indicators, care must be taken as to how they are interpreted. The misinterpretation of a metric or the mistaken belief that a metric is a leading indicator can lead to faulty conclusions.

KPIs are critical components of all earned value measurement systems. Terms such as "cost variance," "schedule variance," "schedule performance index," "cost performance index," and "time/cost at completion" are actually KPIs if used correctly but not always referred to as such. The need for these KPIs is simple: What gets measured gets done! If the goal of a performance measurement system is to improve efficiency and effectiveness, then the KPI must reflect controllable factors. There is no point in measuring an activity if the users cannot change the outcome.

For more than four decades, the only KPIs considered were time and cost or derivatives of time and cost. Today, it is realized that true project status cannot be measured from just time and cost alone. Therefore, the need for additional KPIs has grown. Sometimes several metrics and KPIs can be rolled up into a single KPI. As an example, a customer satisfaction KPI can be a composite of time, cost, quality, and effective customer communications. KPIs are often derived from formulas on how to combine these other metrics into a single KPI that may be specific and/or beneficial to a particular company.

With some metrics and KPIs, the lack of data may be good. An example is safety KPIs. However, if data is needed, then we could use the following:

■ Number of days between accidents
■ Accident rate (i.e., accidents per unit time period)

What is and is not a KPI must be defined by individual decisions unique to that project. Project managers must explain to the stakeholders the differences between metrics and KPIs and why only the KPIs should be reported on dashboards. As an example, metrics focus on the completion of work packages, achievement of milestones, and accomplishment of performance objectives. KPIs focus on future outcomes, and this is the information stakeholders need for decision making. Simply stated, with metrics, users often get bogged down looking at what happened in the past. With KPIs, users figure out how to use this data for decision making in the future. Neither metrics nor KPIs can truly predict that the project will be successful, but KPIs provide more accurate information on what might happen in the future if the existing trends continue. Both metrics and KPIs provide useful information, but neither can tell users what action to take or whether a distressed project can be recovered.

KPIs have been used in a variety of industries and for specialized purposes, such as:

- Construction
- Maintenance
- Risk management
- Safety
- Quality
- Sales
- Marketing
- Information technology (IT)
- Supply chain management
- Nonprofit organizations

TIP The project manager must explain to the dashboard users what is and is not a leading indicator and how the metrics should be interpreted.

TIP Although KPIs reflect controllable factors, not all unfavorable situations can be completely corrected. Stakeholders must be made aware of this fact.

The fastest-growing area for research in metrics appears to be IT. The risk of project failure seems to grow exponentially with the increase in the size of the IT project. Capers Jones, in his book *Assessment and Control of Software Risks* (Yourdon Press, 1994), stated that the majority of project failures in IT seem to be the result of inaccurate metrics and inadequate measurement techniques. More people today, this author included, agree with Jones's observations. Sometimes the only difference between a complete success and a complete failure is whether or not we correctly recognize the early warning signs in the metrics.

In IT, commonly used KPIs include:

- **Code:** Number of lines of code
- **Language understandability:** Language and/or code that is easy to understand and read
- **Movability/immovability:** The ease by which information can be moved
- **Complexity:** Loops, conditional statements, and so on
- **Math complexity:** Time and money needed to execute algorithms
- **Input/output understandability:** How difficult it is to understand the program
- **Antivirus and spyware:** Percentage of systems with the latest updates
- **Repairs:** Mean time to make system repairs

Once the stakeholders understand the need for correct KPIs, other questions must be discussed, including:

- How many KPIs are needed?
- How often should they be measured?

- What should be measured?
- How complex will the KPI become?
- Who will be accountable for the KPI (i.e., who owns the KPI)?
- Will the KPI serve as a benchmark?

Simply because stakeholders are interested in metrics/KPIs and dashboards does not mean that they will understand what they are viewing. It is imperative that stakeholders understand the information on the dashboards and draw the correct conclusions. The risk is that stakeholders may not understand the metric, then draw the wrong conclusion, and lose faith in the metrics concept. Getting back lost faith in a concept may take a great deal of time and become costly. Therefore, during the first few months of using metrics and dashboards, it is imperative that the project manager periodically debriefs each stakeholder to make sure he or she understands what is being viewed and that each is arriving at right conclusions.

We stated previously that what gets measured gets done and that it is through measurement that a true understanding of the information is obtained. If the goal of a metric measurement system is to improve efficiency and effectiveness, then the KPI must reflect controllable factors. There is no point in measuring an activity or a KPI if the users cannot change the outcome. Such KPIs would not be acceptable to stakeholders.

Working with stakeholders is challenging. There are complexities that must be overcome such as:

- Getting stakeholders to agree on the KPIs may be difficult, even if the stakeholders understand KPIs and possess a reasonable level of maturity in project management.
- Before agreeing to provide KPI data to a stakeholder, determine if the KPI data is in the system or needs to be collected.
- The cost, complexity, and timing for obtaining the data must be determined.
- The risks of information system changes and/or obsolescence in some of the organizational process assets that can affect the KPI data collection over the life of the project.
- Some KPIs may not appear until well into the project; over time, the stakeholders may request that additional KPIs be included in the system.

There are two other critical issues that project managers need to consider. First, if the project manager maintains multiple information systems, a measurement can appear and be treated as a KPI in one information system but be recognized as a simple metric in another. As an example, maintaining the project's profit margins might be a simple metric in the project manager's information system but a KPI in the corporate information system. Stakeholders would not be provided with this metric.

TIP KPIs in one industry may not be transferable to another industry. Even in the same industry, KPIs may be used differently in each company.

TIP KPI measurement techniques must be explained to the stakeholders to get their buy-in and approval.

The second issue involves the contractors hired. If the project manager hires consultants and contractors to assist in the management of the project, they may bring with them their own project management methodology, metrics, and KPIs. Project managers must make sure that the information they report is compatible with the project managers' business needs, especially if this information will be presented to the stakeholders as well. The contractor's definition of a KPI may not be the same as the project manager's definition.

4.2 USING THE KPIS

The ultimate purpose of a KPI is to identify what needs to be done to improve performance and keep the strategy on track. If measurements are made over short time blocks, then the team can react quickly to correct mistakes. KPIs are quantifiable measures that are agreed to beforehand that reflect critical success factors (CSFs) of an organization or a project. They measure progress toward organizational goals and strategic importance. It is a waste of precious time selecting and then tracking KPIs that cannot be controlled. As an example, a blood pressure monitor tells if blood pressure is too high (or too low), but it does not tell the doctor what to do to bring it down. KPIs help users reduce uncertainty in order to make better decisions. KPIs lead to proactive project management. Metrics are generic and imply any type of top-down measurement, whereas KPIs are bottom-up measurements. Qualitative measurements of metrics and KPIs can also be used and should be treated as "degrees" of quantitative measurements. Eventually, qualitative measurements become quantitative measurements.

KPIs must be agreed to so that everyone is on the same page. This requires answering the following:

- Are there systems in place for creating the KPIs?
- Are there systems in place for monitoring and reporting the KPIs?
- Are there systems in place for taking action when necessary?
- Will corrective action be done using collaborative problem solving?

Perhaps the most important attribute of a KPI is its relationship to "actionable." KPIs indicate that some action may be necessary to correct a bad situation or to take advantage of an opportunity. Unfortunately, often the people who see the KPIs understand what action should be taken but are not empowered to take action. The solution is to make sure that the KPIs appear on the dashboards of those who have the authority to take the appropriate steps. However, there are always situations where the viewers may fear taking action because of their beliefs and attitudes, and they may then believe that a problem does not exist, at least for now. The result can be a sacrifice of the future for the present. Even if the project manager fails to influence the behavior of current leadership, the KPIs may influence the behavior of future leadership.

The purpose of a KPI is to track performance measures that track changes toward a target. The data in the KPIs can bounce around in each reporting period. Before taking action on an isolated data point, the project manager should examine whether a trend can be identified.

Although most companies use metrics and perform measurements, they seem to have a poor understanding of what constitutes a KPI for projects and how they should be used. Some general principles regarding the use of KPIs are listed next.

- KPIs are agreed to beforehand and reflect the CSFs on the project.
- KPIs indicate how much progress has been made toward the achievement of the project's targets, goals, and objectives.
- KPIs are not performance targets.
- The ultimate purposes of a KPI are the measurement of items directly relevant to performance and the provision of information on controllable factors appropriate for decision making that will lead to positive outcomes.
- Good KPIs drive change but do not prescribe a course of action. They indicate how close a project is to a target but do not indicate what must be done to correct deviations from the target.
- KPIs assist in the establishing of objectives to be targeted with the ultimate purpose of either adding value to the project or achieving the prescribed value.
- KPIs force users to look at the future, whereas metrics alone may allow users to get bogged down looking at history.

Some people argue that the high-level purposes of a KPI are to encourage effective measurement. In this regard, the three high-level purposes are to obtain measurements that lead to:

1. Motivation of the team.
2. Compliance with use of organizational process assets and alignment to business objectives.
3. Performance improvements and the capturing of lessons learned, and best practices.
4. Some companies post KPI information on bulletin boards, in the company cafeteria, on the walls of conference rooms, or in company newsletters as a means of motivating the organization by showing progress toward that target. However, unfavorable KPIs can have an adverse effect on morale.

4.3 THE ANATOMY OF A KPI

Some metrics, such as project profitability, can tell users if things look good or bad but do not necessarily provide meaningful information on what must be done to improve performance. Therefore, a typical KPI must do more than just function as a metric. If KPIs are dissected, the following becomes clear:

- **KEY** = A major contributor to the success or failure of the project. A KPI metric is therefore only "key" when it can make or break the project.
- **PERFORMANCE** = A metric that can be measured, quantified, adjusted, and controlled. The metric must be controllable to improve performance.
- **INDICATOR** = Reasonable representation of present and future performance.

A KPI is part of a measurable objective. Defining and selecting the KPIs are much easier if the CSFs are defined first. KPIs should not be confused with CSFs. CSFs are things that must be in place to achieve an objective. A KPI is not a CSF but may provide a leading indication that the CSF can be met.

Selecting the right KPIs and the right number of KPIs will:

- Allow for better decision making
- Improve performance on the project
- Help identify problem areas faster
- Improve customer-contractor-stakeholder relations

David Parmenter defines three categories of metrics:

1. **Results indicators (RIs):** What have we accomplished?
2. **Performance indicators (PIs):** What must we do to increase or meet performance?
3. **Key performance indicators (KPIs):** What are the critical performance indicators that can drastically increase performance or accomplishment of the objectives?[2]

TIP KPIs can change over the life of a project, but CSFs usually remain the same. Changing CSFs in midstream can be devastating.

Most companies use an inappropriate mix of the three categories and call them all KPIs. Having too many KPIs can slow down projects because of excessive measurements and reporting requirements. Too many can also blur users' vision of actual performance. Too few can likewise cause delays because of the lack of

2 David Parmenter, *Key Performance Indicators* (Hoboken, NJ: John Wiley & Sons, 2007), p. 1.

critical information. Typically, project managers end up with too many rather than too few KPIs.

The number of KPIs can vary from project to project, and that number may be affected by the number of stakeholders. Some people select the number of KPIs based on the Pareto principle, which states that 20% of the total indicators will have an impact on 80% of the project. David Parmenter states that the 10/80/10 rule is usually applied when selecting the number of KPIs:[3]

- **RIs:** 10
- **PIs:** 80
- **KPIs:** 10

Typically, between 6 and 10 KPIs are standard. Factors influencing the number of KPIs include:

- The number of information systems that the project manager uses (i.e., one, two, or three).
- The number of stakeholders and their reporting requirements.
- The ability to measure the information.
- The organizational process assets available to collect the information.
- The cost of measurement and collection.
- Dashboard reporting limitations.

4.4 KPI CHARACTERISTICS

The literature abounds with articles defining the characteristics of metrics and KPIs. All too often, authors use the "SMART" rule as a means of identifying the characteristics:

S = Specific: The KPI is clear and focused toward performance targets or a business purpose.
M = Measurable: The KPI can be expressed quantitatively.
A = Attainable: The targets are reasonable and achievable.
R = Realistic or relevant: The KPI is directly pertinent to the work done on the project.
T = Time-based: The KPI is measurable within a given time period.

TIP Try to educate the stakeholders that there are limits to the number of KPIs that will be reported. This should happen prior to the actual KPI selection process.

The SMART rule was originally developed to establish meaningful objectives for projects and later adapted to the identification of metrics and KPIs. Although the use of the SMART rule does have some merit, its applicability to KPIs is questionable.

---

3 Ibid., p. 9. Pages 130–133 are adapted from Eckerson, *Performance Dashboards* (Hoboken, NJ: John Wiley & Sons, 2006), pp. 201–204.

The most important attribute of a KPI may be that it is actionable. If the trend of the metric is unfavorable, then the users should know what action is necessary to correct the unfavorable trend. The user must be able to control the outcome. This is a weakness when using the SMART rule to select KPIs.

Wayne Eckerson has developed a more sophisticated set of characteristics for KPIs. The list is more appropriate for business-oriented KPIs than project-oriented KPIs, but it can be adapted for project management usage. Table 4.1 shows Eckerson's Twelve Characteristics.

Accountability

An actionable KPI implies that an individual or group exists who "owns" the KPI, is held accountable for its results, and knows what to do when performance declines. Without accountability, measures are meaningless. Thus, it is critical to assign a single business owner to each KPI and make it part of his or her job description and performance review. It is also important to train users to interpret the KPIs and how to respond. Often this training is best done on the job by having veterans transfer their knowledge to newcomers. Some companies attach incentives to metrics, which always underscores the importance of the metric in the minds of individuals. However, just publishing performance scores among peer groups is enough to get most people's competitive juices flowing. It is best to assign accountability to an individual or small group rather than to a large

TABLE 4.1 Twelve Characteristics of Effective KPIs

1. **Strategic.** Performance metrics focuses on the outcome you want to achieve.
2. **Simple.** KPIs should be straightforward and easy to understand, not based on complex indexes that users do not know how to influence directly.
3. **Owned.** Every KPI is "owned" by an individual or group on the business side who is accountable for its outcome.
4. **Actionable.** KPIs are populated with timely, actionable data so users can intervene to improve performance before it is too late.
5. **Timely.** The KPI can be updated frequently so performance can be improved if intervention is needed.
6. **Referenceable:** The users can relate back to the origins of the use of the metric.
7. **Accurate.** The performance metric data can be measured and reported with reasonable accuracy.
8. **Correlated.** The KPI can be used to drive the desired business outcome.
9. **Game-proof.** Frequent testing and analysis on the KPI can be conducted so that the data is realistic and not fudged or circumvented due to laziness.
10. **Aligned.** KPIs are always aligned with corporate strategy and objectives.
11. **Standardized.** Everyone agrees on the definition and meaning of the KPI. KPIs are based on standard definitions, rules, and calculations so they can be integrated across dashboards throughout the organization.
12. **Relevant.** KPIs gradually lose their impact over time, so they must be periodically reviewed and refreshed.

Source: Based on Wayne W. Eckerson 2011.

group, as the sense of ownership and accountability for the metric becomes diffused in large groups.

Empowered

Companies also need to empower individuals to act on the information in a performance dashboard. This seems obvious, but many organizations that deploy performance dashboards hamstring workers by circumscribing the actions they can take to meet goals. Companies with hierarchical cultures often have difficulty here, especially when dealing with frontline workers whose actions they have historically scripted. Performance dashboards require companies to replace scripts with guidelines that give users more leeway to make the right decisions.

Timely

Actionable KPIs require right-time data. The KPI must be updated frequently enough so the responsible individual or group can intervene to improve performance before it is too late. Operational dashboards usually do this by default, but many tactical and strategic dashboards do not. Many of these latter systems contain only lagging indicators of performance and are updated only weekly or monthly. These types of performance management systems are merely electronic versions of monthly operational review meetings, not powerful tools of organizational change.

Some people argue that executives do not need actionable information because they primarily make strategic decisions for which monthly updates are good enough. However, the most powerful change agent in an organization is a top executive armed with an actionable KPI.

Trigger Points

Effective KPIs sit at the nexus of multiple interrelated processes that drive the organization. When activated, these KPIs create a ripple effect throughout the organization and produce stunning gains in performance.

For instance, late planes affect many core metrics and processes at airlines. Costs increase because airlines have to accommodate passengers who miss connecting flights; customer satisfaction declines because customers dislike missing flights; worker morale slips because workers have to deal with unruly customers; and supplier relationships are strained because missed flights disrupt service schedules and lower quality.

When an executive focuses on a single, powerful KPI, it creates a ripple effect throughout the organization and substantially changes the way an organization carries out its core operations. Managers and staff figure out ways to change business processes and behaviors so that they do not receive a career-limiting memo from the chief executive.

Easy to Understand

KPIs must be understandable. Employees must know what is being measured, how it is being calculated, and, more important, what they should do (and should not do) to affect the KPI positively. Complex KPIs that consist of indexes, ratios, or multiple calculations are difficult to understand and, more important, are not clearly actionable.

However, even with straightforward KPIs, many users struggle to understand what the KPIs really mean and how to respond appropriately. It is critical to train individuals whose performance is being tracked and follow up with regular reviews to ensure they understand what the KPIs mean and know the appropriate actions to take. This level of supervision also helps spot individuals who may be cheating the system by exploiting unforeseen loopholes.

It is also important to train people on the targets applied to metrics. For instance, is a high score good or bad? If the metric is customer loyalty, a high score is good, but if the metric is customer churn, a high score is bad. Sometimes a metric can have dual polarity, that is, a high score is good until a certain point and then it turns bad. For instance, a telemarketer who makes 20 calls per hour may be doing exceptionally well, but one who makes 30 calls per hour is cycling through clients too rapidly and possibly failing to establish good rapport with callers.

Accurate

It is difficult to create KPIs that accurately measure an activity. Sometimes, unforeseen variables influence measures. For example, a company may see a jump in worker productivity, but the increase is due more to an uptick in inflation than internal performance improvements. This is because the company calculates worker productivity by dividing revenues by the total number of workers it employs. Thus, a rise in the inflation rate artificially boosts revenues—the numerator in the metric—and increases the worker productivity score even though workers did not become more efficient during this period.

Also, it is easy to create metrics that do not accurately measure the intended objective. For example, many organizations struggle to find a metric to measure employee satisfaction or dissatisfaction. Some use surveys, but some employees do not answer the questions honestly. Others use absenteeism as a sign of dissatisfaction, but these numbers are skewed significantly by employees who miss work to attend a funeral, care for a sick family member, or stay home when daycare is unavailable. Some experts suggest that a better metric, although not a perfect one, might be the number of sick days taken since unhappy employees often take more sick days than satisfied employees.

Relevant

A KPI has a natural life cycle. When first introduced, the KPI energizes the workforce and performance improves. Over time, the KPI loses its impact and must be refreshed, revised, or discarded. Thus, it is imperative that organizations continually review KPI usage.

Performance dashboard teams should track KPI usage automatically, using system logs that capture the number of users and queries for each metric in the system. The team should then present this information to the performance dashboard steering committee, which needs to decide what to do about underused metrics.

Business or financial metrics are usually the results of many factors, and, therefore, it may be difficult to isolate what must be done to implement change. For project-oriented KPIs, the next six characteristics, which are discussed in more depth in Section 4.6, may very well be sufficient:

1. **Predictive:** The KPI is able to predict the future of this trend.
2. **Measurable:** The KPI can be expressed quantitatively.
3. **Actionable:** The KPI triggers changes that may be necessary for corrective action.
4. **Relevant:** The KPI is directly related to the success or failure of the project.
5. **Automated:** Reporting minimizes the chance of human error.
6. **Few in number:** Only what is necessary.

All of these six characteristics are not equal. It may be necessary to prioritize these characteristics based upon the project's requirements and stakeholders' needs.

4.5 CATEGORIES OF KPIS

KPIs can be segmented or clustered per industry. They can also be reported as a group. This is common for business or financial KPIs. Project-based metrics are treated differently because of their inherent differences from financial KPIs, as shown previously in Table 3.1. Unlike financial metrics used for the Balanced Scorecard, project-based metrics can change during each life cycle phase as well as from project to project. Project-based metrics may be highly specific for each project, even in similar industries, and reported individually rather than as a group. Not all KPIs can be grouped. As an example, the KPIs shown next are not easily grouped.

- Percentage of work packages adhering to the schedule.
- Percentage of work packages adhering to the budget.
- Number of assigned resources versus planned resources.
- Percentage of actual versus planned baselines completed to date.
- Percentage of actual versus planned best practices used.
- Project complexity factor.
- Customer satisfaction ratings.
- Number of critical assumptions made.
- Percentage of critical assumptions that have changed.
- Number of cost revisions.
- Number of schedule revisions.
- Number of scope change review meetings.
- Number of critical constraints.
- Percentage of work packages with a critical risk designation.
- Net operating margins.

Sometimes KPIs are categorized according to what they are intended to indicate, similar to the metrics categories discussed in Chapter 3:

- **Quantitative KPIs:** Numerical values
- **Practical KPIs:** Interfacing with company processes
- **Directional KPIs:** Getting better or worse
- **Actionable KPIs:** Effect change
- **Financial KPIs:** Performance measurements

Another means of classification might be leading, lagging, or diagnostic indicators or KPIs:

- Leading KPIs measure drivers for future performance.
- Lagging KPIs measure past performance.
- Diagnostic KPIs measure current performance.

Most dashboards have a combination of leading, lagging, and diagnostic metrics.

4.6 KPI SELECTION

Identifying KPIs or even establishing a KPI library is easy, but selecting the right KPIs can be difficult. Sometimes project managers select a KPI that at first appears to be the perfect metric, then later find out that it is actually a terrible measurement and leads to faulty conclusions by the stakeholders.

KPIs should provide some meaningful information for these four questions usually asked by executives and stakeholders:

1. Where are we today?
2. Where will we end up?

3. Where were we supposed to end up?
4. If necessary, how can we get there in a cost-effective manner without any degradation in the quality of the deliverables or major scope changes?

KPIs are used for information dissemination and must be compatible with dashboard requirements. Some critical factors that can influence the selection process are:

- Size of the dashboard
- Number of dashboards
- Number of KPIs
- Type of audience
- Audience requirements
- Audience project management maturity level

Not all team members may understand the need for KPIs. This is particularly true when using virtual teams that are unfamiliar with KPI measurement practices. Understanding the importance of a KPI is an essential part of the selection process:

- Many things are measurable but not key to the project's success. KPIs are key metrics rather than merely metrics.
- It is important that the number of KPIs be limited so that everyone is focused on the same KPIs and understands them.
- Too many KPIs may distract the project team from what is really important.
- Good metrics are essential for tracking performance toward goals. Poor or inaccurate metrics and indicators lead to bad management decisions.

Without a good understanding of a KPI, project managers may end up with an improper selection process that works as follows:

- Everything that is easy to measure and count is identified.
- Sophisticated dashboards and reporting techniques for everything that is easy to measure and count are developed.
- Then project managers struggle to determine what to do with the information.

Sometimes the selection process may be hindered by factors beyond the control of the project manager. Reasons for this might happen if the project:

TIP Selecting KPIs is easy. Selecting the *right* KPIs is difficult.

- Was bid on at a loss for political reasons.
- Was bid on at a loss with the hope of winning future contracts.

TIP Anything can be measured, but perfect measurements may be unrealistic. Therefore, it may be impossible to select a perfect set of KPIs.

TIP Buy-in by the stakeholders and the team is significantly more important than trying to select the perfect set of KPIs.

- Had its estimated budget slashed by management to win the contract.
- Had an ill-defined statement of work.
- Had a highly optimistic statement of work.

The nature of the project together with the agreed-upon definition of success and the CSFs determine which KPIs to use. Given the potential number of stakeholders, as shown in Figure 4.1, problems can occur if each stakeholder has different needs. Some of these problems are listed next.

- It may be difficult to get customer and stakeholder agreement on the KPIs.
- Project managers must determine if the KPI data is in the system or needs to be collected.
- Project managers must determine the cost, complexity, and timing for obtaining the data.

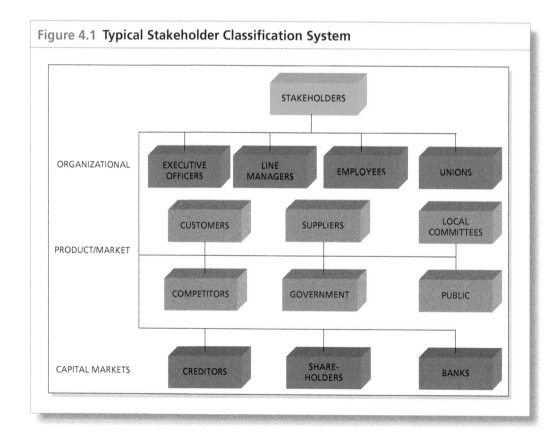

Figure 4.1 Typical Stakeholder Classification System

■ Project managers may have to consider the risks of information system changes and/or obsolescence that can have an impact on KPI data collection over the life of the project.

In a project context, any single metric can be selected as a KPI for a given project because of the relative importance of that KPI to the project manager, client, or stakeholder. For example, the following four metrics can be viewed as KPIs according to who is doing the viewing:

1. Project team morale.
2. Customer satisfaction.
3. Project profitability.
4. Performance trends, such as cost performance index (CPI) and schedule performance index (SPI).

It is possible that a given metric will function as a KPI for one stakeholder but serve as just a simple metric for another. As an example, look at Table 4.2.

The columns in Table 4.2 reflect five of the six criteria for a KPI: predictive, quantifiable, actionable, relevant, and automated. The "yes" entries in the table are subjective entries made by the project manager and possibly the team. As stated previously, the "yes" entries can vary from project to project and for each stakeholder.

The "yes" entries in the table are metrics that may have some of the characteristics of a KPI, but perhaps not all of the characteristics. For example:

■ The number of milestones missed may not be actionable because the project manager may not be able to control this.
■ Likewise, the same holds true for the number or percentage of work packages on budget, and it is unlikely that this can be used as a predictive tool for future work packages.
■ Customer loyalty falls into all categories as long as all of the stakeholders are in agreement and a viable measurement approach is undertaken.
■ Schedule Variance (SV) and Cost Variance (CV) are reasonably good indicators of the present but not as good as the Cost Performance Index (CPI) and the Schedule Performance Index (SPI) for predicting the future.
■ Changes in the risk profile can vary for each project. For a project with a reasonably low risk, this metric may not be used at all.
■ Turnover of key personnel is certainly of interest to the project manager but may or may not be of interest to all of the stakeholders. This metric may function as a KPI for only a selected number of stakeholders.

TABLE 4.2   Converting a Metric to a KPI

METRIC	PREDICTIVE	QUANTIFIABLE	ACTIONABLE	RELEVANT	AUTOMATED
Number of unstaffed hours	Yes	Yes	Yes	Yes	Yes
Number or percentage of milestones missed		Yes		Yes	Yes
Management support hours as percentage of total labor	Yes	Yes			Yes
% of work packages on budget		Yes		Yes	Yes
Number of scope changes		Yes		Yes	Yes
Changes in the risk profile	Yes	Yes	Yes	Yes	Yes
Number or percentage of assumptions that have changed		Yes		Yes	
Customer loyalty	Yes	Yes	Yes	Yes	Yes
Turnover of key personnel, number or percentage		Yes		Yes	
% of overtime labor hours		Yes	Yes		Yes
SV		Yes			Yes
CV		Yes			Yes
SPI	Yes	Yes	Yes	Yes	Yes
CPI	Yes	Yes	Yes	Yes	Yes

Therefore, using the criteria for a KPI stated previously, only 5 of the 14 metrics would be treated as true KPIs and would appear in the dashboard. The other metrics still may be reported but not necessarily through a dashboard reporting system.

Unlike business KPIs, which may remain the same for years, project-based metrics and KPIs can change for a variety of reasons and can have a short life expectancy. Metrics may be treated as KPIs at various stages of a project and replace certain KPIs that may no longer be needed or may be treated as simple metrics for the remainder of the project. When a crisis occurs, the shifting of a metric to a KPI and back may happen. This can also happen when there are changes in the stakeholders.

TABLE 4.3 Possible Viewers for Each metric			
METRIC	**PROJECT MANAGER**	**PROJECT SPONSOR**	**STAKEHOLDERS**
Number of unstaffed hours	Yes	Yes	Yes
Number or percentage of milestones missed	Yes	Yes	Yes
Management support hours as percentage of total labor	Yes	Yes	Yes
Percentage of work packages on budget	Yes	Yes	Yes
Number of scope changes	Yes		Yes
Changes in the risk profile	Yes	Yes	Yes
Number or percentage of assumptions that have changed	Yes		Yes
Customer loyalty	Yes	Yes	Yes
Turnover of key personnel, number, or percentage	Yes		
% of overtime labor hours	Yes		
SV	Yes		
CV	Yes		
SPI	Yes	Yes	Yes
CPI	Yes	Yes	Yes

Previously, it was stated that the project manager may be working with three different information systems. Some of the 14 metrics in Table 4.2 may be treated as KPIs in only one of the information systems, as shown in Table 4.3. All of the metrics, whether they are treated as KPIs or not, are of interest to the project manager. Sponsors and stakeholders may be selective in the metrics that they wish to see in their information system.

Once the KPIs are selected, certain team members must accept ownership for the KPIs they use. Depending on the number of KPIs, it may be advisable to assign a KPI owner to each KPI. However, certain KPIs, such as customer satisfaction, may not be able to be assigned to a single KPI owner.

Many companies today are maintaining KPI libraries. The KPI libraries must take into account the fact that KPIs must evolve over time. If a KPI library exists, then users must ask:

- Should there be a single owner in the organization for each KPI?
- Who, in addition to the KPI's owner, should attend the KPI review meetings?

4.7 KPI MEASUREMENT

KPIs serve no real value if they cannot be measured with any "reasonable" degree of accuracy. As Warren Buffett stated, "It is better to be approximately right than to be precisely wrong." Anything can be measured, but perfect measurements may be unrealistic. Therefore, it may be impossible to select a perfect set of KPIs. KPIs function as rough guides rather than as precise values.

The relatively slow growth in the acceptance of metrics management for projects can be directly related to a poor understanding of metric measurement. Project managers always knew what to measure but not how to measure. Sometimes teams are even allowed to invent their own measurement techniques, and the result is usually chaos. In the past, some of the most important metrics needed for effective governance were considered to be intangibles. Project managers then refused to look for ways to measure intangibles, believing that they were immeasurable. As a result, the wrong projects were selected, the wrong resources were assigned, and there was misleading status reporting and poor decision making. Refusing to measure intangibles can give us a false impression of the true value of the project.

Metrics are needed in order to better understand and, it is hoped, reduce the uncertainties involved in the project. The word "uncertainty" has slightly different meanings according to the field of study. In a project management environment, uncertainty is a state of having limited knowledge such that one may not be able to exactly describe the current or future health of the project. There can be several possible outcomes based on how work is progressing. Effective use of metrics can provide a clearer picture of the project's health and reduce the number of possible outcomes. Associated with each outcome is a risk, and each risk can have a favorable or unfavorable consequence. Metrics cannot reduce the risks, but they can provide the project team, the stakeholders, and the governance group with sufficient information such that the right decisions can be made to reduce or mitigate the risks. Therefore, the better the measurement of the metrics, the more informed the decision makers will be, the better to reduce the possible outcomes and the associated risks.

Measurement can be defined as the quantification of the uncertainties based on some type of observation. Measurement can never totally eliminate uncertainty. The person making the observation must find the proper balance between overconfidence and underconfidence when using the data, especially when reporting the measurements to the stakeholders. There can be a different impression made on stakeholders depending on whether the glass is presented as half full or half empty. Stating that the glass is half full may lead stakeholders to believe that the glass may eventually get full and the liquid level is rising. Stating that the glass is half empty may create the illusion that we have lost half of the liquid in the glass.

Even with sophisticated measurement techniques, uncertainties will still exist. It is unrealistic to believe that there will ever be perfect information and complete certainty for decision making. Even if techniques existed by which perfect information could be obtained, the cost of the measurement would probably be prohibitive. Reality forces us to live with partial information obtained in a cost-effective manner.

The organizational process assets must be capable of capturing the data necessary to make the measurement. Sometimes the method needed to capture the data has to be developed as the project progresses. In this case, all efforts should be made to get a process in place as quickly as possible.

Douglas Hubbard believes that five questions should be asked before KPIs are established for measurement:

1. What is the decision this KPI is supposed to support?
2. What really is the thing being measured by the KPI?
3. Why does this thing and the KPI matter to the decision being asked?
4. What is known about it now?
5. What is the value to measuring it further?[4]

Hubbard also identifies four useful measurement assumptions that should be considered when selecting KPIs:

- The problem in selecting a KPI is not as unique as one thinks.
- One has more data than one thinks.
- One needs less data than one thinks.
- There is a useful measurement that is much simpler than one thinks.[5]

Selecting the right KPIs is essential. On most projects, only a few KPIs are needed. Sometimes too many KPIs are selected, and users end up with some KPIs that provide little or no information value, and the KPI ends up being unnecessary or useless in assisting users in making project decisions.

KPIs are generally defined beforehand but may have to evolve as the project progresses if there are no methods or processes in place to capture the required data initially. When this happens, the result is usually the effect of measurement inversion on the KPI selection process:

- The KPI with the highest information value, especially for decision making, will be avoided or never measured because of the difficulty of data collection.

---

4 Adapted from Douglas W. Hubbard, *How to Measure Anything* (Hoboken, NJ: John Wiley & Sons, 2007), p. 43.

5 Ibid., p. 31.

- KPIs such as time and cost, which are the easiest to measure, will be selected, and often too much time is spent on these variables, which may have the least impact on decision making and the project's final value.

Today, numerous measurement techniques exist. A typical list would include:

- Observations
- Ordinal (i.e., four or five stars) and nominal (i.e., male or female) data tables
- Ranges/sets of value
- Simulation
- Statistics
- Calibration estimates and confidence limits
- Decision models (earned value, expected value of perfect information, etc.)
- Sampling techniques
- Decomposition techniques
- Human judgment
- Rules and formulas (i.e., 50/50 rule, 80/20 rule, 0/100 rule, % complete, etc.)

Regardless of which measurement technique is selected, arguments will always exist over sample size, timing of measurements, duration of measurements, accuracy and precision, who is best qualified to perform the measurements, and others.

The results of KPI measurements can create conflict if employees believe that the information will be:

- Collected on individuals and used against them (e.g. for disciplinary purposes).
- Controlled by management.
- Filtered in both content and distribution (e.g. "They show us information only when it suits their purposes.").
- Used to allocate blame for performance problems.[6]

4.8 KPI INTERDEPENDENCIES

It is almost impossible to determine the status of a project from a single metric or KPI. As shown in Figure 4.2, metrics are interlocked or related. For example, assume that metric #1 is the quality or availability of critical resources, metric #2 is time, and metric #3 is cost. Small changes in the number of qualified resources can have a significant effect on the project's budget and schedule. The effect can be favorable or unfavorable, depending on whether qualified resources are added or removed.

---

6 Parmenter, *Key Performance Indicators*, p. 64.

Figure 4.2 **Metrics Are Related**

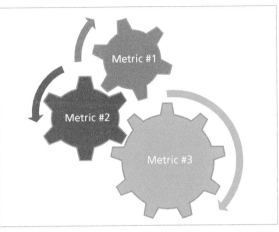

Another important factor is the rate of change of the metrics. This can be seen from the size of the metrics (i.e., wheels) in Figure 4.2. A small change in metric #1 will cause metric #2 to move faster just to keep up. Likewise, metric #3 must move faster than metric #1 and #2 to keep up. In other words, if we lack some critically skilled resources, then the schedule may slip quickly and overtime may be needed at once to produce the deliverables.

KPIs are a set of interrelated performance measures that are necessary to meet the project's CSFs. Looking at the 14 metrics in Table 4.2, the project manager may not be able to determine the actual cause of poor performance or the necessary action to correct the problem. It may be necessary to look at several interrelated metrics. As an example, consider the following two possibilities where a "+" sign is favorable and a "-" sign is unfavorable:

- SV = + and CV = -
- SV = - and CV = +

In the first bulleted item, the project manager may have worked overtime, used higher-salaried workers, or accelerated the schedule. In the second bulleted item, there may have been insufficient resources on the project. In either case, it is hard to tell if performance is good or bad.

Now let's consider another situation:

- SV = -$100,000
- CV = -$250,000

It appears that the situation is bad, but two additional metrics may offer a different picture:

- Number of approved scope changes = 34
- Turnover of critical skilled workers = 9

Combining all four of these KPIs it could be argued that the project might not be in that much trouble, at least for now.

As another example, consider what happens if project managers try to determine the status of the project from just one KPI, namely, CV:

- June: CV = -$10,000
- July: CV = -$20,000

It looks like the situation has gotten worse, since the unfavorable variance has doubled from -$10,000 to -$20,000 in one month. On the surface, this may look bad. But let us assume that in June, EV = $100,000 and in July, EV = $400,000.

If the CV is converted from dollars to percentage using the formula:

$$CV(\%) = CV(\$)/EV (\$)$$

then CV(%) for June was -10%, but CV(%) for July is -5%. In other words, the situation has actually improved even though the magnitude of the variance has increased. Several KPIs may have to be considered and integrated to get an accurate picture of the project's real status.

4.9 KPIS AND TRAINING

Project managers and team members may need to attend training sessions on KPI identification, measurement, control, and reporting. Training sessions must include:

- A comprehensive understanding of KPIs.
- How to identify KPIs.
- How to select the right display for reporting each KPI.
- How to design a project KPI database.
- How to measure each KPI.
- How to decide upon the necessary action, if appropriate, to correct performance.
- How to update the corporate KPI library.

The training should take place prior to the launch of the project. At the project's kick-off meeting, the team will then be briefed on:

- Why KPIs are being introduced.
- How KPIs will be developed.

- How KPIs will be used.
- What KPIs will not be used for.[7]

Care must be taken when launching a training course that the organization is ready for such a change. The organization must recognize:

- The value that the course brings to the organization.
- The need for learning effecting measurement techniques.
- The need for metrics to improve performance.

If the company already has some form of metrics management in place, people may be willing to accept and support the training. Otherwise, the course may be seen as a threat and do more harm than good. Simply stated, project managers must know the organization's maturity level in project management and metrics management before beginning a training course.

4.10 KPI TARGETS

Words such as "customer satisfaction" and "reputation" have no real use as metrics unless they can be measured with some precision. Therefore, we must establish KPI targets, thresholds, and baselines. Targets have the following properties:

- Targets represent a set of values against which measurements will be made.
- Targets must be realistic and not unnecessarily challenging. Otherwise, workers might try to circumvent the targets.
- Targets may require trial-and-error solutions.
- Targets must not be established in a vacuum.

It must be understood that KPIs are not targets. KPIs represent how far an important metric is above or below a predefined target. Typical targets for a KPI might be:

- Simple quantitative targets.
- Time-based targets: measurements made monthly or during a certain time interval.
- At-completion targets: measurements made at the completion of work packages or project completion.
- Stretch targets: become best in class or a target that is greater than specification requirements.
- Visionary targets well into the future: more repeat business from this client.

---

7 Ibid., p. 129.

Stretch targets and visionary targets are often a mix of possible and impossible outcomes, with the impossible outcomes providing encouragement. However, stretch targets can lead to the misinterpretation of information, as discussed in Section 4.11.

Examples of simple quantitative targets include:

- A single value (i.e., completion of 20 tests).
- An upper limit (i.e., < $200,000).
- A lower limit (i.e., > $100,000).
- A range of values (i.e., $400,000 ± 10%).
- A percentage of a specific quantity that may be fixed for the project (i.e., scrap is less than 5% of material costs).
- A percentage of a specific quantity that may change (i.e., planning dollars are less than 35% of total labor dollars).
- Accomplished milestones and deliverables (i.e., must produce and ship at least 10 deliverables each month).
- A percentage of a specific activity that may change over the duration of the project (i.e., planning dollars are not more than 35% of total labor dollars estimated).
- Accomplished milestones and deliverables (i.e., must produce and ship at least 10 deliverables each month).

Figure 4.3 represents a KPI target or boundary box. Normal performance is meeting the target ± 10%. If the company was more than 20% below the target, urgent attention would be required.

It is important to ask stakeholders how much integrity is acceptable for reporting purposes. In Figure 4.3, the threshold or integrity of ± 10% may result from a joint agreement with the client. Typical questions might be:

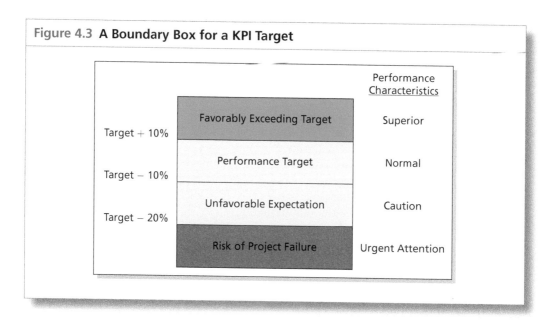

Figure 4.3 A Boundary Box for a KPI Target

- Is the target ± 5% acceptable?
- Is the target ± 10% acceptable?
- Are integrity guidelines established in the project's business case?
- Are integrity guidelines established as part of the enterprise project management system?

Some targets are very difficult to establish such as value targets. Establishing KPIs to identify current and future value is difficult but not impossible. Value-driven project KPIs can be selected by addressing these questions:

- How can I show that the project is creating value for the client?
- How will the client and the stakeholders perceive the value measurements?
- Can I show that the project will also create value for my parent company?

This topic is discussed in more detail in Chapter 5. Some people argue that customer satisfaction surveys are value-reflective KPIs. Figure 4.4 represents a simple customer satisfaction instrument that Mahindra Satyam, its author, refers to as the Customer Delight Index. Companies use this index with the belief that customer satisfaction may lead to repeat business.

The colors are important, as is discussed in Chapter 6. The color green represents something favorable, whereas red indicates something not so good.

When using these types of KPIs, it is the trend that is important rather than a single data point. If the trend shows that the customer satisfaction index is getting worse, then templates or checklists may exist to show what actions the project manager may take to change

Figure 4.4 Mahindra Satyam Customer Delight Index *Source: © 2010Mahindra Satyam. All rights reserved*

SYMBOL	MEANING
○	Data not entered
●	Dissatisfied
△	Satisfied
▢	Delighted

the trend to a more favorable indication. The problem is in determining who or which department has the lead role in the improvement of customer satisfaction.

4.11 UNDERSTANDING STRETCH TARGETS

Performance measurements are an indication of whether a company is moving quickly enough toward a target. However, simply because work is progressing does not mean that work is being accomplished at the desired rate. The tracking information that is monitored would be more meaningful if results or accomplishments are measured rather than the progress along the activities. For example, if workers can produce 10 units in one day and they have a target of 8 units per day, then the workers may be performing their tasks at a slow speed. They are being paid for subpar performance because a poor target was established.

If the target is 11 units per day, the workers may be motivated to see if they can perform better than expected. However, if a target of 15 units per day is established and workers believe that this is unrealistic, they may be demoralized to the extent that they will produce only 9 units per day. If workers try to meet the target, they may ignore unintended consequences, take high-risk shortcuts, violate safety protocols, or even manipulate the numbers to make it appear that the target was reached. Establishing the right target is essential.

Establishing "stretch" targets to encourage continuous improvements in performance is an acceptable approach as long as the correct stretch target is established and the results are reported correctly. The following points must be understood concerning stretch targets:

- An important purpose of any target is to find continuous improvement opportunities. However, stretching too far from reality can be bad.
- Stretch targets are usually established as big, hairy, audacious goals (BHAGs).
- BHAGs move people out of their comfort zones. Some people dislike this even if the BHAG is achievable.
- Failure in achieving the BHAG may lead to lower self-esteem, embarrassment, and frustration.
- In reality, even if the BHAG is not achieved, the actual accomplishment may be the achievement of more than was hoped for otherwise.

Let us look at an example. A customer awards a company a one-year contract to deliver 8 units per month for the first four months and 10 units per month for the remaining eight months of the contract. As the project manager, you are convinced that the project team can easily produce 10 units per month beginning in

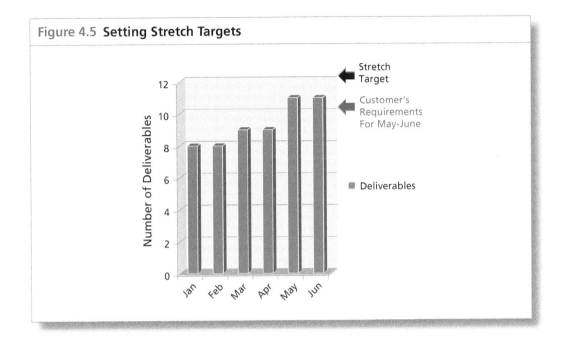

Figure 4.5 **Setting Stretch Targets**

the fifth month, but you establish a BHAG of 12 units per month for the last eight months of the contract, hoping to complete the project ahead of schedule so that the resources can be assigned to other projects.

As seen in Figures 4.4–4.5, you are not achieving the stretch target of 12 units per month beginning in May, but you are exceeding the customer's requirements. How should this be reported as a dashboard metric? If executives see the stretch target, they may believe that you are behind schedule when, in fact, you are actually ahead of schedule. In this case, as seen in Figures 4.4–4.6, it is best to report how far you pull forward from the customer's requirements rather than how far you are behind the BHAG.

4.12 KPI FAILURES

There are several reasons why the use of KPIs often fails on projects. Some of the reasons include:

- People believe that the tracking of a KPI ends at the first line manager level.
- The actions needed to regulate unfavorable indications are beyond the control of the employees doing the monitoring or tracking.
- The KPIs are not related to the actions or work of the employees doing the monitoring.

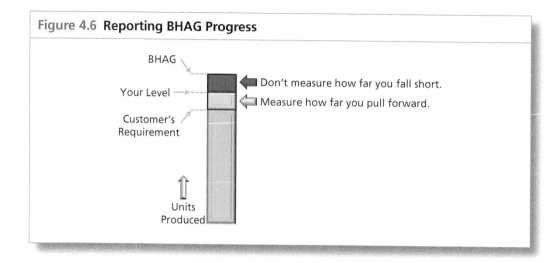

Figure 4.6 Reporting BHAG Progress

- The rate of change of the KPIs is too slow, thus making them unsuitable for managing the daily work of the employees.
- Actions needed to correct unfavorable KPIs take too long.
- Measurement of the KPIs does not provide enough meaning or data to make them useful.
- The company identifies too many KPIs, to the point where confusion reigns among the people doing the measurements.

Years ago, the only metrics that some companies used were those identified as part of the earned value measurement system. The metrics generally focused only on time and cost and neglected metrics related to business success as opposed to project success. Therefore, the measurement metrics were the same on each project and the same for each life cycle phase. Today, metrics can change from phase to phase and from project to project. The hard part is obviously deciding on which metrics to use. Care must be taken that whatever metrics are established do not end up comparing apples and oranges. Fortunately, there are several good books that can assist project managers in identifying proper or meaningful metrics.[8]

Selecting the right KPIs is critical. Since a KPI is a form of measurement, some people believe that KPIs should be assigned only to those elements that are tangible. Therefore, many intangible elements that should be tracked by KPIs never get looked at because someone

---

8 Three books that provide examples of metric identification are Parviz F. Rad and Ginger Levin, *Metrics for Project Management* (Vienna, VA: Management Concepts, 2006); Mel Schnapper and Steven Rollins, *Value-Based Metrics for Improving Results* (Ft. Lauderdale, FL: J. Ross, 2006); and Douglas W. Hubbard, *How to Measure Anything* (3rd ed.) (Hoboken, NJ: John Wiley & Sons, 2014).

believes that measurement is impossible. Anything can be measured regardless of what some people think. According to Douglas Hubbard:

- Measurement is a set of observations that reduces uncertainty where the results are expressed as a quantity.
- A mere reduction, not necessarily elimination, of uncertainty will suffice for a measurement.[9]

Therefore, KPIs can be established even for intangibles like those that are discussed in Chapter 5.

4.13 **KPIS AND INTELLECTUAL CAPITAL**

The growth in IT and the use of a project management office has made it apparent that metrics and KPIs are now being treated as intellectual capital. The curved arrows in Figures 4.4–4.7 represent the 10 knowledge areas in the sixth edition of the *PMBOK® Guide*.[10] The reason why the knowledge areas are in the arrows and connected is because the knowledge areas are related, and each metric and KPI can probably be related to more than one knowledge area. The subjects in the center of the figure represent

Figure 4.7 The PMBOK® Guide and KPIs

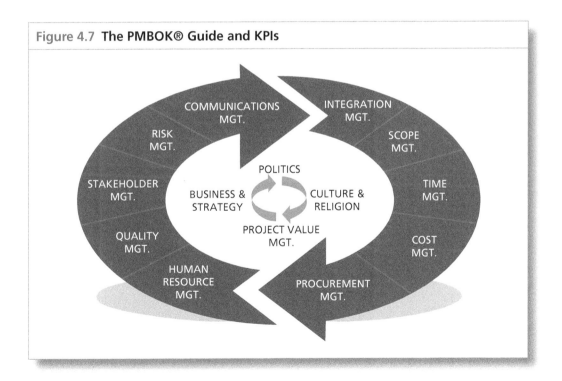

---

9 Hubbard, *How to Measure Anything*, p. 21.
10 PMBOK is a registered mark of the Project Management Institute, Inc.

Figure 4.8 Project Management Knowledge

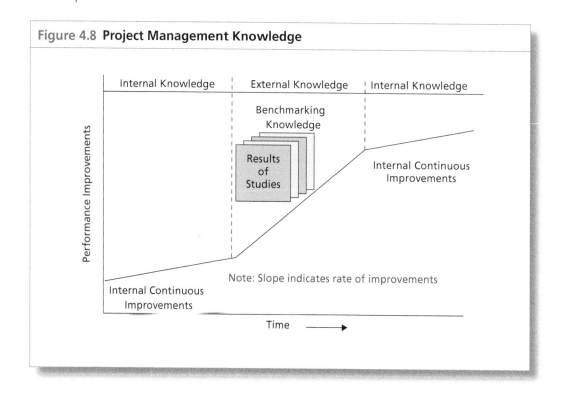

additional knowledge that project managers must learn, especially for managing the more complex global projects. These center subjects also introduce addition metrics that are related to the ten knowledge areas.

Another source of intellectual capital can come from benchmarking. As can be seen in Figures 4.4–4.8, project management benchmarking activities can accelerate the rate of improvements to the project management processes. It is important that metric benchmarking focus on process effectiveness and process maturity considerations rather than the success of an individual project. KPIs are associated with performance improvement initiatives.

The website kpilibrary.com has more than 5000 KPIs in its library. It also performs benchmarking surveys on KPIs using the following categories, among others:

- Operational excellence
- Cost leadership
- Product and service differentiation
- Customer intimacy

Benchmarking studies can support continuous improvement efforts as shown in Table 4.4. The target values in the table were set up as

TABLE 4.4   **Tracking Metrics for Continuous Improvement**

KPI	BENCHMARK STUDIES	CURRENT KPI VALUE	TARGET VALUE
Timing	60 days	55 days	50 days
Cost	$15,000	$14,000	$13,000
Pay grade	Grade 7	Grade 7	Grade 6
Management reserve	$100,000	$90,000	$80,000
Manpower	16 people	15 people	12 people
Quality	2 defects per 3000 units	2 defects per 4000 units	2 defects per 5000 units

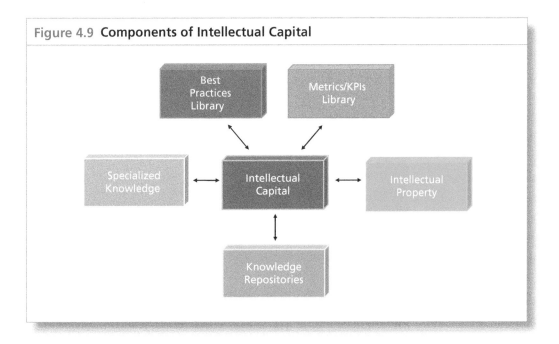

Figure 4.9 **Components of Intellectual Capital**

stretch targets. In Table 4.4, we are comparing the current value of the KPI against the benchmark and the stretch targets. If the current value of the KPI reaches or exceeds the stretch target, then we may be able to assume that continuous improvement has taken place and then raise the bar by inserting new stretch targets.

Figure 4.9 shows the components of intellectual capital. As expected, metrics/KPI libraries are considered corporate intellectual capital. The project management office is the vehicle by which metrics and KPIs are converted into intellectual capital and shared appropriately throughout the company.

4.14 KPI BAD HABITS

Stacey Barr is a globally recognized performance measurement expert who has made a career of challenging many of the long-held beliefs and bad habits people have about how performance measures ought to be chosen, created, and used for organizational performance management. Stacey believes that people share similar struggles with performance measurement. They can't find meaningful performance measures, especially for goals that seem immeasurable. They can't get staff engaged in measuring and improving performance. They don't have measures that drive lasting performance improvement. Stacey once struggled with these challenges, and this became her inspiration for sharing her metric/KPI knowledge with the world. Stacey's research and knowledge of mistakes made in business and strategy performance measurement systems are directly applicable to project management metric management systems.

KPI Bad Habits Causing Your Performance Measurement Struggles

One of the things that Albert Einstein was famous for defining, over and above $E = MC^2$, was insanity: doing the same thing over and over again and expecting different results.

Clearly, if you want to stop struggling to find meaningful performance measures that align to strategy and engage people in improving performance, then you have to change what you're doing.

A few very fruitful changes you can make are to unlearn some limiting habits that you may not even realize are at the root of most performance measurement struggles.

Bad Habit 1: Using Weasel Words to Articulate Your Goals
The problem with strategy in most organizations and companies is that, in its sanitized and word-smithed published form, it's not measurable.

Look at any strategic plan and the chances are astronomically high that you'll see a glut of words like effective, efficient, productive, responsive, sustainable, engaged, quality, flexible, adaptable, well-being, reliable, key, capability, leverage, robust, and accountable.

They are empty words that sound important and fail to say to anything at all, or at least speak of anything that can be verified in the real world, or measured. It's no wonder people with goals or objectives like the following keep asking "How do you measure that?"

Section 4.14 has been provided by Stacey Barr. © 2012 by Stacey Barr. Reproduced by permission. Additional information on Stacey Barr can be found in the following websites: www.staceybarr.com, www.staceybarr.com/measure-up performan cemeasureblueprintonline.com, www.staceybarr.com/products/pumpblueprint workshop.

- "Provide efficient, unique, unbiased and responsive, high-quality support"
- "Strengthen student engagement and learning outcomes by enhancing student support and intervention services"

The new habit to replace the weasel-word habit with is to write goals in clear and simple language that evokes in the minds of people an accurate picture of successful achievement of the goal. If you can't see it in your mind, you won't be able to measure it.

Bad Habit 2: Brainstorming to Find KPIs or Performance Measures

For the most part, people are not that conscious or aware of the approach they take to find or choose performance measures. Brainstorming is the most common approach, however. People try to select performance measures for a particular CSF or key result area or objective or goal, or for their function or process, by asking a simple question: "So, what measures could we use?" Everyone sits around and randomly suggests potential performance measures for that particular CSF or key result area or objective or goal or for their function or process. They might produce a list that looks something like this if they were brainstorming measures for staff engagement:

- Turnover
- Sick days
- Retention rate
- Introduction of talent management
- Overtime
- Staff survey
- Engagement index
- Staff satisfaction with their job
- Leadership development
- Performance management

Brainstorming can generate lots of ideas for measures quite rapidly, it's easy to do, no special knowledge or skill is required, it's familiar so it won't be distracting, and it can be very collaborative and engage people in being part of the measure selection process. But brainstorming rarely produces good measures.

The truth is you're not really finished after the brainstorming is over because you still have to work out how to get a final selection of measures from that long list. And in all honesty, voting or ranking the ideas and skimming the few that rise to the top is not the answer. Ideas for potential measures need to be vetted or tested to weed out the ideas on the list that are not measures at all ("introduction of talent management"), that aren't really relevant to the goal ("overtime"), and that are not feasible to implement.

Instead of habitually brainstorming to find performance measures, think about listing potential measures that are the best observable evidence of the successful achievement of your goal, and then choose the measures that are most relevant balanced with being feasible to implement.

Bad Habit 3: Getting People to Sign Off on Their Acceptance of the Measures

Performance measurement has such a terrible stigma. Many people associate it with the inane drudgery of data collection, with pointing fingers and with big sticks that come beating down on them when things go wrong. They associate it with the embarrassment of being compared with whoever is performing best this month. The emotions people typically feel about performance measurement are frustration, cynicism, defensiveness, anxiety, stress, and fear.

What we really want is for performance measurement to be seen as a natural and essential part of work. We want people to associate it with learning more about what works and what doesn't, with valuable feedback that keeps us on the right track, with continuous improvement of business success. We want people to feel curiosity, pride, confidence, anticipation, and excitement through using performance measures of performance results that matter. This is buy-in, not sign-off.

Buy-in is a natural product of showing people that measurement is about feedback, not judgment; of giving them tools that make measurement easy and fun; of allowing them to decide the measures most useful for their goals.

Of course, you need to stay sensitive to the fact that measurement of performance is an organization-wide system, and each team is only a part of that system. But the trade-off should be biased more toward their buy-in than it is toward sophistication of the measures. You can improve the sophistication of measures on a foundation of buy-in more easily than you can get buy-in to a suite of sophisticated measures.

Bad Habit 4: Assuming Everyone Knows the Right Way to Implement Your Measures

In general, a lot of effort is wasted in bringing performance measures to life. The waste is in the time spent to select measures that are never brought to life or in the time spent bringing measures to life in the wrong way. Thus the labor of bringing many measures into the world is far more excessive and painful than it needs to be. This, as you no doubt have experienced, breeds cynicism, a feeling of being overwhelmed, and disdain for and disengagement from the process of measuring anything, let alone measuring what matters.

People argue about data or measure validity instead of making decisions about how to improve performance. Measures misinform and mislead decisions because of the wrong calculation or analysis being used. Too many conflicting versions of the same measure result from duplication and lack of discipline in performance reporting processes, and causes confusion and cynicism about the value of measures.

This was the case for Martin, a manager in a freight company, who was receiving 12 different versions of a measure of the cycle time of coal trains from a range of business analysts throughout his department. Because no two of these 12 different measures matched, he had no idea which one was the true and accurate measure of cycle time. Martin had 12 measures and no information.

Defining a performance measure means fleshing out the specifics of its calculation, presentation, interpretation, and ownership. This is the new habit to learn to avoid making assumptions that result in waste and misinformation in performance measurement.

Bad Habit 5: Using Performance Reports

If your performance reports are stacking up in a pile, unread and unused, then they're obviously not stacking up well as sources of invaluable insight to guide performance improvement.

It's an emotional thing, performance reporting. Executives give up the precious little time they have for their families and nine holes of golf to instead paw through piles of strategic reports often more than an inch thick. Or they leave the pile of reports on their desk and make decisions from their gut instead.

Managers earnestly trawl through operational reports to check if anything needs a bit more positive light thrown on it. Supervisors and teams cynically scoff about the volumes of time and effort they waste reporting tables of statistics that track their daily activities to audiences they never see or hear from.

Performance reports need to provide the content that truly matters most, and provide that content so it is fast and easy for managers to digest. But most performance reports are just the opposite:

- They are thrown together in an ad hoc way, making it very hard to navigate to the information most relevant or urgent.
- They are cluttered and cumbersome with too much detail that drowns out the important signals with trivial and unactionable distractions.
- Their information is displayed poorly, using indigestible tables and silly graphs that are designed with entertainment in mind and unwittingly result in dangerous misinterpretation of the information.
- The layout is messy and unprofessional, wasting visual real estate, detracting from the report's importance, and disengaging users before they find the insights they can use.

The habit of using performance reports to justify our existence has to stop. Rather, we ought to get comfortable with making performance reports answer only three simple questions: What's performance doing? Why is it doing that? And what response, if any, should we take?

Bad Habit 6: Comparing This Month's Performance to Last Month's, to the Same Month's Last Year, and to Target

Managers at one of the major sawmills in a timber company thought their performance dashboard was the duck's guts. It tracked a multitude of various performance measures about how the sawmill operations were going, and the data update for many of these measures was almost live, so their dashboard could be updated very regularly.

Traffic lights—red, green, and yellow visual flags that summarize if performance is bad, good, or heading toward unacceptable—were also updated each time the data feed refreshed. The managers and supervisors would react to these traffic lights with unnecessarily large interventions, like changing the settings on timber processing equipment or altering work procedures.

These traffic lights were changing according to data, often from a small sample, like a day. Long-term trends and natural variability in the data were ignored. Rather than looking for signals in their data, managers and supervisors were reacting to the noise, reacting to any variation in the data at all.

Everyone was so busy reacting to data, the key elements that did need changing were left unattended; therefore, the overall performance worsened.

Performance will *always* vary up and down over time—you can safely assume that comparing any two points of performance data will always reveal a difference of *some* magnitude. Drawing a conclusion about whether performance has changed by comparing this month to last month is tantamount to making things up as you go along.

The insightful conclusions—which will lead you to act when you should and not act when you shouldn't—come from the patterns in your performance data. Insights do not come from comparisons between the points of your performance data. Unlearn that bad habit of making point-to-point comparisons and performance will improve.

Bad Habit 7: Treating Performance Measurement Separately from Planning, Reporting, and Strategy Execution

Performance measurement is a process that weaves through your existing management processes. It doesn't stand alone and apart from them.

The steps of selecting performance measures for goals or objectives need to be performed as part of the strategic and operational planning processes, or you end up with goals and objectives that are vague and immeasurable.

Reporting processes, including the business intelligence systems, data warehouses, and information dashboards that support them, need to quickly refocus on the data and information for new performance measures and their cause analysis.

Strategy execution needs to be informed by the current performance measures and their targets, and the strategies themselves need to be changed when the measures show they aren't working.

Performance measurement isn't something you do after your strategic plan is cast in stone and just in time for the annual review. It's not something we do for bureaucratic reasons. We do it because it provides the feedback that about how well we're achieving the endeavors we chose to pursue. If those endeavors are important enough to pursue, they are too important not to measure, and measure well.

4.15 BRIGHTPOINT CONSULTING, INC.—DASHBOARD DESIGN: KEY PERFORMANCE INDICATORS AND METRICS

Introduction

The most important aspect of designing an effective dashboard is making sure you have selected the right metrics and KPIs to display. No matter how stunning or clever a visual design is, if it isn't displaying meaningful insights and data relevant to its audience it will just end up being a pretty display that no one looks at.

This article will show you some proven techniques that will help you to collect and define appropriate performance indicators for executive and operational dashboards. While the techniques discussed here are focused on dashboard design, these same principles can be used across many different business intelligence requirements gathering efforts.

Material in section 4.15 has been taken from BrightPoint Consulting white paper, "Dashboard Design: Key Performance Indicators and Metrics," by Tom Gonzalez, managing director, BrightPoint Consulting, Inc., © 2005 by BrightPoint Consulting, Inc. Reproduced by permission. All rights reserved. Mr. Gonzalez is the founder and managing director of BrightPoint Consulting, Inc., serving as a consultant to both Fortune 500 companies and small to medium businesses alike. With over 20 years of experience in developing business software applications, Mr. Gonzalez is a recognized expert in the fields of business intelligence and enterprise application integration within the Microsoft technology stack. BrightPoint Consulting, Inc. is a leading technology services firm that delivers corporate dashboard and business intelligence solutions to organizations across the world. BrightPoint Consulting leverages best-of-breed technologies in data visualization, business intelligence and application integration to deliver powerful dashboard and business performance solutions that allow executives and managers to monitor and manage their business with precision and agility. For further company information, visit BrightPoint's Web site at www.brightpointinc.com. To contact Mr. Gonzalez, email him at tgonzalez@brightpointinc.com.

With the prevalence of dashboard tools and displays seen in widely varied software offerings, people have different understandings of what a dashboard, metric, and key performance indicator (KPI) consist of. To make sure we are all speaking the same language, I will define a set of terms that will form the basis of our design discussion.

Metrics and Key Performance Indicators

Metrics and KPIs are the building blocks of many dashboard visualizations, as they are the most effective means of alerting users as to where they are in relationship to their objectives. The definitions that follow form the basic building blocks for dashboard information design, and they build upon themselves, so it is important that you fully understand each definition and the concepts discussed before moving on to the next definition.

Metrics and KPIs are the building blocks of many dashboard visualizations as they are the most effective means of alerting users as to where they are in relationship to their objectives. The definitions that follow form the basic building blocks for dashboard information design, and they build upon themselves, so it is important that you fully understand each definition and the concepts discussed before moving on to the next definition.

Measure: A measure is numerically quantifiable piece of data. Sales, Profit, Retention Rate, are all examples of specific measures.

Dimension: A dimension represents different facets of a given measure. For instance, *time* is often used as a dimension to analyze different measures. Some other common dimensions are *region, product, division, market segment*, etc.

Hierarchy: Dimensions can be further broken down into hierarchies. For instance, the time dimension may also form a hierarchy such as *Year > Quarter > Month > Day*.

Grain: Each of level within the hierarchy is referred to as the *grain* of the dimension. For example, if you were looking at a Geography dimension, the individual grains (levels) might be *Region > Country > State/Province > City > Postal Code*

Metric: A metric, which is the type of data we are often displaying in a dashboard, is a *measure* that represents a piece of data in relationship to one or more specific *dimensions* and associated *grains*.

Gross Sales by Week *is an example of a metric. In this case, the* measure *would be dollars (gross sales) and the* dimension *would be time with the associated* grain *(week.)*

Looking at a measure across more than one dimension such as gross sales by territory and time is called multi-dimensional analysis.

Designing dashboards for multi-dimensional analysis can add significant complexity to your dashboard design and user experience due to the fact we are using a two-dimensional surface (device screen) to represent multiple dimensions. While there are effective layout and interaction patterns to handle this type of analysis, it is important to make note of that prior to your design phase as these types of designs require more sophisticated approaches.

Key Performance Indicator (KPI): A KPI is simply a *metric* that is tied to a target. Most often a KPI represents how far a metric is above or below a predetermined target. KPIs usually are shown as a ratio of actual to target and are designed to instantly let a business user know if they are on or off their plan without the end user having to consciously focus on the metrics being represented. For instance, we might decide that in order to hit our quarterly sales target, we need to be selling $10,000 of widgets per week. The metric would be *widget sales per week;* the target would be $10,000. If we used a percentage gauge visualization to represent this KPI and we had sold $8,000 in widgets by Wednesday, the user would instantly see that they were at 80% of their goal.

When selecting targets for your KPIs, you need to remember that a target will have to exist for every *grain* you want to view within a metric. Having a dashboard that displays a KPI for gross sales by day, week, and month will require that you have identified targets for each of these associated grains.

Scorecards, Dashboards, and Reports

The difference between a scorecard, dashboard, and report can be one of fine distinctions. Each of these tools can combine elements of the other, but at a high level they all target distinct and separate levels of the business decision-making process.

Scorecards: Starting at the highest, most strategic level of the business decision-making spectrum, we have scorecards. Scorecards are primarily used to help align operational execution with business strategy. The goal of a scorecard is to keep the business focused on a common strategic plan by monitoring real-world execution and mapping the results of that execution back to a specific strategy. The primary measurement used in a scorecard is the key performance indicator. These KPIs are often a composite of several metrics or other KPIs that measure the organization's ability to execute a strategic objective. An example of a scorecard KPI is an indicator named "Profitable Sales Growth" that combines several weighted measures such as: new customer acquisition, sales volume, and gross profitability into one final score.

Dashboards: A dashboard falls one level down in the business decision-making process from a scorecard, as it is less focused on a strategic objective and more tied to specific operational goals. An operational goal may directly contribute to one or more higher-level strategic objectives. Within a dashboard, execution of the operational goal itself becomes the focus, not the higher-level strategy. The purpose of a dashboard is to provide the user with actionable business information in a format that is both intuitive and insightful. Dashboards leverage operational data primarily in the form of metrics and KPIs.

Reports: Probably the most prevalent BI tool seen in business today is the traditional report. Reports can be very simple and static in nature, such as a list of sales transactions for a given time period, to more sophisticated cross-tab reports with nested grouping, rolling summaries, and dynamic drill-through or linking. Reports are best used when the user needs to look at raw data in an easy to read format. When combined with scorecards and dashboards, reports offer a tremendous way to allow users to analyze the specific data underlying their metrics and KPIs.

Gathering KPI and Metric Requirements for a Dashboard

Traditional BI projects will often use a bottom-up approach in determining requirements, where the focus is on the domain of data and the relationships that exist within that data. When collecting metrics and KPIs for your dashboard project, you will want to take a top-down approach. A top-down approach starts with the business decisions that need to be made first and then works its way down into the data needed to support those decisions.

In order to take a top-down approach, you *must* involve the actual business users who will be utilizing these dashboards, as these are the only people who can determine the relevancy of specific business data to their decision-making process.

When interviewing business users or stakeholders, the goal is to uncover the metrics and KPIs that lead the user to a specific decision or action. Sometimes users will have a very detailed understanding of what data is important to them, and sometimes they will only have a high-level set of goals. By following the practices outlined in the text that follows, you will be able to distill the information provided to you by the user into a specific set of KPIs and metrics for your dashboards.

Interviewing Business Users

In our experience working directly with clients and gathering requirements for executive and operational dashboard projects in a variety of industries, we have found that the interview process revolves around two simple questions: "What business questions do you need answers to, and once you have those answers what action would you take or what decision would you make?"

Question 1: What Business Questions Do You Need Answers To?

The purpose here is to help business users define their requirements in a way that allows us to get to the data behind their question. For instance, a VP of sales might have the question: "Which salespeople are my top producers?" or "Are we on target for the month?" In the case of the question "Which sales people are my top producers?," we might then follow up with a couple of questions for the VP and ask her "Would this measure be based on gross sales? Would you like to see this daily, weekly, or monthly?"

We want to identify the specific data components that will make up the KPI or metric. So, we need to spend enough time with the user discussing the question that we clearly understand the *measure, dimension, grain,* and *target* (in the case of a KPI) that will be represented.

Question 2: Based on the Answer to Question 1, What Other Questions Would This Raise or What Action Would You Take?

Once we understand the metric or KPI that is needed to answer the user's question, we then need to find out if the user wants to perform further analysis based on that answer, or if he or she would be able to take an action or make a decision. The goal is to have users keep breaking down the question until they have enough information to take action or make a decision. This process of drilling deeper into the question is analogous to peeling back the layers of an onion; we want to keep going deeper until we have gotten to the core, which in this case is the user's ability to make a decision or take action.

As a result of this iterative two-part question process, we are going to quickly filter out the metrics and KPIs that could be considered just interesting from the ones that are truly critical to the user's decision-making process.

Putting It All Together—The KPI Wheel

In order to help with this requirements' interview process, BrightPoint Consulting has created a tool called the KPI Wheel (see Figures 4.4– 4.10). The interview process is very rarely a structured linear conversation and more often is an organic free-flowing exchange of ideas and questions. The KPI Wheel allows us to have a naturally flowing conversation with the end user, while at the same time keeping us focused on the goal of gathering specific requirements.

The KPI Wheel is a tool that can be used to collect all the specific information that will go into defining and visualizing a metric or KPI. We will use this tool to collect the following information:

Figure 4.10 KPI Wheel

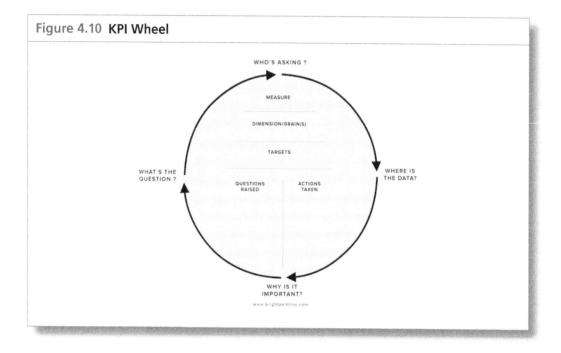

1. The business question that we are trying to help the user answer.
2. Which business users this question would apply to.
3. Why the question is important.
4. Where data resides to answer this question.
5. What further questions this metric or KPI could raise.
6. What actions or decisions could be taken with this information.

Start Anywhere, but Go Everywhere

The KPI Wheel is designed as a circle because it embodies the concept that you can start anywhere but go everywhere, thus covering all relevant areas. In the course of an interview session, you will want to refer to the wheel to make sure you are filling in each area as it is discussed. As your conversation flows, you can simply jot down notes in the appropriate section, and you can make sure to follow up with more questions if some areas remain unfilled. The beauty behind this approach is that a user can start out very high level—"I want to see how sales are doing"—or at a very low level—"I need to see product sales broken down by region, time, and gross margins." In either scenario, you are able to start at whatever point the user feels comfortable and then move around the wheel filling in the needed details.

Question 1: What's the Question?

This area of the wheel refers to the basic "What business question do you need an answer to?" We can often start the interview with this question, or we can circle back to it when the user starts off with a specific metric in mind by asking them "What business question would that metric answer for you?" This segment of the wheel drives the overall context and relevance of the whole metric or KPI.

Question 2: Who's Asking?

For a given metric, we want to know who will be using this information to make decisions and take action. It is important to understand the various users within the organization who may be viewing this metric. We can either take note of specific individuals or just refer to a general group of people who would all have similar business needs.

Question 3: Why Is It Important?

Because a truly effective dashboard can become a tool that is used every day, we want to validate the importance of each metric and KPI that is displayed. Oftentimes, in going through this requirements-gathering process, we will collect a long list of potential metrics and KPIs, and at some point, the user will have to make a choice about what data is truly the most important for them to see on a regular basis. We suggest using a 1–10 scale in conjunction with a description of why the metric is important, so when you begin your dashboard prototyping, you will have context as to the importance of this metric.

Question 4: Data Sources

For a given metric or KPI, we also want to identify where the supporting data will come from. Sometimes, in order to calculate a metric along one or more dimensions, we need to aggregate data from several different sources. In the case of the metric "Top selling products by gross margins," we may need to pull data from both a customer relations management system and an enterprise resource planning system. At this stage it is good enough to simply indicate the business system that holds the data; it is unnecessary to dive into actual table/field name descriptions at this point.

Upper Half: Measures, Dimensions, and Targets

We want to make sure that we have captured the three main attributes that create a metric or KPI, and have the user validate the grains of any given dimensions. If we are unable to pin down the measure and dimension for a metric and/or the target for a KPI, then we will be unable to collect and visualize that data when it comes time to design our dashboard.

Lower Half: Questions Raised

In this section of the KPI Wheel, we want to list any other questions that may be raised when we have answered our primary question. This list can serve as the basis for the creation of subsequent KPI Wheels that are used for definition of further metrics and KPIs.

Lower Half: Actions to Be Taken

For any given metric or KPI, we want to understand what types of decisions can be made or what types of actions will be taken depending on the state of the measurement. By filling out this section, we are also able to help validate the importance of the metric and separate the "must-have" KPI from the "nice-to-have" KPI.

Wheels Generate Other Wheels

In filling out a KPI Wheel, the process will often generate the need for several more KPIs and metrics. This is one of the purposes of doing an initial analysis in the first place: to bring all of the user's needs up to the surface. As you work through this requirements-gathering effort, you will find that there is no right path to getting your answers, questions will raise other questions, and you will end up circling back and covering ground already discussed in a new light. It is important to be patient and keep an open mind as this is a process of discovery. The goal is to have a concrete understanding of how you can empower the user through the use of good metrics and KPIs.

As you start to collect a thick stack of KPI Wheels, you will begin to see relationships between the KPIs you have collected. When you feel that you have reached a saturation point and neither you nor the user can think of any more meaningful measurements, you will then want to review all the KPI Wheels in context with each other. It is a good practice to aggregate the KPIs and create logical groupings and hierarchies, so you clearly understand the relationships that exist between various metrics. Once these steps have been accomplished, you will have a solid foundation to start your dashboard visualization and design process upon.

A Word about Gathering Requirements and Business Users

Spending the needed time with a formal requirements-gathering process is often something not well understood by business users, especially senior executives. This process will sometimes be viewed as a lot of unnecessary busywork that interrupts the user's already hectic day. It is important to remember that the decisions you are making now about what data is and is not relevant will have to be done at some point, and the only one who can make this determination is the user himself. The question is whether you spend the time to make those fundamental decisions now, while you are simply moving around ideas, or later, after you have painstakingly designed the dashboards and built complex data integration services around them.

As with all software development projects, the cost of change grows exponentially as you move through each stage of the development cycle. A great analogy is the one used for home construction. What is the cost to move a wall when it is a line on a drawing versus the cost to move it after you have hung a picture on it?

Wrapping It All Up

While this article touches upon some of the fundamental building blocks that can be used in gathering requirements for a dashboard project, it is by no means a comprehensive methodology. Every BI architect has a set of best practices and design patterns they use when creating a new solution. It is hoped that some of the processes mentioned here can be adapted and used to supplement current best practices for a variety of solutions that leverage dashboard technologies.

5

VALUE-BASED PROJECT MANAGEMENT METRICS

CHAPTER OVERVIEW	For some stakeholders, value is positioned at the top of the priority list. Establishing value metrics is now a necessity. However, there are shortcomings and pitfalls that must be addressed.
CHAPTER OBJECTIVES	■ To understand what is meant by value ■ To understand the need for measurements of value ■ To understand the shortcomings with value measurement ■ To understand how value has changed the way we manage projects ■ To understand how to create a value-based metric ■ To understand the need for creating a value baseline
KEY WORDS	■ Boundary box ■ Value ■ Value baseline ■ Value conflicts ■ Value measurements ■ Value metrics ■ Value-driven projects

5.0 INTRODUCTION

For years, the traditional view of project management was that, if a project was completed and it adhered to the triple constraints of time, cost, and performance (or scope), the project was successful. Perhaps in the eyes of the project manager, the project appeared to be a success. In the eyes of the customer or the stakeholders, however, the project might be regarded as a failure.

As stated in Chapter 1, project managers are now becoming more business oriented. Projects are being viewed as part of a business for the purpose of providing value to both the ultimate customer and the parent corporation. Project managers are expected to understand business operations more so today than in the past. As the project managers become more business oriented, the definition of success

Project Management Metrics, KPIs, and Dashboards: A Guide to Measuring and Monitoring Project Performance, Fourth Edition. Harold Kerzner.
© 2023 John Wiley & Sons, Inc. Published 2023 by John Wiley & Sons, Inc.
Companion Website: www.wiley.com/go/kerznermetrics4e

on a project now includes both a business and a value component. The business component may be directly related to value.

SITUATION: The information technology (IT) group at a large public utility always serviced all IT requests without question. All requests were added to the queue and would eventually get done. The utility implemented a project management office (PMO) that was assigned to develop a template for establishing a business case for the request, clearly indicating the value to the company if the project were completed. In the first year of using the business case template, one third of all of the projects in the queue were tossed out.

Projects must provide some degree of value when completed and must meet the competing constraints. Perhaps the project manager's belief is that meeting the competing constraints provides value, but that is not always the case. Why should a company work on projects that provide no near-term or long-term value? Too many companies either are working on the wrong projects or simply have a poor project portfolio selection process, and no real value appears at the completion of the projects even though the competing constraints have been met.

Assigning resources that have critical skills that are in demand on other projects to projects that provide no appreciable value is an example of truly inept management and poor decision making. Yet selecting projects that will guarantee value or an acceptable return on investment (ROI) is very challenging because some of today's projects do not provide the targeted value until years into the future. This is particularly true for research and development and new product development, where as many as 50 or more ideas must be explored to generate one commercially successful product. Predicting the value at the start and tracking the value during execution is difficult. In the pharmaceutical industry, the cost of developing a new drug could run about $850 million, take 3000 days to go from exploration to commercialization, and provide no meaningful ROI. In that industry, less than 3% of the research and development projects are ever viewed as a commercial success and generate more that $400 million per year in revenue.

There are multiple definitions of value. For the most part, value is like beauty; it is in the eyes of the beholder. In other words, value may be viewed as a perception at project selection and initiation based on data available at the time. At project completion, however, the actual value becomes a reality that may not meet the expectations that had initially been perceived.

Another problem is that the achieved value of a project may not satisfy all of the stakeholders, since each stakeholder may have had a different perception of value as it relates to his or her business function. Because of the money invested in some projects, establishing value-based metrics is essential. The definition of value, along with the metrics, can be industry-specific, company-specific, or even dependent on the size, nature, and business base of the firm. Some stakeholders may view value as job security or profitability. Others might view value as image, reputation, or the creation of intellectual property. Satisfying all stakeholders is a formidable task often difficult to achieve. In some cases, it may simply be impossible. In any event, value-based metrics must be established along with traditional metrics. Value metrics show whether value is being created or destroyed.

TIP Do not make promises to stakeholders about final value unless there are metrics that will confirm that their expectations can or will be met.

5.1 VALUE OVER THE YEARS

Before discussing value-based metrics, it is important to understand how the necessity for value identification has evolved. Surprisingly enough, much research on value has taken place over the past 15 years. Some of the items covered in the research include:

- Value dynamics
- Value gap analysis
- Intellectual capital valuation
- Human capital valuation
- Economic value-based analysis
- Intangible value streams
- Customer value management/mapping
- Competitive value matrix
- Value chain analysis
- Valuation of IT projects
- Balanced Scorecard

Following are some of the models that have developed over the past 15 years as a result of the research:

- Intellectual capital valuation
- Intellectual property scoring
- Balanced scorecard
- Future Value Management™
- Intellectual Capital Rating™
- Intangible value stream modeling
- Inclusive Value Measurement™
- Value performance framework
- Value measurement methodology (VMM)

The reason why these models have become so popular in recent years is because we have developed techniques for the measurement and determination of value. This is essential in order to have value metrics on projects.

TABLE 5.1 Application of VPF to Project Management	
VPF ELEMENT	PROJECT MANAGEMENT APPLICATION
Understand key principles of valuation	Working with the project's stakeholders to define value
Identification of key value drivers for the company	Identification of key value drivers for the project
Assessing performance on critical business processes and measures through evaluation and external benchmarking	Assessing performance of the enterprise project management methodology and continuous improvement using the PMO
Creating a link between shareholder value and critical business processes and employee activities	Creating a link among project values, stakeholder values, and team member values
Aligning employee and corporate goals	Aligning employee, project, and corporate goals
Identification of key "pressure points" (high-leverage improvement opportunities) and estimating potential impact on value	Capturing lessons learned and best practices that can be used for continuous improvement activities
Implementation of a performance management system to improve visibility and accountability in critical activities	Establishing and implementing a series of project-based dashboards for customer and stakeholder visibility of KPIs
Development of performance dashboards with high-level visual impact	Development of performance dashboards for stakeholder, team, and senior management visibility

Source: Column 1 is from Jack Alexander, Performance Dashboards and Analysis for Value Creation *(Hoboken, NJ: John Wiley & Sons, 2007), p. 6.*

There is some commonality among many of these models such that they can be applied to project management. For example, Jack Alexander created a model called the Value Performance Framework (VPF).[1] The model focuses on building shareholder value and is heavily biased toward financial key performance indicators (KPIs). However, the key elements of VPF can be applied to project management, as shown in Table 5.1. The first column contains the key elements of VPF from Alexander's book and the second column illustrates the application to project management.

5.2 VALUES AND LEADERSHIP

The importance of value can have a significant impact on the leadership style of project managers even though value leadership metrics are not always created. Historically, project management leadership was perceived as the inevitable conflict between individual values and organizational values. Today, companies are looking for ways to get employees to align their personal values with the organization's values.

---

1 Jack Alexander, *Performance Dashboards and Analysis for Value Creation* (Hoboken, NJ: John Wiley & Sons, 2007), p. 5.

TABLE 5.2 Changing Values	
MOVING AWAY FROM: INEFFECTIVE VALUES	**MOVING TOWARD: EFFECTIVE VALUES**
Mistrust	Trust
Job descriptions	Competency models
Power and authority	Teamwork
Internal focus	Stakeholder focus
Security	Taking risks
Conformity	Innovation
Predictability	Flexibility
Internal competition	Internal collaboration
Reactive management	Proactive management
Bureaucracy	Boundaryless
Traditional education	Lifelong education
Hierarchical leadership	Multidirectional leadership
Tactical thinking	Strategic thinking
Compliance	Commitment
Meeting standards	Continuous improvements

Source: *Ken Hultman and Bill Gellerman*, Balancing Individual and Organizational Values (San Francisco: Jossey-Bass, 2002), pp. 105–106.

Several books have been written on this subject, and the best one, in this author's opinion, is *Balancing Individual and Organizational Values* by Ken Hultman and Bill Gellerman.[2] Table 5.2 shows how our concept of value has changed over the years. Close examination of the table shows that the changing values affect more than just individual versus organizational values. Instead, it is more likely to be a conflict among four groups, as shown in Figure 5.1. The needs of each group might be as listed next.

Project Manager

- Accomplishment of objectives
- Demonstration of creativity
- Demonstration of innovation

Team Members

- Achievement
- Advancement

2 Ken Hultman and Bill Gellerman, *Balancing Individual and Organizational Values* (San Francisco: Jossey-Bass, 2002).

Figure 5.1 Project Management Value Conflicts

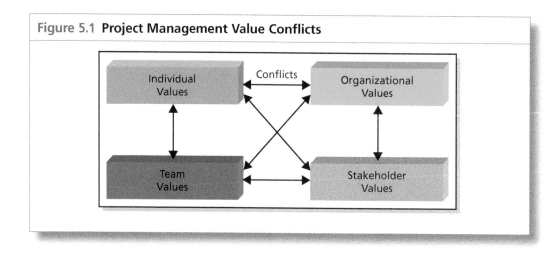

- Ambition
- Credentials
- Recognition

Organization

- Continuous improvement
- Learning
- Quality
- Strategic focus
- Morality and ethics
- Profitability
- Recognition and image

Stakeholders

- Organizational stakeholders: Job security
- Product/market stakeholders: Quality performance and product usefulness
- Capital markets: Financial growth

There are several reasons why the role of the project manager and the accompanying leadership style have changed. Some reasons include:

- Project managers are now managing businesses as if they were a series of projects.
- Project management is now viewed as a full-time profession.
- Project managers are now viewed as both business managers and project managers and are expected to make decisions in both areas.
- The value of a project is measured more in business terms than just in technical terms.
- Project management is now being applied to parts of the business that traditionally have not used project management.

5.3 **COMBINING SUCCESS AND VALUE**

Based on some of the value models discussed previously, such as the Balanced Scorecard model, a classification system for projects can be identified. The types of projects, which can be identified and tracked according to their business alignment and value, can be classified into four categories:

1. **Enhancement or internal projects:** These are projects designed to update processes, improve efficiency and effectiveness, and possibly improve morale.
2. **Financial projects:** Companies require some form of cash flow for survival. These are projects for clients external to the firm and have an assigned profit margin.
3. **Future-related projects:** These are long-term projects to produce a future stream of products or services capable of generating a future cash flow. These projects may be an enormous cash drain for years with no guarantee of success.
4. **Customer-related projects:** Some projects may be performed, even at a financial loss, to maintain or build a customer relationship. However, performing too many of these projects can lead to financial disaster.

Today, these types of projects focus more on value than on the competing constraints. Value-driven constraints emphasize stake-holder satisfaction and decisions and the value that is expected on the project. In other words, success is when the value is obtained, it is hoped within the triple or competing constraints. As a result, we can define the four cornerstones of success using Figure 5.2.

Figure 5.2 **Four Cornerstones of Success**

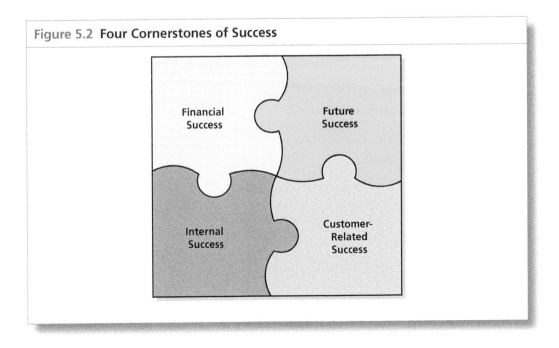

Very few projects are completed without some trade-offs. Metrics provide some of the necessary information needed for decisions on tradeoffs. This holds true for both the traditional projects and those that are based on value components and metrics. Traditional trade-offs result in a lengthening of the schedule and an increase in the budget. The same holds true for the value-driven projects, but the major difference is with performance. With traditional trade-offs, performance tends to be reduced to satisfy other requirements. With value-driven projects, performance tends to be increased in hopes of providing added value, and this tends to cause much larger cost over-runs and schedule slippages than with traditional trade-offs. The amount of additional time and funding that the stakeholders will allow is dependent on the tracking of the metrics.

Project managers generally do not have the sole authority for scope or performance increases or decreases. For traditional trade-offs, the project manager and the project sponsor, working together, may have the authority to make trade-off decisions.

However, for value-driven projects, all or most of the stakeholders may need to be involved. This can create additional issues such as:

- It may not be possible to get all of the stakeholders to agree on a value target during project initiation.
- It may not be possible to get all of the stakeholders to agree on the metrics or KPIs.
- Getting agreement on scope changes, extra costs, and schedule extensions is significantly more difficult the farther along the project is.
- Stakeholders must be informed of the value expected from the project at project initiation and continuously briefed as the project progresses; that is, no surprises!

Conflicts among the stakeholders may occur. As an example:

- During project initiation, conflicts among stakeholder are usually resolved in favor of the largest financial contributors.
- During execution, conflicts over future value become more complex, especially if major contributors threaten to pull out of the project.

For projects that have a large number of stakeholders, a single sponsor may not provide effective project sponsorship. Therefore, committee sponsorship may be necessary. Committee members may include, among others:

- Perhaps a representative from all stakeholder groups
- Influential executives
- Critical strategic partners and contractors

Responsibilities for the sponsorship committee may include:

- Taking a lead role in the definition of the targeted value
- Taking a lead role in the acceptance of the actual value
- Ability to provide additional funding
- Ability to assess changes in the enterprise environment factors
- Ability to validate and revalidate the assumptions

Sponsorship committees may have significantly more expertise than the project manager in defining and evaluating the value in a project.

Each of the quadrants in Figure 5.2 can have its own unique set of critical success factors (CSFs) and likewise its own unique metrics and KPIs. Following are typical CSFs for each quadrant.

Internal Success

- Adherence to schedule, budget, and quality/scope (triple constraint)
- Mutually agreed-upon scope change control process
- Does not disturb the main flow of work
- Clear understanding of the objectives (end user involvement)
- Maintaining the timing of sign-offs
- Execution without disturbing the corporate culture
- Building lasting internal working relationships
- Consistently respecting each other's opinions
- Searching for value-added opportunities

Financial Success

- Integrating program and project success into one definition
- Maintaining ethical conduct
- Adherence to regulatory agency requirements
- Adherence to health, safety, and environmental laws
- Maintaining or increasing market share
- Maintaining or improving ROI, net present value, internal rate of return, payback period, etc.
- Maintaining or improving net operating margins

Future Success

- Improving the processes needed for commercialization
- Emphasizing follow-on opportunities
- Maintaining technical superiority
- Protecting the company image and reputation
- Maintaining a knowledge repository
- Retaining presale and post-sale knowledge
- Aligning projects with long-term strategic objectives
- Informing the teams about strategic plans
- Willingness of team members to work with this project manager again

Figure 5.3 Categories of Success Metrics

Financial Success	Future Success
• Quantitative • Directional • Financial	• Quantitative • Financial
Internal Success	**Customer-Related Success**
• Quantitative • Practical • Directional • Actionable • Milestone	• Directional • End Result

Customer-related Success

- Keeping promises made to the customers over and over again
- Maintaining customer contact and interfacing continuously
- Focusing on customer satisfaction from start to finish
- Improving customer satisfaction ratings on a continuous basis
- Using every customer's name as a reference
- Measuring variances against customer-promised best practices
- Maintaining or improving on customer delivery requirements
- Building long-term relationships between organizations

Chapter 3 identified the different type of metrics. The type of metric that is most suitable for each success quadrant is shown in Figure 5.3.

5.4 RECOGNIZING THE NEED FOR VALUE METRICS

The importance of the value component in the definition of success cannot be overstated. Consider the following eight postulates:

Postulate #1: Completing a project on time and within budget does not guarantee success if you were working on the wrong project.

Postulate #2: Completing a project on time and within budget is not necessarily success.

Postulate #3: Completing a project within the triple constraints does not guarantee that the necessary business value will be there at project completion.

Postulate #4: Having the greatest enterprise project management methodology (EPMM) in the world cannot guarantee that value will be there at the end of the project.

Postulate #5: "Price is what you pay. Value is what you get" (Warren Buffett).

Postulate #6: Business value is what customers believe is worth paying for.

Postulate #7: Success is when business value is achieved.

Postulate #8: Following a project plan to conclusion is not always success if business-related changes were necessary but never implemented.

These eight postulates suggest that value may become the dominating factor in the selection of a project portfolio. Project requestors must now clearly articulate the value component in the project's business case or run the risk that the project will not be considered. If the project is approved, then value metrics must be established and tracked. However, it is important to understand that value may be looked at differently during the portfolio selection of projects because the trade-offs that take place are among projects rather than the value attributes of a single project.

Postulate #1 shows what happens when management makes poor decisions during project selection, establishment of a project portfolio, and when managing project portfolios. Companies end up working on the wrong project or projects. What is unfortunate about this scenario is that the deliverable that was requested can be produced but:

- There is no market for the product.
- The product cannot be manufactured as engineered.
- The assumptions may have changed.
- The marketplace may have changed.
- Valuable resources were wasted on the wrong project.
- Stakeholders may be displeased with management's performance.
- The project selection and portfolio management process is flawed and needs to be improved.
- Organizational morale has diminished.

Postulate #2 is the corollary to Postulate #1. Completing a project on time and on budget does not guarantee:

- A satisfied client/customer
- That the customer will accept the product/service
- That performance expectations will be met
- That value exists in the deliverable
- Marketplace acceptance
- Follow-on work
- Success

SITUATION: During the initiation of the project, the project manager and the stakeholders defined project success and established metrics for each of the competing constraints. When it became obvious that all of the constraints could not be met, the project manager concluded that the best alternative was a trade-off on value. The stakeholders became irate upon hearing the news and decided to prioritize the competing constraints themselves. This took time and delayed the project.

Postulate #3 focuses on value. Simply because the deliverable is provided according to a set of constraints is no guarantee that the client will perceive value in the deliverable. The ultimate objective of all projects should be to produce a deliverable that meets expectations and achieves the desired value. Although the importance of the competing constraints always seems to be emphasized when defining the project, very little time is spent in defining the value characteristics and resulting metrics that are expected in the final deliverable. The value component or definition must be agreed upon jointly by the customer and the contractor (buyer/seller) during the initiation stage of the project.

Most companies today have some type of project management methodology in place. Unfortunately, all too often there is a mistaken belief that the methodology will guarantee project success. Methodologies:

- Cannot guarantee success
- Cannot guarantee value in the deliverable
- Cannot guarantee that the time constraint will be adhered to
- Cannot guarantee that the quality constraint will be met
- Cannot guarantee any level of performance
- Are not a substitute for effective planning
- Are not the ultimate panacea to cure all project ills
- Are not a replacement for effective human behavior

TIP The definition of value must be aligned with the strategic objectives of both the customer and the contractor.

Methodologies can improve the chances for success but cannot guarantee success. Methodologies are tools and, as such, do not manage projects. Projects are managed by people, and, likewise, tools are managed by people. Methodologies do not replace the people component in project management. They are designed to enhance the performance of people.

In Postulate #5, the quote from Warren Buffett emphasizes the difference between price and perceived value. Most people believe that customers pay for deliverables. This is not necessarily true. Customers pay for the value they expect to receive from the deliverable. If the deliverable has not achieved value or has limited value, the result is a dissatisfied customer.

Some people believe that a customer's greatest interest is quality. In other words, "Quality comes first." Although that may seem true on the surface, the customer generally does not expect to pay an extraordinary amount of money just for high quality. Quality is just one component in the value equation. Value is significantly more than just quality.

When a customer agrees to a contract with a contractor/supplier for a deliverable, the customer is actually looking for the value in the deliverable. The customer's definition of success is "value achieved."

Unfortunately, unpleasant things can happen when the project manager's definition of success is the achievement of the deliverable (and possibly the triple constraint) and the customer's definition of success is value. This is particularly true when customers want value and the project manager, as the contractor, focuses on the profit margins of the projects.

Postulate #7 is a summation of Postulates #1 through #6. Perhaps the standard definition of success using just the triple constraints should be modified to include a business component, such as value, or even be replaced by a more specific definition of value. Other competing constraints may need to be included.

Sometimes the value of a project can change over time, and the project manager may not recognize that these changes have occurred. Failure to establish value expectations or lack of value in a deliverable can result from:

- Market unpredictability.
- Market demand that has changed, thus changing constraints and assumptions.
- Technology advances or inability to achieve functionality.
- Critical resources were that were not available or resources who lacked the necessary skills.

Establishing value metrics early on can allow a project manager to identify if a project should be canceled. The earlier the project is canceled, the quicker the project's resources can be assigned to those projects that have a higher perceived value and probability of success. Unfortunately, early warning signs are not always present to indicate that the value will not be achieved. The most difficult metrics to establish are value-driven metrics.

TIP Degradation in value metrics is a clear indication that the project is in trouble and that it may be canceled.

5.5 **THE NEED FOR EFFECTIVE MEASUREMENT TECHNIQUES**

Selecting metrics and KPIs is not that difficult, provided they can be measured. This is the major obstacle with the value-driven metrics. On the surface, they look easy to measure, but there are complexities. However, even though value appears in the present and in the future, it does not mean that the value outcome cannot be quantified.

TABLE 5.3 Measuring Value	
VALUE METRIC	**MEASUREMENT**
Profitability	Easy
Customer satisfaction	Hard
Goodwill	Hard
Penetrate new markets	Easy
Develop new technology	Moderate
Technology transfer	Moderate
Reputation	Hard
Stabilize workforce	Easy
Utilize unused capacity	Easy

Table 5.3 illustrates some of the metrics that are often treated as value-driven KPIs.

Traditionally, business plans identified the benefits expected from the project, and the benefits were the criteria for project selection. Today, portfolio management techniques require identification of the value as well as the benefits. However, conversion from benefits to value is not easy.[3]

Shortcomings in the conversion process can make the conversion difficult. Figure 5.4 illustrates several common shortcomings.

There are other shortcomings with the measurement of KPIs. KPIs are metrics for assessing value. With traditional project management, metrics are established by the EPMM and fixed for the duration of the project's life cycle. With value-driven project management, however, metrics, can change from project to project, during a life cycle phase and over time because of:

- The way the customer and contractor jointly define success and value at project initiation
- The way the customer and contractor come to an agreement at project initiation regarding what metrics should be used on a given project
- The way the company defines value
- New or updated versions of tracking software
- Improvements to the EPMM and the accompanying project management information system
- Changes in the enterprise environmental factors

---

3 For additional information on the complexities of conversion, see Jack J. Phillips, Timothy W. Bothell, and G. Lynne Snead, *The Project Management Scorecard* (Oxford, UK: Butterworth Heinemann, 2002), Chapter 13.

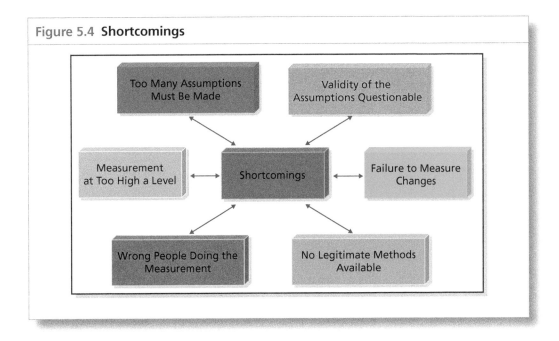

Figure 5.4 **Shortcomings**

Even with the best possible metrics, measuring value can be difficult. Benefits less costs indicate the value and determine if the project should be done. The challenge is that not all costs are quantifiable. Some values are easy to measure, while others are more difficult. The value metrics that are easy to measure are often called soft or are established by the EPMM and fixed for the project's life cycle. With value-driven project management, however, metrics can change from project to project, during a life cycle phase and over time, because of tangible value metrics, whereas the hard values are often considered intangible value metrics. Table 5.4 illustrates some of the easy and hard value metrics to measure. Table 5.5 shows some of the problems associated with measuring both hard and soft value metrics.

Today, some consider the intangible elements to be more important than the tangible elements. This appears to be happening on IT projects where executives are giving significantly more attention to intangible values. The critical issue with intangible values is not necessarily in the end result but in the way that the intangibles were calculated.[4]

Tangible values are usually expressed quantitatively, whereas intangible values may be expressed through a qualitative assessment. There are three schools of thought regarding value measurement:

School #1: The only thing that is important is ROI.

---

4 For additional information on the complexities of measuring intangibles, see ibid., Chapter 10. The authors emphasize that the true impact on a business must be measured in business units.

TABLE 5.4   Typical Financial Value Metrics

EASY (SOFT/TANGIBLE) VALUE METRICS	HARD (INTANGIBLE) VALUE METRICS
ROI calculators	Stockholder satisfaction
Net present value	Stakeholder satisfaction
Internal rate of return	Customer satisfaction
Cash flow	Employee retention
Payback period	Brand loyalty
Profitability	Time to market
Market share	Business relationships
	Safety
	Reliability
	Goodwill
	Image

TABLE 5.5   Problems with Measuring Value Metrics

EASY (SOFT/TANGIBLE) VALUE METRICS	HARD (INTANGIBLE) VALUE METRICS
Assumptions are often not fully disclosed and can affect decision making.	Value is almost always based on subjective attributes of the person doing the measurement.
Measurement is very generic.	The perceived value may be more of an art than a science.
Measurement never captures the correct data in a meaningful way.	Limited models are available to perform measurement.

School #2: ROI can never be calculated effectively; only the intangibles are important.

School #3: If something cannot be measured, then it does not matter.

The three schools of thought appear to be all-or-nothing approaches where value is either 100% quantitative or 100% qualitative. The best approach is most likely a compromise between a quantitative and qualitative assessment of value. It may be necessary to establish an effective range, as shown in Figure 5.5, which is a compromise among the three schools of thought. The effective range can expand or contract.

The timing of value measurement is absolutely critical. During the life cycle of a project, it may be necessary to switch back and forth from qualitative to quantitative assessment of the metric and, as stated previously, the actual metrics or KPIs may then be subject to change. Certain critical questions must be addressed:

- When or how far along the project life cycle can concrete metrics be established, assuming this can be done at all?

Figure 5.5 Quantitative versus Qualitative Assessment

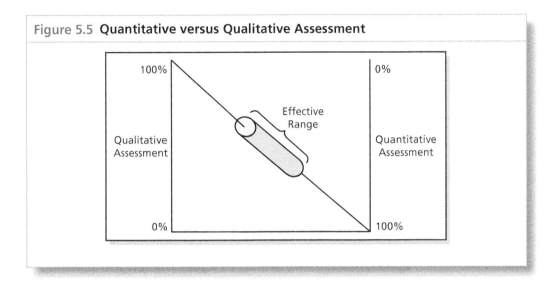

- Can value simply be perceived and, therefore, no value metrics are required?
- Even if there are value metrics, are they concrete enough to reasonably predict actual value?
- Will project managers be forced to use value-driven project management metrics on all projects, or are there some projects where this approach is not necessary? Examples might include projects that are:
- Well defined versus ill defined
- Strategic versus tactical
- Internal versus external
- Can a criterion be developed for when to use value-driven project management, or should it be used on all projects but at lower intensity levels?

For some projects, using metrics to assess value at project closure may be difficult. Project managers must establish a time frame for how long to wait to measure the final or real value or benefits from a project. This is particularly important if the actual value cannot be identified until sometime after the project has been completed. Therefore, it may not be possible to appraise the success of a project at closure if its true economic value cannot be realized until sometime in the future.

Some practitioners of value measurement question whether value measurement is better using boundary boxes instead of life cycle phases. Potential problems with life cycle phase metrics for value-driven projects are listed next.

- Metrics can change between phases and even during a phase.
- One may not be able to account for changes in the enterprise environmental factors.

- The focus may be on the value at the end of the phase rather than the value at the end of the project.
- Team members may get frustrated not being able to calculate value quantitatively.

Boundary boxes, as shown in Figure 5.6, have some similarity to statistical process control charts and can assist in metric measurements. Upper and lower strategic targets for the value of the metrics are established. As long as the KPIs indicate that the project is still within the upper and lower value targets, the project's objectives and deliverables may not undergo any scope changes or trade-offs.

Value-driven projects must undergo value health checks to confirm that the project will make a contribution of value to the company. Value metrics, such as KPIs, indicate the current value. What is also needed is an extrapolation from the present into the future. Using traditional project management combined with the traditional EPMM, the time at completion and the cost at completion can be calculated. These are common terms that are part of earned value measurement systems (EVMSs). However, as stated earlier, being on time and within budget is no guarantee that the perceived value will be there at project completion.

Therefore, instead of using an EPMM, which focuses on earned value measurement, project managers may need to create a value management methodology (VMM), which stresses the value variables. With VMM, time to complete and cost to complete are still used, but a new term, "value (or benefits) at completion," is introduced. Determination of value at completion must be done periodically throughout the project. However, periodic reevaluation of benefits and value at completion may be difficult for these reasons:

- There may be no reexamination process.
- Management is not committed and believes that the reexamination process is meaningless.
- Management is overoptimistic and complacent about existing performance.

Figure 5.6 **Boundary Box**

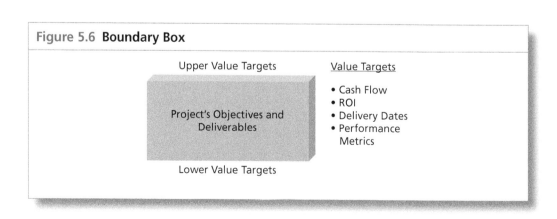

Upper Value Targets

Value Targets

Project's Objectives and Deliverables

- Cash Flow
- ROI
- Delivery Dates
- Performance Metrics

Lower Value Targets

- Management is blinded by unusually high profits on other projects (misinterpretation).
- Management believes that the past is an indication of the future.

5.6 CUSTOMER/STAKEHOLDER IMPACT ON VALUE METRICS

For years, customers and contractors have been working toward different definitions of project success. The project manager's definition of success was profitability and was tracked through financial metrics. The customer's definition of success was usually the quality of the deliverables. Unfortunately, quality was measured at the closure of the project because it was difficult to track throughout the project. Yet quality was often considered the only measurement of success.

Today, clients and stakeholders appear to be more interested in the value they will receive at the end of the project. If we were to ask 10 people, including project personnel, the meaning of value, we would probably get 10 different answers. Likewise, if we were to ask which CSF has the greatest impact on value, we would get different answers. Each answer would be related to the individual's work environment and industry. Today, companies seem to have more of an interest in value than in quality. This does not mean that companies are giving up on quality. Quality is part of value. Some people believe that value is simply quality divided by the cost of obtaining that quality. In other words, the less a company pays for obtaining the customer's desired level of quality, the greater the value to the customer.

The problem with this argument is that it assumes that quality may be the only attribute of value that is important to the client and, therefore, that we need to determine better ways of measuring and predicting just quality.[5]

Unfortunately, there are other attributes of value, and many of these other attributes are equally as difficult to measure and predict. Customers can consider many attributes to represent value, but not all of the value attributes are equal in importance.

Unlike the use of quality as the solitary parameter, value allows a company to better measure the degree to which the project will satisfy its objectives. Quality can be regarded as an attribute of value along with other attributes. Today, every company has quality and produces quality in some form. This is necessary for survival. What differentiates one company from another, however, are the other attributes, components, or factors used to define value. Some of these

---

5 Throughout this chapter, the "client" can be internal to the company, external to the company, or the customers of an external client. Stakeholders also can be considered clients.

attributes might include price, timing, image, reputation, customer service, and sustainability.

In today's world, customers make decisions to hire a contractor based on the value they expect to receive and the price they must pay to receive this value. Actually, the value may be more of a "perceived" value that may be based on trade-offs on the attributes selected as part of the client's definition of value. *The client may perceive the value of the project to be used internally in his or her company or pass it on to customers through the client's customer value management (CVM) program.* If the project manager's organization does not or cannot offer recognized value to clients and stakeholders, then the project manager will not be able to extract value (i.e., loyalty) from them in return. Over time, they will defect to other contractors.

The importance of value is clear. According to a study by the American Productivity and Quality Center:

> Although customer satisfaction is still measured and used in decision making, the majority of partner organizations [used in this study] have shifted their focus from customer satisfaction to customer value.[6]

Project managers of the future must consider themselves as the creators of value. In the courses I teach, I define a "project" as "a set of values scheduled for sustainable realization." Project managers therefore must establish the correct metrics so that the client and stakeholders can track the value that will be created. Measuring and reporting customer value throughout the project is now a competitive necessity. If it is done correctly, it will build emotional bonds with your clients.

5.7 CUSTOMER VALUE MANAGEMENT

For decades, many companies believed that they had an endless supply of potential customers. Companies called this the "doorknob" approach, whereby they expected to find a potential customer behind every door. Under this approach, customer loyalty was nice to have but not a necessity.

Customers were plentiful and often had little regard for the quality of the deliverables. Those days may be gone.

As the quality movement began to take hold during the 1980s, so did the need for effective customer relations management (CRM), as shown in Figure 5.7. Most CRM programs focused on three things:

1. Finding the right customers
2. Developing the right relationships with these customers
3. Retaining customers

---

6 American Productivity and Quality Center, "Customer Value Measurement: Gaining Strategic Advantage," 1999, p. 8.

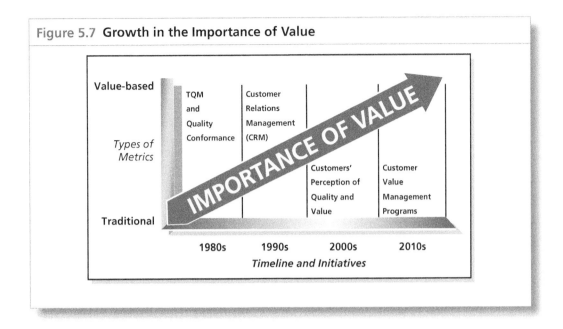

Figure 5.7 **Growth in the Importance of Value**

This included stakeholder relations management and seeking out ways to maintain customer loyalty.

Historically, sales and marketing were responsible for CRM activities. Today, project managers are doing more than simply managing a project; they are managing part of a business. Therefore, they are expected to make business decisions as well as project decisions, and this includes managing activities related to CRM. Project managers soon found themselves managing projects that required effective stakeholder relations management as well as customer relations management. Satisfying the needs of both the client and various stakeholders was difficult.

As CRM began to evolve, companies soon found that there were different perceptions among their client base regarding the meaning of quality and value. In order to resolve these issues, companies created customer value management (CVM) programs. CVM programs address the critical question: *Why should customers purchase from you rather than from the competition?* The answer was most often the value provided through products and deliverables. Loyal customers appear to be more value sensitive than price sensitive. Loyal customers are today a scarce resource and also a source of value for project managers and their organizations. Value breeds loyalty.

There are other items, such as trust and intangibles, that customers may see as a form of value. As stated by a technical consultant:

> The business between the vendor and the customer is critical. It's situational but in technical consulting for instance the customer may really value only the technical prowess of a vendor's team; project management is expected to be competent in this case. If project

management itself is adding value then isn't that really a matter of the customer's view of the project manager providing services above and beyond the normal view of functional responsibility? That would come down to the relationship with the customer. Ask any customer what they truly value in a vendor and they will tell you it is trust because there is a reliance on the customer's business strategy succeeding based on how well the vendor executes.

For example, in answer to the question, Why do you value vendor X?, you could imagine the following answers from customers: "So and so always delivers for me, and I can count on them to deliver a quality (defined however) product on time and at the agreed price," or "Vendor X really helped me be successful with my management by pulling in the schedule by x weeks," or "I really appreciated a recent project done by vendor Y because they handled our unexpected design changes with professionalism and competence."

Now most project management is within the sphere of operations in a vendor's organization. The customer-facing business relationship is handled by some company representative in most cases. The project manager would be brought in once the work is under way, and typically the direct reporting is to someone underneath the person who authorized the work; so in this case the project manager has the opportunity to build value with the underling but not the executive.

We discount the personality of the project managers as though this isn't an issue. It is a major issue and people need to realize that it is. Understanding one's own personality and the personality of the customer is vital to getting a label of value added from a customer. If the project manager isn't flexible in this area, then creating value with the customer becomes more difficult.

Anyway, there are as many variations to this theme as there are projects since personality and other interpersonal relationship nuances are involved. However, so much of successful project management is all about these intangibles.

CVM today focuses on maximizing customer value, regardless of the form. In some cases, CVM must measure and increase the lifetime value of the deliverables of the project for each customer and stakeholder. By doing this, the project manager is helping customers manage their profitability as well.

Performed correctly, CVM can and will lead to profitability, but being profitable does not mean that a project manager is performing CVM correctly. Sales and marketing personnel rather than project managers historically managed CVM activities and viewed their role as handling customer complaints, enforcing warranty promises, and looking for ways to sell more units to a customer. Today, CVM activities are involving project managers and focus on determining the customer's definition of value-in-use and establishing metrics that can support this definition. Determining the correct value-in-use metrics may require that the project managers observe how the customers are using the products and services. Each company may have a different set of metrics to satisfy customer needs.

There are benefits to implementing CVM effectively, as shown in Table 5.6. CVM is the leveraging of customer and stakeholder business relationships throughout the project. Because each project will have different customers and different stakeholders, CVM must be custom fit to each organization and possibly each project.

If CVM is to be effective, then presenting the right information to the client and stakeholders becomes critical. CVM introduces a value mindset into decision making. Many CVM programs fail because of poor metrics and not measuring the right things. Just focusing on the end result does not indicate if what the project manager is doing is right or wrong. Having the correct value-based metrics is essential. Value is in the eyes of the beholder, which is why there can be different value metrics.

CVM relies heavily on customer value assessment. Traditional CVM models are light on data and heavy on assumptions. For CVM to work effectively, it must be heavy on data and light on assumptions. Most successful CVM programs perform data mining, where the correct attributes of value are found along with data that supports the use of those attributes. However, valuable resources should not be wasted calculating value metrics unless the client perceives the value of using the metric. Project management success in the future will be measured by how well the project manager provides superior customer value. To do this, the project manager must know what motivates the customers and the customers' customers.

Executives generally have a better understanding of the customers' needs than do the workers at the bottom of the organizational chart where the work takes place. Therefore, the workers may not see, understand, or appreciate the customer's need for value. Without the use of value metrics, project managers focus on the results of the process rather than the process itself and miss opportunities to add value. Only when a crisis occurs do is the process put under a microscope. Value metrics provide workers with a better understanding of the customer's definition of value.

TABLE 5.6   Before and after CVM Implementation

FACTOR	BEFORE CVM	AFTER CVM (WITH METRICS)
Stakeholder communications	Loosely structured	Structured using a network of metrics
Decision making	Based on partial information	Value-based informed decision making
Priorities	Partial agreements	Common agreements using metrics
Trade-offs	Less structured	Structured around value contributions
Resource allocation	Less structured	Structured around value contributions
Business objectives	Projects poorly aligned to business	Better alignment to business strategy
Competitiveness	Market underperformer	Market outperformer

Understanding the customer's perception of value means looking at the disconnects or activities that are not value-added work. Value is created when non-value-added work can be eliminated rather than looking at ways to streamline project management processes. As an example, consider the following situation.

SITUATION: A company had a project management methodology that mandated that a risk management plan be developed on each and every project regardless of the magnitude of the risk. The risk management plan was clearly defined as a line item in the work breakdown structure and the clients eventually paid the cost of this. On one project, the project manager created a value metric and concluded that the risks on the project were so low that risk should not be included as one of the attributes of the value metric. The client agreed with this, and the time and money needed to develop a risk management plan was eliminated from the project plan. On this project, the risk management plan was seen as a disconnect and eliminated so that added value could be provided to the client. The company recognized that the risk management plans might be a disconnect on some projects and made the risk management plan optional at the discretion of the project manager and the clients.
This situation is a clear example that value is created when outputs are produced using fewer inputs or more outputs are created with the same number of inputs. However, care must be taken to make sure that the right disconnects are targeted for elimination.

Project managers must work closely with the customers for CVM to be effective. This includes:

- Understanding the customer's definition of satisfaction and effective performance.
- Knowing how the customer perceives the project manager's price/value relationship (some clients still believe that value is simply quality divided by price).
- Making sure that the customers understand that value can be expressed in both nonfinancial and financial terms.
- Ensuring client understands the project manager's distinctive competencies and determining if those competencies are appropriate candidates for value attributes.
- Being prepared to debrief the customer and stakeholders on a regular basis for potential improvements and best practices.
- Validating that the client is currently using or is willing to use the value metrics for its own informed decision making.
- Understanding which value attributes are most important to customers.
- Building a customer project management value model or framework possibly unique for each customer.
- Making sure that the project manager's model or framework fits the customer's internal business model.
- Designing metrics that interface with the customer's business model.

- Recognizing that CVM can maximize lifetime profitability with each customer.
- Changing from product-centric or service-centric marketing to project management-centric marketing.
- Maximizing the economics of customer loyalty.

5.8 THE RELATIONSHIP BETWEEN PROJECT MANAGEMENT AND VALUE

Companies today are trying to link quality, value, and loyalty. These initiatives, which many call CVM initiatives, first appeared as business initiatives performed by marketing and sales personnel rather than project management initiatives. Today, however, project managers are slowly becoming more involved in business decisions, and value has become extremely important.

Quality and customer value initiatives are part of CVM activities and are a necessity if a company wishes to obtain a competitive advantage. Competitive advantages in project management do not come just from being on time and within budget at the completion of each project. Offering something that competitors do not offer may help. However, true competitive advantage is found when the project manager's efforts are directly linked to the customer's value initiatives, and whatever means by which this can be shown will give the project manager a step up on the competition. Project managers must develop value-creating strategies and must also know how to create a project value baseline, as is discussed in Section 5.21.

Customers today have become more demanding and are requiring the contractor to accept the customer's definition of value according to the attributes selected by the customer. Each customer can, therefore, have a different definition of value. Contractors may wish to establish their own approach for obtaining this value based on their company's organizational process assets rather than having the customer dictate it. If a project manager establishes his or her own approach to obtaining the desired value, it should not be assumed that customers will understand the approach. They may need to be educated. Customers who recognize and understand the value that the project manager is providing are more likely to want a long-term relationship with the project manager's firm.

To understand the complexities with introducing value to project management activities, let us assume that companies develop and commercialize products according to the phases in Figure 5.8. Once the project management phase is completed, the deliverables are turned over to someone in marketing and sales responsible for program management and ultimately commercialization of the product. Program management and commercialization may be done for products developed internally for one's own company or they may be done for the client or even for the client's own customers. In any event, it is during or after commercialization when companies survey their customers as to feedback on customer satisfaction and the value of the product. If customers are unhappy with the final value, it is an expensive process to go back to the project stage and repeat the project in order to try to improve the value for the end-of-line customers.

The failures from Figure 5.8 can be attributed to:

- Project managers are allowed to make only project decisions rather than both project and business decisions.
- Project managers are not informed as to the client's business plans as they relate to what the project manager is developing.
- Customers do not clearly articulate to the project manager, either verbally or through documented requirements, the exact value that they expect.
- Customers fund projects without fully understanding the value needed at the completion of the project.
- Project managers interface with the wrong people on the project.
- No value-based metrics are established in the project management phase whereby informed decision making can take place to improve the final value.
- Trade-offs are made without considering the impact on the final value.
- Quality and value are considered as synonymous; quality is considered the only value attribute.

Returning to Figure 5.7, it can be seen that as CVM activities are approached, traditional metrics are replaced with value-based metrics. The value metric should not be used to replace other metrics that people may be comfortable using for tracking project performance. Instead, it should be used to support other metrics being used. As stated by a healthcare IT consulting company:

> [The need for a value metric] "hits directly on the point" that so many of us consultants encounter on a regular basis surrounding the fact that often the metrics that are utilized to ensure success of projects and programs are too complex, don't really demonstrate the key factors of what the sponsors, stakeholders, investors, client, etc., "need"

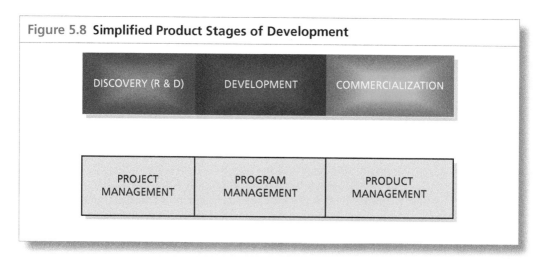

Figure 5.8 Simplified Product Stages of Development

DISCOVERY (R & D)	DEVELOPMENT	COMMERCIALIZATION
PROJECT MANAGEMENT	PROGRAM MANAGEMENT	PRODUCT MANAGEMENT

in order to demonstrate value and success in a project and do not always "circle back" to the mission, vision, and goals of the project at the initial kick-off. What we have found is that the majority of our clients are constantly in search of something simple, streamlined, and transparent (truly just as simple as that). In addition, they generally request a "pretty picture" format; something that is clear, concise and easy and quick to glance at in order to determine areas that require immediate attention. They crave for the red, yellow, green. They yearn for a crisp measure to state what the "current state" truly is [and whether the desired value is being achieved]. They are tired and weary of the "circles," they are exhausted of the mounds of paperwork, and they want to take it back to the straight facts.

Most companies are now in the infancy stage of determining how to define and measure value. For internal projects, we are struggling in determining the right value metrics to assist us in project portfolio management with the selection of one project over another. For external projects, the picture is more complex. Unlike traditional metrics used in the past, value-based metrics are different for each client and each stakeholder. Figure 5.9 shows the three dimensions of values: the parent company's values, the client's values, the client's customers' values, and a fourth dimension could be added, namely stakeholder values. It should be understood that the value that the completion of a project brings to the project manager's organization may not be as important as the total value that the project brings to the client's organization and the client's customer base.

For some companies, the use of value metrics will create additional challenges. As stated by a global IT consulting company:

Figure 5.9 Dimensions of Value

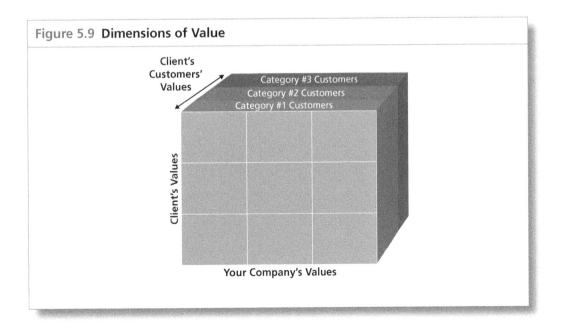

This will be a cultural change for us and for the customer. Both sides will need to have staff competent in identifying the right metrics to use and weightings; and then be able to explain in layman's language what the value metric is about.

The necessity for such value initiatives is clear. As stated by a senior manager:

I fully agree on the need for such value initiatives and also the need to make the importance clear to senior management. If we do not work in that direction, it will be very difficult for the companies to clearly state if they are working efficiently enough, providing value to the customer and the stockholders, and in consequence being sufficiently predictive with regard to its future in the market.

The window into the future now seems to be getting clearer. As a guess as to what might happen, consider the following:

- Clients will perform CVM activities with their own clients — their customers—to discover what value attributes are considered as important. Their success is achieved by providing superior value to their customers.
- These attributes of quality will be presented to the project manager at the initiation of the project so that value-based metrics can be created for the project using these attributes if possible. The project manager must interact with clients to understand their value dimensions.
- The project manager must then create value metrics and be prepared to educate the client on the use of the metrics. It is a mistake to believe that clients will fully understand the value metrics approach. Interact closely with your client to make sure you are fully aware of any changes in the value attributes they are finding in their CVM efforts.
- Since value creation is a series of key and informed decisions, be prepared for value attribute trade-offs and changes to your value metrics.

Value is now being introduced into project management practices. Value management practices have been around for several decades and have been hidden under the radar in many companies. Some companies performed these practices in value engineering (VE) departments. Today the primary processes of value management, as shown in Figure 5.10, are becoming readily apparent.[7]

---

7 For an excellent book discussing the evolution of value management, see Michel Thiry, *Value Management Practice* (Newtown Square, PA: The Project Management Institute, 1997). Another good reference, which discusses how value can be used in making project decisions, is Thomas G. Lechler and John C. Byrne, *The Mindset for Creating Project Value* (Newtown Square, PA: The Project Management Institute, 2010). Mel Schnapper and Steven Rollins created an excellent book, *Value-Based Metrics for Improving Results; An Enterprise Project Management Toolkit* (Ft. Lauderdale, FL: J. Ross, 2006). The authors discuss how this technique was applied at 3 M Corporation. Chapters 17 to 25 directly relate value-based metrics to the areas of knowledge in the *PMBOK" Guide.* PMBOK is a registered mark of the Project Management Institute, Inc.

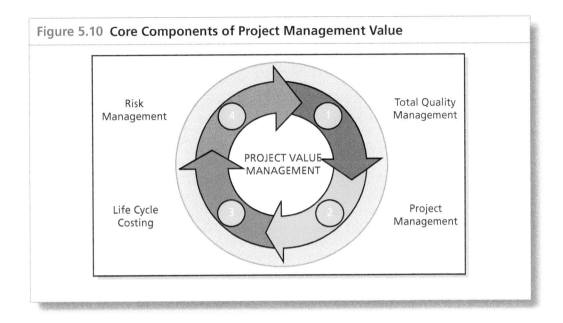

Figure 5.10 **Core Components of Project Management Value**

Although only four processes appear in Figure 5.10, it could be argued that all of the areas of knowledge in the *PMBOK* Guide* are part of project value management.

5.9 **BACKGROUND OF METRICS**

You have been managing a project for the past several months. During that time, the client appeared quite happy with your performance, especially because all of the status reports indicated that your performance was within time and cost. You patted yourself on the back for doing a good job, you received accolades from your team and management, and then reality set in as the end of the project neared; the client was unhappy with the end result and did not believe that they were receiving the value that they expected when the project was initiated. The client even commented that they probably should have canceled the project before wasting all of this money. What went wrong?

The problem can be addressed in one word, *metrics*, or perhaps we should say using the wrong metrics or the lack of metrics that could have projected or demonstrated value throughout the project. Having some metrics is certainly better than having no metrics at all. However, having the right metrics, especially the inclusion of metrics that can in some way describe and communicate the value of the project, is best. Metrics must fully communicate what was needed by the customer.

When projects fail, project managers conduct debriefing sessions that focus more on blame-laying and finger-pointing than identifying the root cause of the failure. Sometimes project managers go through meticulous pain to identify every possible reason for a failure without identifying the most critical and real causes of the failure. As an example, in some industries, such as IT, surveys disclose a multitude of causes of failure. Well-known IT surveys include:

- The Standish Group Chaos Reports (1995–2010)
- The OASIG Study (1995)
- The KPMG Canada Survey (1997)
- The Bull Survey (1998)

Yet in each of these surveys, very little effort is expended on the fact that the wrong metrics may have been used. Also, many of the causes for failure could have been prevented if metrics had been established to track the potential causes of failure. Then people wonder why these surveys show that less than 30% of the IT projects are completed successfully.

Every year surveys are published that show basically the same reasons for the failures of IT projects. If the same causes appear year after year, why do project managers refuse to identify new metrics to track these causes of failure? Just imagine how many billions of dollars could have been saved on IT projects if project managers started with the right metrics.

Measuring performance on a project is more than looking at just time and cost. For more than half a century, project managers were taught that the "holy grail" of performance metrics was getting the job done within time and cost. Unfortunately, achieving time and cost does not guarantee that business value will be there at the end of the project nor does it guarantee that the project manager has been working on the right project. Instead, it simply means that project managers spent what they thought they would spend, and it took as long as they thought. It does not mean that the project was or will be a success.

Redefining Success

For decades, the definition of project success was meeting the proverbial triple constraints, as shown in Figure 5.11. Time and cost were two of the three sides to the triangle, and the third side was scope, technology, performance, or quality, depending on who was defining success. Whenever other constraints appeared, such as risk, business value, image reputation, safety, and sustainability, they were inserted into the center of the triangle with the belief that they either lengthened or compressed the boundary triple constraints.

Today, project management practitioners agree that there are more than three constraints on most projects, referred to as *competing constraints*. The old adage of working with just the triple constraints has gone by the wayside. The competing constraints also are prioritized. As an example of prioritization, the design and development of new attractions at Disney World and Disneyland were characterized

Figure 5.11 Traditional Triple Constraints

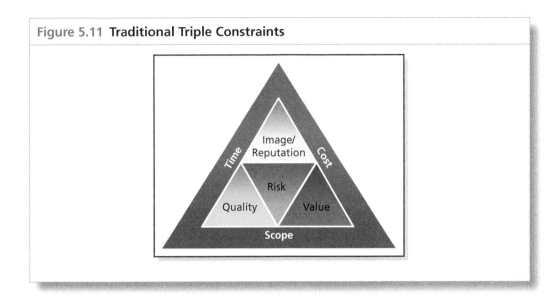

by six constraints; time, cost, scope, safety, aesthetic value, and quality. During my consulting with Disney, it was apparent the three priority constraints were safety, aesthetic value, and quality. If and when trade-offs were required, the only options were trade-offs on time, cost, and scope. Safety, aesthetic value, and quality constraints were considered untouchable.

One of the first steps on a project is the agreement between the contractor and the client/stakeholders on the requirements of the project and the various limitations or constraints. For years, contractors, clients, and stakeholders each had their own definitions of success, which created havoc with performance reporting. Every project can have different constraints and a different definition of success, but the final definition of project success must be agreed to jointly by the contractor and the client/stakeholders. Otherwise, confusion will reign as to what is or is not important.

Today's need for more than just three constraints to define success is quite apparent. Unfortunately, as with most changes to the way business is done, new headaches appear. For example, new constraints require new and more sophisticated metrics. For years many companies abided by the rule of inversion, which stated that time and cost were the only two metrics that needed to be tracked because they were the easiest to measure. The more difficult metrics, which could provide an indication of the project's true value, were omitted because they could not be measured. All project management software packages track time and cost, and many of the packages track only time and cost as the core metrics. However, as stated previously, time and cost alone cannot accurately determine performance and the success or value of a project. They tell only part of the story.

As project managers become more experienced in the use of project management and undertake more complex projects, the necessity of using additional constraints to validate performance and success is recognized. This is a necessity to create a project value baseline and is discussed in Section 5.21. For every competing constraint that appears on projects today, there must be one or more metrics to track that constraint. This need for value metrics will require major enhancements to many of the project management software packages currently in the marketplace.

Growth in the Use of Metrics

For more than two decades, project managers reluctantly recognized and understood the cost of paperwork on projects and yet did nothing about it. Some companies estimated that 8–10 hours were needed for each page of a report or handout given to the customer. This included organizing the report, typing, proofing, editing, retyping, graphic arts, approvals, reproduction, distribution, classification, storage, and destruction. Based on the cost per billable hour, it was not uncommon for a company to spend $1200–$2000 per page.

The solution to this problem was quite simple: go to paperless project management. But in order to minimize the cost of paperwork, other methods of conveying information must be used, such as dashboards. The purpose of a dashboard is to convert raw data into meaningful information that can be easily understood and used for informed decision making. Thus, the need existed for effective metrics that could appear in dashboard reporting systems. The value of dashboards was now quite clear:

- Reduction or consolidation of reports.
- Less time wasted in preparing and reading reports.
- Reduction in time needed for project monitoring and control.
- Informed decision making based on current or real-time data.
- More time available for important project management work.

Unfortunately, today more artwork than needed is added to the dashboards, part of the trend over recent decades toward what has come to be called "infographics." Some of the problems associated with the growth of infographics include:

- There is a heavy focus on designs, colors, images, and text rather than on the quality of the information being presented.
- A decline in the quality of the information makes it difficult for stakeholders to use the data properly.
- There are too many pretty graphics that can be misleading and hard to understand.
- Dashboards have been converted from a project management performance tool to a marketing/sales tool.
- Some graphic artists do not understand or utilize information visualization best practices.
- Political factors sometimes dictate dashboard design.

■ Rushing into often complex technology before understanding the needs of the user group.
■ Metric scope creep occurs because of a lack of a formalized metric selection process.
■ Failing to understand which metrics are:
  ■ Must have—fundamental to the requirements
  ■ Nice to have—but not a necessity (partial value)
  ■ Can wait—does not add value
  ■ Not needed—forget it (no value at all)

Better and clearer representation of the metrics selected is necessary.

Companies have been using metrics for business applications and developing business strategies for some time. However, the application of business metrics to project management is difficult to achieve because of differences shown in Table 5.7. Business and project metrics both have important use, but in different contexts.

For project management applications, companies used the rule of inversion (i.e., selecting only metrics that were easy to measure and track) and created a list of six core metrics, as shown in Table 5.8. Some companies use additional or supporting metrics, such as:

■ **Deliverables (schedule):** Late versus on time
■ **Deliverables (quality):** Accepted versus rejected
■ **Management reserve:** Amount available versus amount used
■ **Risks:** Number of risks in each core metric category
■ **Action items:** Action items in each core category
■ **Action items aging:** How many action items not completed are over one month, two months, three or more months

The core metrics are often represented by a dashboard labeled "Health Metrics," as shown in Figure 5.12. Most upper management and customer organizations look for a single, simple metric that will indicate to them if a given project is on the road to success. If a single

TABLE 5.7   Business versus Project Metrics/KPIs

VARIABLE	BUSINESS/FINANCIAL	PROJECT
Focus	Financial measurement	Project performance
Intent	Meeting strategic goals	Meeting project objectives, milestones, and deliverables
Reporting	Monthly or quarterly	Real-time data
Items to be looked at	Profitability, market share, repeat business, number of new customers, etc.	Adherence to competing constraints, validation, and verification of performance
Length of use	Decades or even longer	Life of the project
Use of data	Information flow and changes to strategy	Corrective action to maintain baselines
Target audience	Executive management	Stakeholders and working levels

TABLE 5.8 The Core Metrics	
MEASURE	**INDICATOR**
Time	Schedule performance index (SPI)
Cost	Cost performance index (CPI)
Resources	Quality and number of actual versus planned staff
Scope	Number of change requests
Quality	Number of defects against user acceptance criteria
Action Items	Number of action items behind schedule

value-based metric does not exist, then they usually accept some or all of the core metrics shown in Figure 5.12. A simple signal light is all they may be looking for:

- **Green:** On the road to success (No action required)
- **Yellow:** Problems have surfaced that can derail the progress of the project. (Upper management and the customer will want to know what is being done to solve these problems.)
- **Red:** Progress has stopped because the project is derailed. (Upper management and the customer want to know what resources they should apply to get the project healthy or whether it is beyond repair so that it has to be abandoned. *Is it no longer of value?*)

This is the management/customer value metric in its simplest form. According to an aerospace firm:

> The challenge is identifying what the intrinsic value of the project is. This can only be defined by a dedicated process to obtain consensus

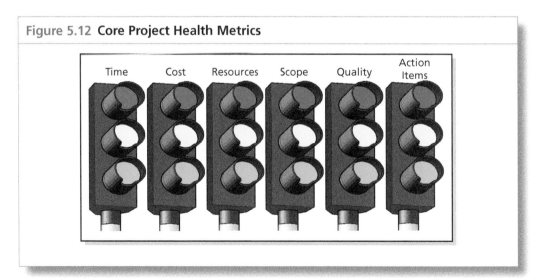

Figure 5.12 **Core Project Health Metrics**

from management and the customer. Once this is defined and agreed to, then the project manager can select the minimum set of metrics that will allow him to gauge the success of his efforts. The set of metrics must be at a minimum so that the tracking of metrics does not in itself become a burden to the successful progress of the project.

Unfortunately, there are problems with using just the core metrics to explain the health of a project:

- The core metrics are usually interdependent and must be considered together to get an accurate picture of status;
- Additional metrics may have to be added specific to the project at hand;
- Explaining the core metrics by the colors of red, yellow, and green can be confusing.
- Core metrics are similar to vital signs taken when visiting a doctor's office. Doctors always take the same core metrics: height, weight, temperature, and blood pressure. From these core metrics alone, the doctor usually cannot diagnose the problem or prescribe a corrective course of action. Additional metrics, such as blood work and x-rays, may be required.

The importance of using metrics other than just time and cost, or even the core metrics, has been known for some time. Knowing it and doing something about it are two different things, however. In the past, project managers avoided using more metrics because they did not know how to measure them. Today, books on the marketplace promote the concept that "anything can be measured."

Techniques have been developed by which image, reputation, goodwill, and customer satisfaction, just to name a few, can be measured. Some of the measurement techniques include:[8]

- Observations
- Ordinal (e.g., four or five stars) and nominal (e.g., male or female) data tables
- Ranges, sets of value, number, headcount, percentages
- Simulation
- Statistical measurement
- Calibration estimates and confidence limits
- Decision models (earned value, expected value of perfect information, etc.)
- Sampling techniques
- Decomposition techniques
- Direct versus indirect measurement
- Human judgment

---

8 For a description of several of these techniques, see Douglas W. Hubbard, *How to Measure Anything* (Hoboken, NJ: John Wiley & Sons, 2007).

5.10 SELECTING THE RIGHT METRICS

Because of these measurement techniques, companies are now tracking a dozen or more metrics on projects. Although this sounds good, it has created the additional problem of potential information overload. Having too many performance metrics may provide viewers with more information than they actually need. Therefore, they may not be able to discern the true status of the project or what information is really important. It may be hard to ascertain what is important and what is not, especially if decisions must be made. Providing too few metrics can make it difficult for viewers to make informed decisions. There is also a cost associated with metric measurement, and it must be determined if the benefits of using this many metrics outweigh the costs of measurement. Cost is important because users tend to select more metrics than actually are needed.

There are three categories of metrics:

1. **Traditional metrics:** These metrics are used for measuring the performance of the applied project management discipline more than the results of the project and how well project managers are managing according to the predetermined baselines. (e.g., cost variance and schedule variance)
2. **Key performance indicators (KPIs):** These are the few selected metrics that can be used to track and predict whether the project will be a success. These KPIs are used to validate that the CSFs defined at the initiation of the project are being met (e.g., time at completion, cost at completion, and customer satisfaction surveys).
3. **Value (or value reflective) metrics:** These are special metrics that are used to indicate whether the stakeholders' expectations of project value are or will be met. Value metrics can be a combination of traditional metrics and KPIs (value at completion and time to achieve full value).

Each type of metric has a primary audience, as shown in Table 5.9. There can be three information systems on a project:

1. One for the project manager
2. One for the project manager's superior or parent company
3. One for the stakeholders and the client

TABLE 5.9 Audiences for Various Metrics

TYPE OF METRIC	AUDIENCE
Traditional metrics	Primarily the project manager and the team, but may include the internal sponsor(s) as well
Key performance indicators	Some internal usage but mainly used for status reporting for the client and the stakeholders
Value metrics	Can be useful for everyone but primarily for the client

There can be a different set of metrics and KPIs for each of these information systems.

Traditional metrics, such baselines as the cost, scope, and schedule, track and can provide information on how well project managers are performing according to the processes in each Knowledge Area or Domain Area in the *PMBOK* Guide*. Project managers must be careful not to micromanage their projects and establish 40 to 50 metrics.

Typical metrics may include:

- Number of assigned versus planned resources
- Quality of assigned versus planned resources
- Project complexity factor
- Customer satisfaction rating
- Number of critical constraints
- Number of cost revisions
- Number of critical assumptions
- Number of unstaffed hours
- Percentage of total overtime labor hours
- Cost variance
- SPI
- CPI

This is obviously not an all-inclusive list. These metrics may have some importance for the project manager but not necessarily the same degree of importance for the client and the stakeholders.

Clients and stakeholders are interested in critical metrics or KPIs. These chosen few metrics are reported to the client and stakeholders and provide an indication of whether success is possible; however, they do not necessarily identify if the desired value will be achieved. The number of KPIs is usually determined by the amount of real estate on a computer screen. Most dashboards can display between 6 and 10 icons or images where the information can be readily seen with reasonable ease.

Understanding what a KPI means requires a dissection of each of the terms:

Key: A major contributor to success or failure
Performance: Measurable, quantifiable, adjustable, and controllable elements
Indicator: Reasonable representation of present and future performance

Obviously, not all metrics are KPIs. There are six attributes of a KPI, and these attributes are important when identifying and selecting the KPIs.

1. **Predictive:** Able to predict the future of this trend
2. **Measurable:** Can be expressed quantitatively
3. **Actionable:** Triggers changes that may be necessary

4. **Relevant:** The KPI is directly related to project success or failure
5. **Automated:** Reporting minimizes the chance of human error
6. **Few in number:** Only what is necessary

Applying these six attributes to traditional metrics is highly subjective and will be based on the agreed-upon definition of success, the CSFs that were selected, and possibly the whims of the stakeholders. There can be a different set of KPIs for each stakeholder based on each stakeholder's definition of project success and final project value. The use of these six attributes could significantly increase the costs of measurement and reporting, especially if each stakeholder requires a different dashboard with different metrics.

Previously 12 possible metrics that could be used on projects were identified, but how many of those 12 are actually regarded as a KPI? If the first five of the six KPI attributes just identified are applied, the result could be the representation as shown in Table 5.10.

Only 6 of the 12 metrics (1, 3, 4, 8, 11, and 12) may be regarded as a KPI and, once again, this is often a highly subjective selection

TABLE 5.10   Selecting the KPIs

	PREDICTIVE	MEASURABLE	ACTIONABLE	RELEVANT	AUTOMATED
1. # of assigned versus planned resources	Yes	Yes	Yes	Yes	Yes
2. Quality of assigned versus planned resources		Yes		Yes	Yes
3. Project complexity factor	Yes	Yes	Yes	Yes	Yes
4. Customer satisfaction rating	Yes	Yes	Yes	Yes	Yes
5. Number of critical constraints		Yes	Yes	Yes	
6. Number of cost revisions		Yes	Yes	Yes	Yes
7. Number of critical assumptions		Yes	Yes	Yes	
8. Number of unstaffed hours	Yes	Yes	Yes	Yes	Yes
9. Percentage of overtime labor hours		Yes	Yes		Yes
10. Cost variance		Yes			Yes
11. SPI	Yes	Yes	Yes	Yes	Yes
12. CPI	Yes	Yes	Yes	Yes	Yes

process. In this example, these would be the critical metrics that would be shown on the project dashboard and could be necessary for informed decision making. The other metrics can still be used, but users might need to drill down on the screens to get access to the traditional metrics.

5.11 THE FAILURE OF TRADITIONAL METRICS AND KPIS

Although some people swear by metrics and KPIs, there are probably more failures than success stories. Typical causes of metric failure include:

- Performance is expressed in traditional or financial terms only.
- Measurement inversion is used; using the wrong metrics or selecting only those metrics that are easy to measure and report.
- Performance metrics are not linked to requirements, objectives, and success criteria.
- No determination of whether the customer was satisfied.
- Lack of understanding as to which metrics indicate project value.

Metrics used for business purposes tend to express all information in financial terms. Project management metrics cannot always be expressed in financial terms. Also, in project management, metrics that cannot effectively predict project success and/or failure and are not linked to the customer's requirements often are identified.

Perhaps the biggest issue today is in which part of the value chain metrics are used. Michael Porter, in his book *Competitive Advantage*, used the term "value chain" to illustrate how companies interact with upstream suppliers, the internal infrastructure, downstream distributors and end-of-the-line customers.[9] Although metrics can be established for all aspects of the value chain, most companies do not establish metrics for how the end-of-the-line customer perceives the value of the deliverable. Those companies that have developed metrics for this part of the value chain are more likely doing better than those that have not. These metrics are identified as *customer-related* value metrics.

5.12 THE NEED FOR VALUE METRICS

In project management, it is now essential to create metrics that focus not only on business (internal) performance but also on performance toward customer satisfaction. If the customer cannot see the value in the project, then the project may be canceled and repeat business will not be forthcoming. Good value metrics can also result in less customer and stakeholder interference and meddling in the project.

---

9 Michael Porter, *Competitive Advantage* (New York: Free Press, 1985).

Figure 5.13 **Typical Steps in the Performance Metrics Process**

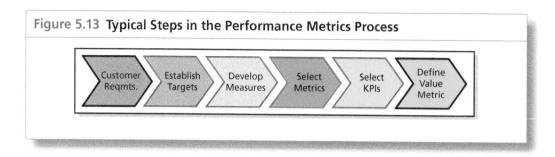

The performance metrics process for project management is shown in Figure 5.13.

The need for an effective metrics management program that focuses on value-based metrics is clear:

- There must be a customer/contractor/stakeholders' agreement on how a set of metrics will be used to define success or failure; otherwise, there are just best guesses. Value metrics will allow for a better agreement on the definitions of success and failure.
- Metrics selection must cover the reality of the entire project; this can be accomplished with a set of core metrics supported by a value metric.
- A failure in effective metrics management, especially value metrics, can lead to stakeholder challenges and a loss of credibility.

Project managers need to develop value-based metrics that can forecast stakeholder value, possibly shareholder value, and most certainly project value. Most models for creating this metric are highly subjective and are based on assumptions that must be agreed upon up front by all parties. Traditional value-based models that are used as part of a business intelligence application are derivatives of the quality, cost, delivery (QCD) model.

5.13 CREATING A VALUE METRIC

The ideal situation would be the creation of a single value metric that stakeholders can use to make sure that the project is meeting or exceeding their expectations of value. The value metric can be a combination of traditional metrics and KPIs. Discussing the meaning of a single value metric may be more meaningful than discussing the individual components; the whole is often greater than the sum of the parts.

There must be support for the concept of creating a value metric. According to a global IT consulting company:

> There has to be buy-in from both sides on the importance and substance of a value metric; it can't be the latest fad—it has to be understood as a way of tracking the value of the project.

The next list specifies typical criteria for a value metric.

- Every project will have at least one value metric or value KPI. In some industries, it may not be possible to use just one value metric.
- There may be a limit, such as five, for the number of value attributes that are part of the value metric. As companies mature in the use of value metrics, the number of attributes can grow or be reduced. Not all desired attributes will be appropriate or practical.
- Weighting factors will be assigned to each component.
- A project value baseline (discussed in Section 5.21) will be tracked using the value metrics.
- The weighting factors and the component measurement techniques will be established by the project manager and the stakeholders at the outset of the project. There may be company policies on assigning the weighting factors.
- The target boundary boxes for the metrics will be established by the project manager and possibly the PMO. If a PMO does not exist, then a project management committee may take responsibility for accomplishing this, or it may be established by the funding organization.

To illustrate how this might work, let us assume that, for the IT projects the project manager performs for stakeholders, the attributes of the value metric will be:

- Quality (of the final software package)
- Cost (of development)
- Safety protocols (for security of information)
- Features (functionality)
- Schedule or timing (for delivery and implementation)

These attributes are agreed to by the project manager, the client, and the stakeholders at the outset of the project. The attributes may come from the metric/KPI library or may be new attributes. Care must be taken to make sure that the organizational process assets can track, measure, and report on each attribute. Otherwise, additional costs may be incurred, and these costs must be addressed up front so that they can be included in the contract price.

Time and cost are generally attributes of every value metric. However, there may be special situations where time and cost, or both, are not value metric attributes:

- The project must be completed by law, such as environmental projects, where failure to perform could result in stiff penalties.
- The project is in trouble, but necessary, and whatever value that can be salvaged must be.
- A new product must be introduced to keep up with the competition regardless of the cost.

- Safety, aesthetic value, and quality are more important than time, cost, or scope.

Other attributes are almost always included in the value metric to support time and cost.

The next step is to set up targets with thresholds for each attribute or component. This is shown in Figure 5.14. If the attribute is cost, then perhaps performing within ± 10% of the cost baseline is normal performance. Performing at greater than 20% over budget could be disastrous, whereas performing at more than 20% below budget is superior performance. However, there are cases where a + 20% variance could be good and a –20% variance could be bad.

The exact definition or range of the performance characteristics could be established by the PMO if company standardization is necessary or through an agreement with the client and the stakeholders. In any event, targets and thresholds must be established.

The next step is to assign value points for each of the cells in Figure 5.14, as shown in Figure 5.15. In this case, two value points were assigned to the cell labeled "Performance Target." The standard approach is to then assign points in a linear manner above and below the target cell. Nonlinear applications are also possible, especially when thresholds are exceeded.

In Table 5.11, weighting factors are assigned to each of the attributes of the value metric. As before, the weighting percentages could be established by the PMO or through an agreement with the client (i.e., funding organization) and the stakeholders. The PMO might be

Figure 5.14 Value Metric/KPI Boundary Box

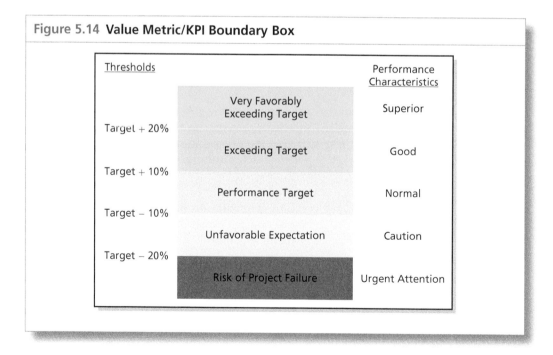

Figure 5.15 Value Points for a Boundary Box

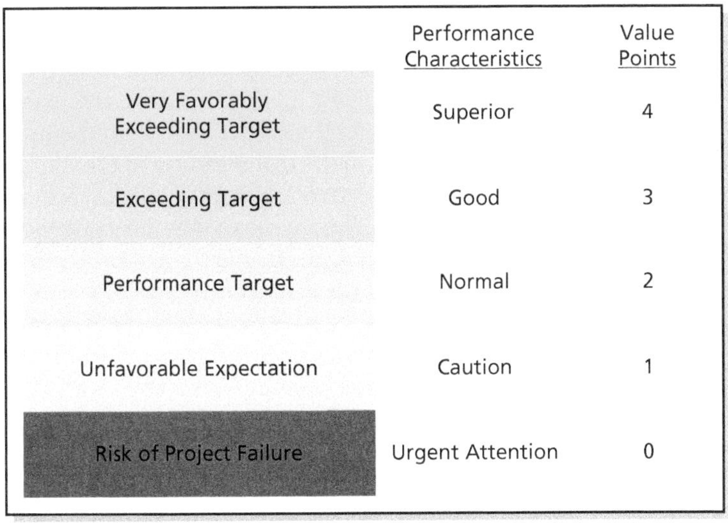

	Performance Characteristics	Value Points
Very Favorably Exceeding Target	Superior	4
Exceeding Target	Good	3
Performance Target	Normal	2
Unfavorable Expectation	Caution	1
Risk of Project Failure	Urgent Attention	0

used for company standardization on the weighting factors. However, allowing the weighting factors are allowed to change indiscriminately sets a dangerous precedent.

Now the weighting factors can be multiplied by the value points and summed to get the total value contribution. If all of the value measurements indicated that performance targets were being met, then 2.0 would be the worth of the value metric. However, in this case, performance is being exceeded with regard to quality, safety and schedule, and therefore the final worth of the value metric is 2.7. This metric implies that stakeholders are receiving additional value that is most likely meeting or exceeding expectations.

TABLE 5.11 Value Metric Measurement

VALUE COMPONENT	WEIGHTING FACTOR	VALUE MEASUREMENT	VALUE CONTRIBUTION
Quality	10%	3	0.3
Cost	20%	2	0.4
Safety	20%	4	0.8
Features	30%	2	0.6
Schedule	20%	3	0.6
			TOTAL = 2.7

Several issues still must be considered when using this technique:

- Normal performance must be clearly defined. The users must understand what this means. Is this level actually the target level, or is it the minimal acceptable level for the client? If it is the target level, then having a value below 2.0 might still be acceptable to the client if the target were greater than what the requirements asked for.
- Users must understand the real meaning of the value metric. When the metric goes from 2.0 to 2.1, how significant is that? Statistically, this is a 5% increase. Does it mean that the value increased by 5%? How can project managers explain to laypeople the significance of such an increase and the impact on value?

Value metrics generally focus on the present and/or future value of the project and may not provide sufficient information regarding other factors that may affect project health. As an example, let us assume that the value metric is quantitatively assessed at 2.7. From the customers' perspective, they are receiving more value than they anticipated. But other metrics may indicate that the project should be considered for termination. For example,

- The value metric is 2.7, but the remaining cost of development is so high that the product may be overpriced for the market.
- The value metric is 2.7, but the time to market will be too late.
- The value metric is 2.7, but a large portion of the remaining work packages have a very high critical risk designation.
- The value metric is 2.7, but significantly more critical assumptions are being introduced.
- The value metric is 2.7, but the project no longer satisfies the client's needs.
- The value metric is 2.7, but competitors have introduced a product with a higher value and quality.

In Table 5.12, the number of features in the deliverable was reduced, which allowed the project manager to improve quality

TABLE 5.12   Value Metric with a Reduction in Features

VALUE COMPONENT	WEIGHTING FACTOR	VALUE MEASUREMENT	VALUE CONTRIBUTION
Quality	10%	3	0.3
Cost	20%	2	0.4
Safety	20%	4	0.8
Features	30%	1	0.3
Schedule	20%	3	0.6
			TOTAL = 2.4

and safety as well as accelerate the schedule. Since the worth of the value metric is 2.4, additional value still is being provided to the stakeholders.

In Table 5.13, additional features were added, and quality and safety were improved. However, to do this, schedule slippage and a cost overrun have been incurred. The worth of the value metric is now 2.7, which implies that the stakeholders are still receiving added value. The stakeholders may be willing to incur the added cost and schedule slippage because of the added value.

Whenever it appears that the project may be over budget or behind schedule, project managers can change the weighting factors and over-weight those components that are in trouble. As an example, Table 5.14 shows how the weighting factors can be adjusted. Now, if the overall worth of the value metric exceeds 2.0 with the adjusted weighting factors, the stakeholders still may consider continuing the project. Sometimes companies identify minimum and maximum weights for each component, as shown in Table 5.15. However, there is a risk that management may not be able to adjust to and accept weighting factors that can change from project to project or even during a project. Also, standardization and repeatability of the solution may disappear with changing weighting factors.

TABLE 5.13   A Value Metric with Improved Quality, Features, and Safety

VALUE COMPONENT	WEIGHTING FACTOR	VALUE MEASUREMENT	VALUE CONTRIBUTION
Quality	10%	3	0.3
Cost	20%	1	0.2
Safety	20%	4	0.8
Features	30%	4	1.2
Schedule	20%	1	0.2
			TOTAL = 2.7

TABLE 5.14   Changing the Weighting Factors

VALUE COMPONENT	NORMAL WEIGHTING FACTOR	WEIGHTING FACTORS WITH SIGNIFICANT SCHEDULE SLIPPAGE	WEIGHTING FACTORS WITH SIGNIFICANT COST OVERRUN
Quality	10%	10%	10%
Cost	20%	20%	40%
Safety	20%	10%	10%
Features	30%	20%	20%
Schedule	20%	40%	20%

TABLE 5.15	Weighting Factor Ranges		
VALUE COMPONENT	MINIMAL WEIGHTING VALUE	MAXIMUM WEIGHTING VALUE	NOMINAL WEIGHTING VALUE
Quality	10%	40%	20%
Cost	10%	50%	20%
Safety	10%	40%	20%
Features	20%	40%	30%
Schedule	10%	50%	20%

Companies generally are reluctant to allow project managers to change weighting factors once a project is under way, and they may establish policies to prevent unwanted changes from occurring. The fear is that the project manager may change the weighting factors just to make a project look good. However, there are situations where a change may be necessary:

- Customers and stakeholders are demanding a change in weighting factors possibly to justify the continuation of project funding.
- The risks of the project have changed in the downstream life cycle phases, and a change in weighting factors is necessary.
- As the project progresses, new value attributes are added to the value metric.
- As the project progresses, some value attributes no longer apply and must be removed from the value metric.
- The enterprise environment factors have changed, requiring a change in the weighting factors.
- The assumptions have changed over time.
- The number of critical constraints has changed over time.

It must be remembered that project management metrics and KPIs can change over the life of a project. Therefore, the weighting factors for the value metric also may be susceptible to changes.

Sometimes, because of the subjectivity of this approach, when the information is presented to the client, which measurement technique was used for each target should be included. This is shown in Table 5.16. The measurement techniques may be subject to negotiations at the beginning of the project.

The use of metrics and KPIs has been with us for decades, but the use of a value metric is relatively new. Therefore, failures in the use of this technique are still common and may include:

- Not being forward looking; the value metric focuses on the present rather than the future.

TABLE 5.16	Weighting Factors and Measurement Techniques			
VALUE COMPONENT	WEIGHTING FACTOR	MEASUREMENT TECHNIQUE	VALUE MEASUREMENT	VALUE CONTRIBUTION
Quality	10%	Sampling techniques	3	0.3
Cost	20%	Direct measurement	2	0.4
Safety	20%	Simulation	4	0.8
Features	30%	Observation	2	0.6
Schedule	20%	Direct measurement	3	0.6

- Not going beyond financial metrics and, thus, failing to consider the value in knowledge gained, organizational capability, customer satisfaction, and political impacts.
- Believing that value metrics (and the results) that other companies use will be the same for another company.
- Not considering how the client and stakeholders define value.
- Allowing the weighting factors to change too often, to make the project's results look better.

As with any new technique, additional issues always arise. Some typical questions that project managers are now trying to answer in regard to the use of a value metric are listed next.

- What if only three of the five components can be measured, for example, in the early life cycle phases of a project?
- If only some components can be measured, should the weighting factors be changed or normalized to 100% or left alone?
- Should the project be a certain percentage complete before the value metric has any real meaning?
- Who will make decisions regarding changes in the weighting factors as the project progresses through its life cycle phases?
- Can the measurement technique for a given component change over each life cycle phase, or must it be the same throughout the project?
- Can the subjectivity of the process be reduced?

5.14 PRESENTING THE VALUE METRIC IN A DASHBOARD

Figure 5.16 illustrates how the value metric may appear on a dashboard. The value attributes and ratings in the table in the upper-right corner reflect the values in the month of April. In January, the magnitude of the value metric was about 1.7. In April, the magnitude is 2.7.

Figure 5.16 Project Value Attributes

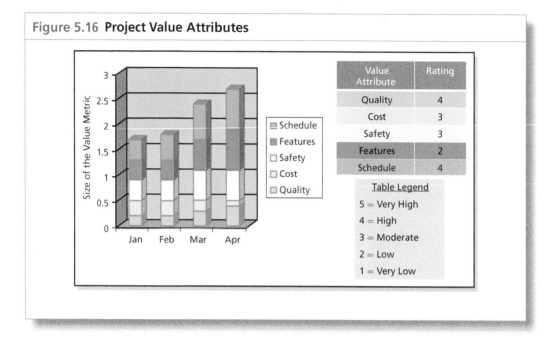

Stakeholders can easily see the growth in value over the past four months. They can also see that four of the five attributes have increased in value over this time period, whereas the cost attribute appears to have diminished in value.

Eventually, as project managers become more knowledgeable in the use of value metrics, they may end up with a single value metric that can be obtained through an objective and automated process. In the near term, however, the value metric process will be to be more qualitative than quantitative and highly subjective based on the value attributes that were selected.

5.15 INDUSTRY EXAMPLES OF VALUE METRICS

This section provides examples of how various companies use value metrics. The number of companies using value metrics is still quite small. Some of the companies surveyed could not provide accurate weighting factors because the factors can change for each project. Other companies differentiated between project success and product success and stated that the value attributes and weighting factors were different for each as well. (Note: Company names were withheld at the request of the companies.)

In several of the examples, there are descriptions of the attributes. In most cases, the attributes of the final value metric are a composite of various KPIs as discussed previously in this chapter.

Aerospace and Defense: Company 1

- Schedule: 25% (based on objective data via the project's EVMS)
- Cost: 25% (based on objective data via the project's EVMS)

- Technical factor: 30% (based on technical performance measures established at the beginning of the project)
- Quality factor: 10% (based on an ongoing audit of adherence to established quality standards and procedures)
- Risk factor: 10% (based on how well risk mitigation plans are implemented and followed)

These percentages are for a generic project. Depending on the nature of the project and the constraints imposed by the client, the percentages would be adjusted. The actual percentages would be coordinated with the client so that the client would know and agree to the weighting to be used. The projects that this method would apply to would result in a product or system, or both. Most of the clients in aerospace and defense are extremely sensitive to cost and schedule, which is why the percentages for those areas are as high as they are.

Aerospace and Defense: Company 2

- Quality: 35%
- Delivery: 25%
- Cost: 20%
- Technology: 5%
- Responsiveness: 10%
- General management: 5%

Capital Projects: Company 2

- Revenue growth/generation: 30% (Our primary focus has been on growing revenue through leveraging alternative options.)
- Cost efficiencies: 30% (This has a direct bottom-line impact and seen almost equal to generating revenue.)
- Handle/market share growth: 20% (There is a revenue impact here as well.)
- Project schedule: 10% (We have many time-constrained projects because of our core business operating model. We have the natural trade-off decisions; however, the first three factors are weighted more heavily.)
- Project cost: 10% (Generally tracked to ensure increasing costs do not overwhelm expected benefits)

IT Consulting (External Clients): Company 1 (No Percentages Provided)

- Risks
- Scope
- Resources
- Quality
- Schedule
- Overall status

IT Consulting (External Clients): Company 2

- Quality: 40% (as determined by feedback from internal engagement leaders, review of project deliverables before they are submitted to the client, number of iterations before the deliverables are client-ready, and, eventually, by the client's satisfaction with the provided deliverables and the manner in which client interactions and expectations were managed throughout the engagement; i.e., did we make and keep our promises, did we offer the client a "no-surprise" experience, and did we demonstrate professionalism through effective interaction and communications?)
- Talent: 20% (as determined by feedback from firm and engagement leaders as well as satisfaction of project team members in terms of how the project was planned and delivered; i.e., was a positive working environment created; were the views and opinions of individual team members valued and considered; were the roles and contributions of individual team members well defined, properly communicated, and understood by all involved; did the project provide an opportunity for personal and professional growth and development?; etc.)
- Marketplace: 10% (as determined by the evaluation of the firm and engagement leadership as to the extent to which the given project demonstrated understanding of the client and the client's industry as well as the extent to which the project contributed to establishing or supporting the firm's preeminence within the given service line or industry)
- Financial: 30% (as determined by on-time and on-budget delivery of the project, profitability of the engagement, achieved recovery rate or the level of applied discount to standard rates, and ability to submit invoices and collect payment within a reasonable amount of time)

IT Consulting (External Clients): Company 3

- Customer satisfaction: 30% (their perception on how well the project is going and satisfaction with the team and solution)
- Budget: 20%
- Schedule: 10%
- Solution deployed: 20% (solution is being used by the customer in production and providing value)
- Support Issues: 10%
- Opportunity generation: 10% (a successful project for us would also generate additional opportunities with the customer—difficult to measure until well after project completion)

These factors would apply to all industries and solutions. The only differences are that the solution deployed is not applicable in certain project types (such as a health check) and opportunity generation is applicable only when looking back in time after a project is completed.

IT Consulting (External Clients): Company 4

- Customer satisfaction/conditions of satisfaction: 30%
- Manage expectations/communication: 20%
- Usability/performance: 20%
- Quality: 20%
- Cost: 10%

It is less about time and cost now and more about can/will the product/service be used and is the client happy with the delivery of the final product/service?

IT Consulting (External Clients): Company 5

These value metric attributes are primarily for consulting in healthcare.

- Quality: 25%
- Cost: 20%
- Durations/timeliness: 15%
- Resource utilization: 10%
- Incorporation of processes (clinical, technical, business focused): 30%

IT Consulting (External Clients): Company 6

Similar to Company 5, this company also provided IT consulting services to healthcare but did not provide a breakdown of the value metric. The company believes that the value metric (whether specifically outlined on a project or not) is the true reason why these projects continue to be funded beyond their schedule and budget.

IT Consulting (Internal): Company 1

- Scope: 25%
- Project client satisfaction: 22.5%
- Schedule: 17.5%
- Budget: 17.5%
- Quality: 17.5%

Software Development: (Internal) (No Percentages Provided)

- Code: number of lines of code
- Language understandability: Language and/or code is easy to understand and read
- Movability/immovability: The ease with which information can be moved
- Complexity: Loops, conditional statements, etc.
- Math complexity: Time and money to execute algorithms
- Input/output understandability: How difficult is it to understand the program?

Telecommunications: Company 1

- Financial: 35%
- Quality/customer satisfaction: 35%
- Process adherence: 15%
- Teamwork: 15%

Telecommunications: Company 2 (No Percentages Provided)

- Customer Satisfaction
- Employee Satisfaction
- Quality
- Financial
- Cost

New Product Development

- Features/Functions: 35% (This is where the company believes it can differentiate itself from its competitors.)
- Time to Market: 25%
- Quality: 25%
- Cost: 15%

Automotive Suppliers

- Quality: 100%
- Cost: 100%
- Safety: 100%
- Timing: 100%

It is interesting to note that in this company, there were four value metrics rather than just one, and the company believed that the four value metrics could not be combined into a single metric. This is why 100% is assigned to each. In the auto industry, being less than 100% on each value metric could delay the launch of a product and create financial problems for the client and all of the suppliers.

Global Consulting: Company 1 (Not Industry Specific and No Weights)

This company believes that soft skills and personal attributes are some key factors affecting the outcome of projects and so far very little association has been made between these key factors and project failure. This is why some of the soft skills are part of the company's value metric. The challenge is in quantifying these soft skills factors for CVM. This will vary for each project, which is why percentages are not provided.

- **Management:** Consideration given to quick resolution of issues, minimal time wasted, minimal recycling of ideas, timely escalations.

- **Communications/Relationship Building:** Verbal and written status reports, weekly meetings, and the like. Agreed to at the start of the project.
- **Competency:** Impact and influence as well as performance and knowledge of project management.
- **Flexibility and Commitment:** Balancing clients' requests and requirements with that of suppliers and third-party contractors—ability to monitor and control to bring optimum results.
- **Quality:** The quality definition would vary based on project deliverables. This would be influenced by the industry.
- **Usability:** How much of what is implemented can give immediate added value—this is based on the premise that few organizations progress beyond what has been implemented within the first two years by which time the original users have moved on and their successors would not have had the vision to move the systems to the next level. On the contrary, the reverse happens, as less functionality is utilized by the inheritors of the systems.
- **Delivery Strategy:** This pertains directly to the strategy or approach for implementation of the solution offered to the client—the solution on paper could have seemed to meet the client's need, but if it is not executed optimally, it can cause added pain rather than gains. This speaks directly to governance approaches, production of specified deliverables, performing tasks on schedule, and the like.
- **Customer Focus:** To what extent does the project management team seek the interest of the customer/client?
- **HSE (health, safety, and environmental) Policies:** The satisfaction level with the project team's compliance with health, safety, and environmental policies while performing duties at the customer's site.

Global Consulting: Company 2 (Not Industry-specific and No Weights)

- Profitability
- Schedule
- Impact/Result
- Customer Satisfaction
- Safety
- Erosion

Erosion is the value of work performed in excess of what was originally estimated in terms of effort or duration and not recovered through project change management procedures. In simple terms, erosion is the difference between billable work (that which has been estimated, proposed, and accepted by the customer) and non-billable work (that which has not been estimated, proposed, and accepted by the customer).

5.16 USE OF CRISIS DASHBOARDS FOR OUT-OF-RANGE VALUE ATTRIBUTES

Today, most companies that use dashboards to communicate with clients and stakeholders include drill-down capability on the dashboards. The top dashboard contains the KPIs and the value metric. If additional information is required, the drill-down capability allows more detailed information to appear on the screen.

Some companies have taken this concept a step further. Rather than requiring the viewer to drill down to find the cause of a potential problem, companies are creating a crisis dashboard that is similar to an exception report. All KPIs and value metric attributes that exceed the minimum threshold limits unfavorably will appear in the crisis dashboard. According to one of the contributing companies:

What we have developed is a kind of algorithm that summarizes the status of the project. It is based on the following:

> There are indicators or semaphores associated to several aspects of the project that we believe are contributors to value (cost, schedule, margin variation, risk, issues, pending invoicing, pending payments, milestones, etc.)
>
> The indicators get a green, yellow, or red colors, depending on thresholds and variation percentages that are characteristic of each business unit.
>
> Colors have an allocated value (0,1,4).
>
> Then a new overall indicator or metric is created called **"CLOA" (calculated level of attention)** that goes from "very low" to "very high" or even to "requiring intensive care." The value of CLOA is assigned through a formula that takes into account the colors of the indicators and some of the absolute figures. The intent is not to create unnecessary alarms if the amounts involved are not significant.
>
> This single indicator is mainly used (until now) by the finance controllers, the local PMOs, the QA departments, portfolio managers, etc., to claim their attention on specific projects. The goal is, of course, to provide an **early detection** mechanism of failing projects and to establish the corresponding corrective actions.
>
> The algorithm is not perfect, but it is becoming more and more useful as we refine it. In any case, the "owners" of this indicator are the finance controllers of the business unit, not the project managers. Only the finance controllers are able to modify its value and to report the reasons for modification. The report is then stored in the system for historical backup.

Rather than using a crisis dashboard, some companies simply use alerts. Alerts or attention-getters are indications that the KPI's threshold boundary condition or integrity level has been reached. In general, alerts serve as early warning signs that some action must be taken before the situation gets worse. The situation might actually be worse than it appears and stakeholders may assume the worst possible scenario. Alerts are similar to exception reports.

Not all KPIs will trigger alerts. Also, not all alerts indicate unfavorable trends. One company established an alert trigger if the company became too efficient and produced more units than the warehouse could handle. Shadan Malik describes several types of alerts that may be appropriate for enterprise dashboards:

- Critical alert
- Important alert

- Informational alert
- Public alert
- Private alert
- Unread alert[10]

In a project management environment, there may be three types of alerts:

1. Project team alert
2. Management alert
3. Stakeholder alert

5.17 ESTABLISHING A METRICS MANAGEMENT PROGRAM

The future of project management must include metrics management. Certain facts about metrics management can be identified:

- Project managers cannot effectively promise deliverables to a stakeholder unless they can also identify measurable metrics.
- Good metrics allow project managers to catch mistakes before they lead to other mistakes.
- Unless project managers identify a metrics program that can be understood and used, the project managers are destined to fail.
- Metrics programs may require change, and people tend to dislike change.
- Good metrics are rallying points for the project management team and the stakeholders.

There are also significant challenges facing organizations in the establishment of value-based metrics:

- Project risks and uncertainties may make it difficult for the project team to identify the right attributes and perform effective measurement of the value attributes.

10 Shadan Malik, *Enterprise Dashboards* (Hoboken, NJ: John Wiley & Sons, 2005), p. 66.

- The more complex the project, the harder it is to establish a single value metric.
- Competition and conflicting priorities among projects can lead to havoc in creating a value metrics program.
- Added pressure by management and the stakeholders to reduce the budget and shorten the schedule may have a serious impact on the value metrics.

Metric management programs must be cultivated. Here are some facts to consider in establishing such a program:

- There must be an institutional belief in the value of a metrics management program.
- The belief must be visibly supported by senior management.
- The metrics must be used for informed decision making.
- The metrics must be aligned with corporate objectives as well as project objectives.
- People must be open and receptive to change.
- The organization must be open to using metrics to identify areas of and for performance improvement.
- The organization must be willing to support the identification, collection, measurement, and reporting of metrics.

Best practices and benefits can be identified as a result of using metrics management correctly and effectively. Some of the best practices include:

- Confidence in metrics management can be built using success stories.
- Displaying a "wall" of metrics for employees to see is a motivational force.
- Senior management support is essential.
- People must not overreact if the wrong metrics are occasionally chosen.
- Specialized metrics generally provide more meaningful results than generic or core metrics.
- The minimization of the bias in metrics measurement is essential.
- Companies must be able to differentiate between long-term, short-term, and lifetime value.

The benefits of metrics include:

- Companies that support metrics management generally outperform those that do not.
- Companies that establish value-based metrics are able to link the value metrics to employee satisfaction and better business performance.

5.18 **USING VALUE METRICS FOR FORECASTING**

Performance reporting is essential for effective decision making to take place. In general, there are three types of performance reports:

1. **Progress reports:** These reports describe the work accomplished to date. This includes:
   - The planned amount work completed up to the timeline of the report.
   - The actual amount of work accomplished up to the timeline.
   - The actual cost accumulated up to the timeline.
2. **Status reports:** These reports indicate the status by comparing the progress to the baselines and determining the variances. This includes:
   - The schedule variance up to the timeline of the report.
   - The cost variance up to the timeline.
3. **Forecast reports:** Progress reports and status reports are snapshots of where the project is today. The forecast reports, which are usually of significant importance to the stakeholders, indicate where the project will end up. This includes:
   - The expected cost at the completion of the project.
   - The expected time duration or date at completion of the project.

There are other items that can be included in these reports. However, the main concern here is with the forecast reports and the metrics used to make forecasts.

Traditional forecast reports provide information on the time and cost expected at the completion of the project. This data can be calculated from extrapolation of trends or formulas or from projections of the metrics and KPIs. Unfortunately, this data may not be sufficient to provide management with the necessary information to make effective business decisions and to decide whether to continue with the project or consider termination.

Two additional pieces of information may be necessary: the expected benefits at completion and the expected value at completion. Most EVMSs in use today do not report these two additional pieces of information, probably because there are no standard formulas for them. Also, the value and benefits at completion may not be known until months after the project is completed.

The benefits and value at completion must be calculated periodically. However, depending on which life cycle phase a project is in, there may be insufficient data to perform the calculation quantitatively. In such cases, a qualitative assessment of benefits and value at completion may be necessary, assuming, of course, that information exists to support the assessment. Expected benefits and value are more appropriate for business decision making and usually provide a strong basis for continuation or cancellation of the project.

TABLE 5.17   A Comparison of EVMS, EPMS, and VMM

VARIABLE	EVMS	EPMS	VMM
Time	Yes	Yes	Yes
Cost	Yes	Yes	Yes
Quality		Yes	Yes
Scope		Yes	Yes
Risks		Yes	Yes
Tangibles			Yes
Intangibles			Yes
Benefits			Yes
Value			Yes
Trade-offs			Yes

Using value metrics, an assessment of value at completion can indicate whether value trade-offs are necessary. Reasons for value trade-offs include:

- Changes in the enterprise environmental factors
- Changes in the assumptions
- Better approaches have been found, possibly with less risk
- Availability of highly skilled labor
- A breakthrough in technology

As stated earlier, most value trade-offs are accompanied by schedule extensions. Two critical factors that must be considered before schedule extensions take place are:

1. Extending a project for the desired or added value may incur risks.
2. Extending a project consumes resources that may have already been committed to other projects in the portfolio.

Traditional tools and techniques may not work well on value-driven projects. The creation of a VMM may be necessary to achieve the desired results. A VMM can include the features of EVMSs and enterprise project management systems (EPMSs), as shown in Table 5.17. However, additional variables must be included for the capturing, measuring, and reporting of value and possibly value metrics.

5.19 METRICS AND JOB DESCRIPTIONS

Because project managers are now expected to be knowledgeable regarding metrics, job descriptions for project managers are being revised to include a level of knowledge in metrics management. Table 5.18 illustrates some of the expectations for a company that has five levels of job descriptions for project managers.

TABLE 5.18	Placing Metrics Knowledge into Job Descriptions
GRADE LEVEL	**COMPETENCY**
1	Understand project metrics and key performance indicators
2	Be able to identify and create project-specific metrics
3	Be able to track and report metrics on a project
4	Be able to measure and evaluate metrics
5	Be able to extract best practices from metrics

The growth of measurement techniques has accelerated the importance metrics, and this includes both tangible and intangible forms of value metrics. Project management is slowly becoming metrics-driven project management. The traditional metrics used for decades no longer satisfy the needs of clients and stakeholders. Value-based metrics will become critical in stakeholder relations management. In addition, metrics management will lead project managers to a better understanding of the necessity of developing more sophisticated knowledge management techniques.

5.20 GRAPHICAL REPRESENTATION OF METRICS

The old adage "a picture is worth one thousand words" certainly holds true when graphically displaying metrics. Figure 5.17 shows the resources that are assigned versus the planned resources. For Work Package #1, five people were scheduled to be assigned, but only four

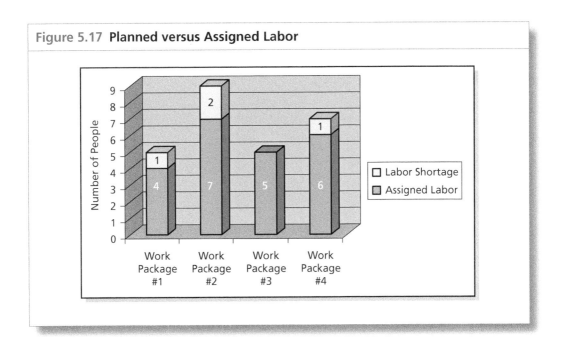

Figure 5.17 **Planned versus Assigned Labor**

people are currently working on this work package. The metric can also be used to show if excess labor is assigned to a work package.

Figure 5.17 shows a shortage of resources, but that may not be bad if the workers assigned have higher skills than originally planned for. This is shown in Figure 5.18. From this figure, the assigned resources are pay grades 6, 7, and 8. If using only pay grades 5 and 6 was planned, then this may be good. However, if some pay grade 9 workers were anticipated, then there may be a problem.

In Figure 5.19 depicts regular time, overtime, and unstaffed hours. In January, people worked 600 hours on regular shift and

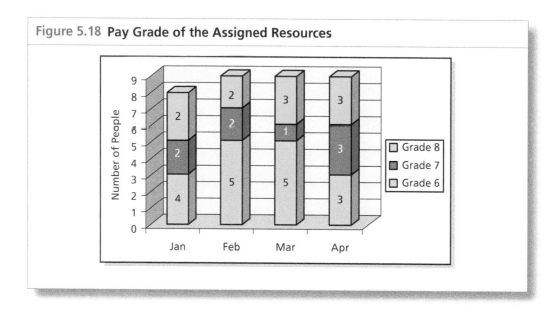

Figure 5.18 Pay Grade of the Assigned Resources

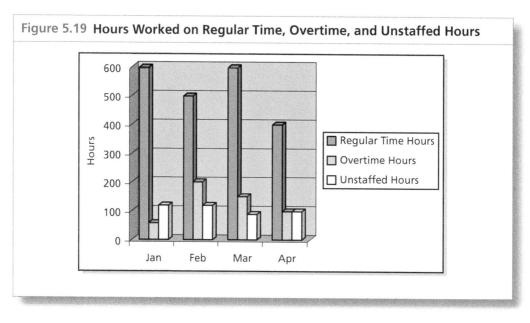

Figure 5.19 Hours Worked on Regular Time, Overtime, and Unstaffed Hours

50 hours on overtime. The project was still short some 100 hours of work that needed to be done. This could lead to a schedule slippage.

Stakeholders often want to know what percentage of the work packages scheduled have been completed and how many are still open or possibly late. This is shown in Figure 5.20. It is most likely a good sign that the work packages late are shrinking each month. This representation could also be used to illustrate a percentage of all of the work packages, regardless of whether they have started or not.

Figure 5.20 **Work Packages Scheduled for Completion, Including Those Completed and Those Still Open**

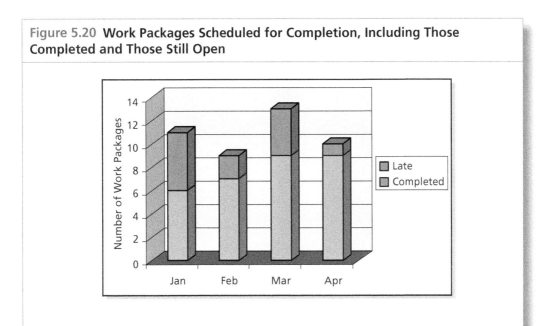

Figure 5.21 **Work Packages with a Critical Risk Designation**

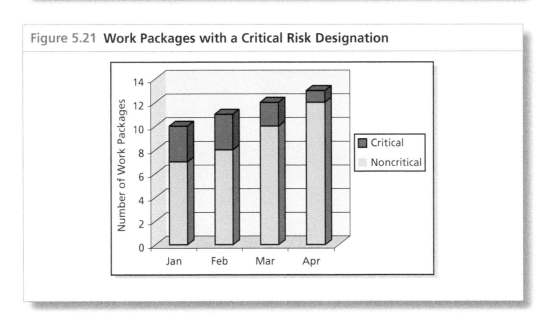

Some companies provide critical risk designations to each work package. In Figure 5.21, the number of work packages with critical risk designations is diminishing over time. This is a good sign.

Figure 5.22 depicts the number of work packages adhering to the cost baseline. Since the number is increasing over time, costs seem under control. When costs, scope, and schedules need to be revised, baseline changes result. The number of baseline changes is usually an indication of the quality of the up-front planning process and/or the company's estimating capability. This is seen in Figure 5.23. This type

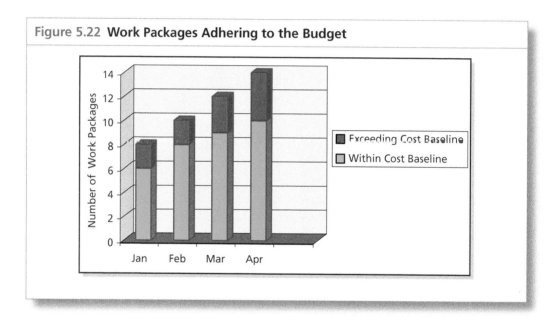

Figure 5.22 Work Packages Adhering to the Budget

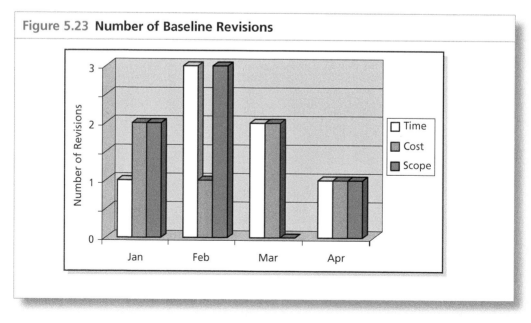

Figure 5.23 Number of Baseline Revisions

of metric is important because it illustrates the rate of change in the requirements over time. The number of baseline revisions can be the result of scope changes being made throughout the project. This is shown with the metric displayed in Figure 5.24.

Cost and schedule slippages can be the result of poor governance on a project and the inability to close out action items in a timely manner. This is shown in Figure 5.25. Action items that remain open for two or three months can have a serious impact on the final deliverables of the project as well as on stakeholder satisfaction. However,

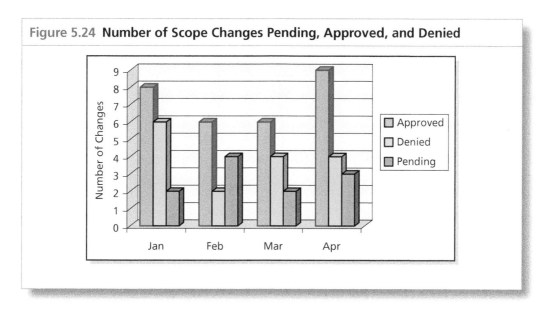

Figure 5.24 **Number of Scope Changes Pending, Approved, and Denied**

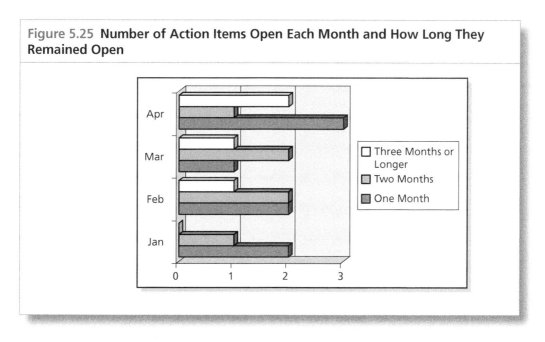

Figure 5.25 **Number of Action Items Open Each Month and How Long They Remained Open**

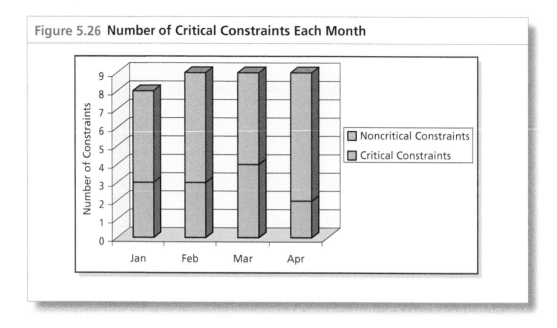

Figure 5.26 **Number of Critical Constraints Each Month**

there are situations where action items may need to be open for an extended period of time. Two such examples include taking advantage of a business opportunity and looking at risk mitigation techniques.

Not all constraints are equal. Some companies designate their constraints as critical and noncritical. Both critical and noncritical constraints, as shown in Figure 5.26, must be tracked closely throughout the project. Changes in the importance of the constraints can have a serious impact on the final value the stakeholders expect to receive.

Assumptions can also change during a project. The longer the project, the greater the likelihood that the number of assumptions will change. The assumptions must be tracked closely. Significant changes in the assumptions can cause a project to be canceled. This is shown in Figure 5.27. In February, eight of the assumptions were the same as in January. Two of the assumptions changed. From the table in the right of the figure, one of the two assumptions changed was new, and the second assumption was modified.

Most companies today maintain a best practices library. When bidding on a contract, companies often make promises to the client that all of the best practices in the library that relate to this project will be used. In order to validate that these promises are being kept, the project team can track the number of best practices that are being used as compared to what was promised. This is shown in Figure 5.28. In this figure, 10 best practices were promised to the customer. Seven of the best practices have already been used, two will be used in the future, and one best practice will not be used.

Some companies assign a project complexity factor to each project based on the aggregate risks. This is shown in Figure 5.29. In this example, a complexity factor of 15 would indicate a very serious risk. Risks are normally the greatest at the beginning of a project,

Figure 5.27 Number of Critical Assumptions That Are New or Have Been Changed

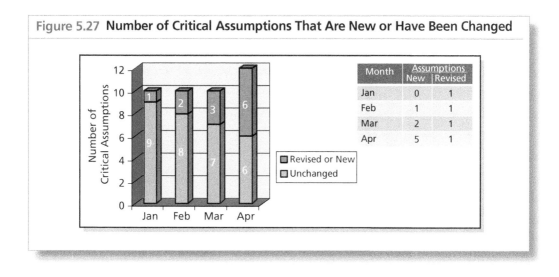

Figure 5.28 Actual versus Promised Best Practices Used

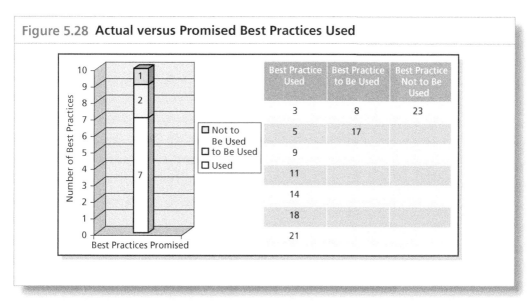

when the least amount of information is known. Since the complexity factor appears to be diminishing over time, things appear to be proceeding well.

Sometimes companies that maintain a metrics library identify the artwork along with a description of the metric. This is shown in Figure 5.30. Figure 5.30 shows how Figure 5.29, the project complexity factor, would appear in the metric library. The column on the right in Figure 5.30 shows some of the information used to describe the metric.

On large projects, stakeholders are often interested in the total manpower assigned to the project. Figure 5.31 shows this in the form of a metric. Another important metric is the rate at which the management

Figure 5.29 Project Complexity Factor

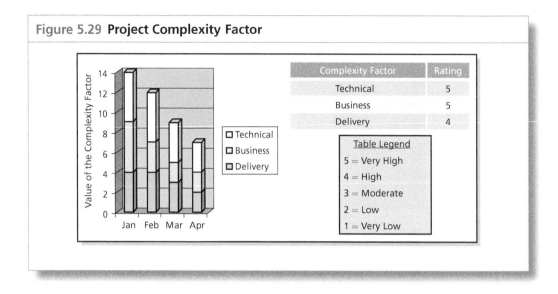

Figure 5.30 Project Complexity Factor Appearing in the Metric Library

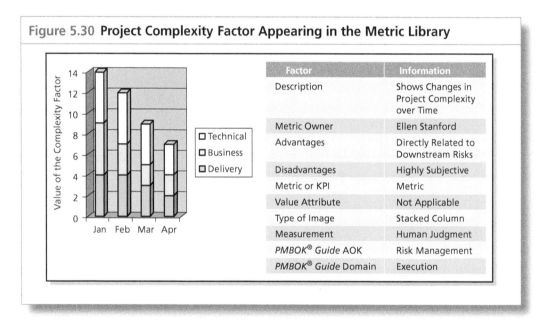

reserve is being used up. This is shown in Figure 5.32. In January, a management reserve of $100,000 was established. By the end of March, $60,000 had been used and $40,000 remained. In April, another $30,000 was added to the reserve fund and a total of $80,000 had been used.

Figure 5.33 shows the number of deliverables that were delivered on time or late each month. This metric usually is accompanied by another metric, as shown in Figure 5.34, which shows how many of the deliverables were accepted and rejected.

Figure 5.35 shows the trend on the CPI and the SPI. Since they represent trends, it is better to show them on the trend chart on the

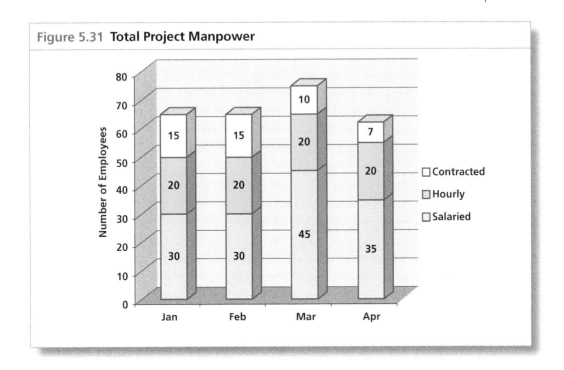

Figure 5.31 **Total Project Manpower**

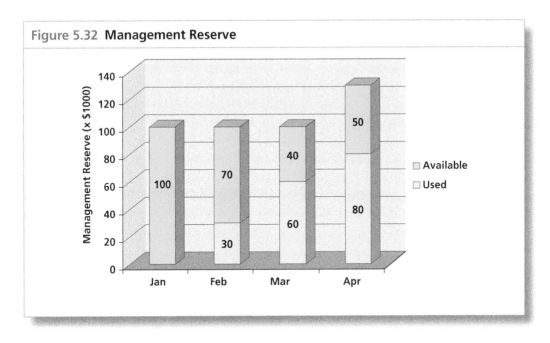

Figure 5.32 **Management Reserve**

left rather than using a gauge that shows the average value. Under normal performance, a value of 1.0 for SPI would indicate meeting the schedule., but the gauge shows an "average" SPI value. The trend chart on the left shows that the trend for SPI is unfavorable even though the average value is 1.0. Care must be taken to use the correct

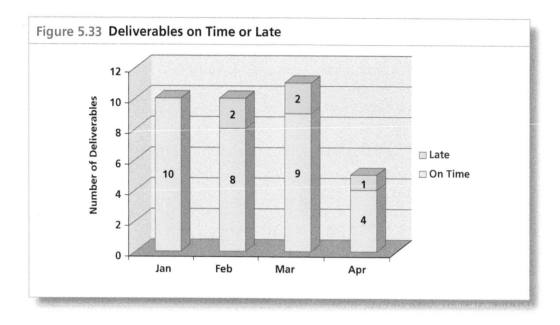

Figure 5.33 **Deliverables on Time or Late**

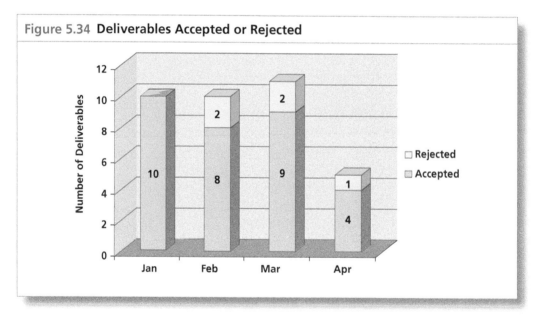

Figure 5.34 **Deliverables Accepted or Rejected**

image when displaying the metric. Since this chart is used to show trends, often it is not placed on the dashboard until there exist a sufficient number of data points to indicate a trend.

The CPI generally is more meaningful than the cost variance because it shows trends. However, if the cost variance is to be provided in metric form, it may be best to show the difference between last month and this month. This is shown in Figure 5.36. Work package #8 is significantly worse than last month. Work package #1 is still unfavorable, but there has been a favorable decrease in the

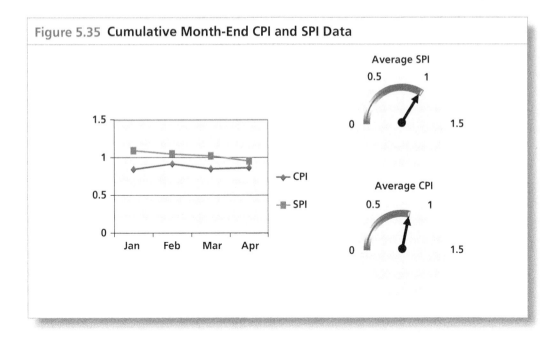

Figure 5.35 **Cumulative Month-End CPI and SPI Data**

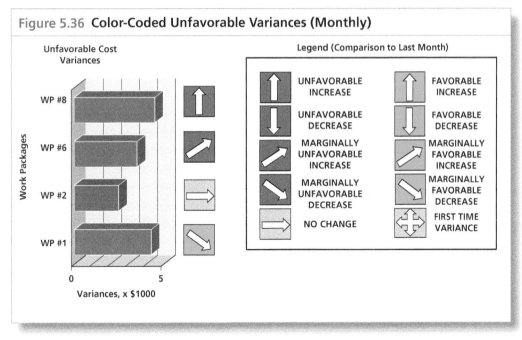

Figure 5.36 **Color-Coded Unfavorable Variances (Monthly)**

magnitude of the unfavorable variance. In other words, there can be improvements in the magnitude of a negative variance. Likewise, in Figure 5.37. there can be unfavorable decreases in the magnitude of a positive variance.

Trends are normally used to predict the estimated cost at completion (EAC). This is shown in Figure 5.38. The original budget is

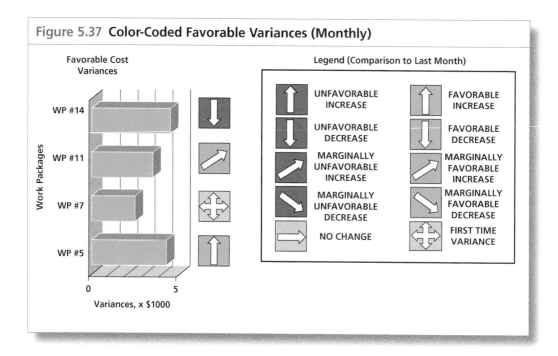

Figure 5.37 Color-Coded Favorable Variances (Monthly)

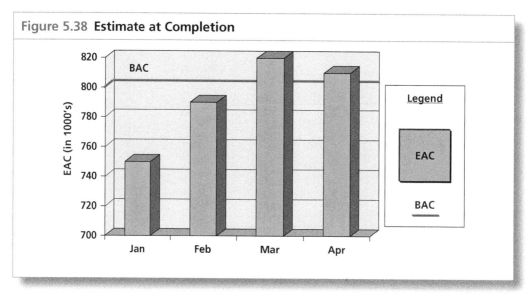

Figure 5.38 Estimate at Completion

$800,000, and the fluctuations in EAC can be seen. Several formulas can be used for EAC, and sometimes the fluctuations occur when the formulas being used are changed.

Sometimes the risks on a project are chronic; they simply do not go away even with mitigation attempts. Figure 5.39 shows how a metric can be used to track the aging of a risk. One metric alone cannot always show the status of the project. A combination of metrics may be needed.

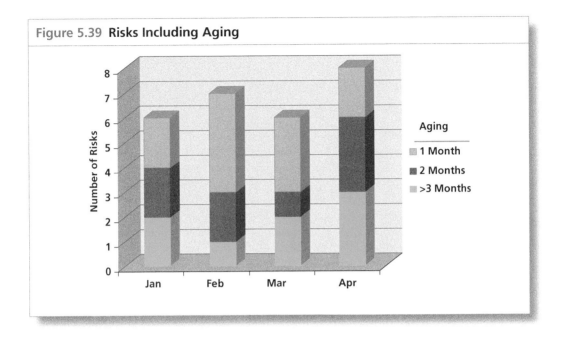

Figure 5.39 **Risks Including Aging**

5.21 CREATING A PROJECT VALUE BASELINE

All of the information covered in the chapter thus far makes it clear that a baseline for measuring value is needed. Executives and clients expect project managers to monitor and control projects effectively. As part of monitoring and control, project managers must prepare progress, status, and forecast reports that clearly articulate the performance of the project. But to measure performance, a reference point or baseline from which measurements can be made is needed. The necessity for a baseline is clear:

- Without a baseline, performance cannot be measured.
- If performance cannot be measure, it cannot be managed.
- Performance that can be measured gets watched.
- What gets watched gets done.

For a project to be able to be controlled, it must be organized as a closed system. This requires that baselines be established for perhaps all constraints, including scope, time, and cost at a minimum. Without such baselines, a project is considered out of control. It may be impossible to track what has changed without knowing where the project began. It may also be impossible to identify value-added opportunities.

The Performance Measurement Baseline

The reference point for measuring performance is the performance measurement baseline (PMB). The PMB serves as the metric benchmark against which performance can be measured in terms of the triple constraints of time, cost, and scope. It can also be used as the basis for business value tracking, provided that value metrics and value constraints be established.

The primary reasons for establishing, approving, controlling, and documenting the PMB were to:

- Ensure achievement of project objectives.
- Manage and monitor progress during project execution.
- Ensure accurate information on the accomplishment of the deliverables and requirements.
- Establish performance measurement criteria.

The PMB is finalized at the end of the planning phase once the requirements have been defined, the initial costs have been developed and approved, and the schedule has been set. Once established, the PMB serves as the benchmark from which to measure and gauge the project's progress. The baseline is then used to measure how actual progress compares to planned performance. Performance measurement may be meaningless without an accurate baseline as a starting point. Unfortunately, project managers in the past created baselines based on just those elements of work they felt were important, and this may or may not have been in full alignment with customer requirements or the customer's need for identification of value. *The baseline was what the project manager planned to do, not necessarily what the customer had asked for.*

Project Value Management

The problem with the way the PMB was used traditionally is that it did not account for the value that was expected from the project by people outside of the company. The importance of value has been known for some time. For years, many companies established value engineering (VE) departments that focused on internal value achieved in engineering and manufacturing activities. Later VE was expanded and called value analysis (VA). It included consideration of internal business value. Today there exists combined VE and VA, called value management (VM), where VM = VE + VA.

Today project managers need to understand the importance of project VM and its relationship to both the customer's and their consumers' understanding of value. Project VM is a mind-set, and its principles should be employed at the outset of the project and used throughout the project's full life cycle. Project VM is a combination of attributes, people, requirements, enterprise environmental factors, and circumstances. The attributes of project value may include time, cost, quality, risk, and other such factors. Project value is more than just obtaining customer satisfaction at the lowest price or with a minimum amount of expended resources. Project value cannot be determined exclusively in terms of the traditional triple constraints.

Focusing on the triple constraints is not a mind-set to create project value. Establishing good value-reflective metrics allows the definition of project success to be made in terms of factors other than the traditional triple constraints. For example, say a project manager works in a company that manufactures a component for a customer who, in turn, uses the component in a product it sells to consumers (i.e., its customers). Each can have a completely different definition of success, such as:

- **Consumer:** A deliverable or solution that removes obstacles from or improves the consumer's way of life.
- **Customer:** A deliverable or solution that is aligned to the customer's corporate strategic goals and objectives.
- **Project manager:** Providing sustainable business value within the competing constraints.

Simply stated, VOC is no longer just the voice of your customer, who may not be the final user of your products and services; it is no longer just the voice of the customer. VOC can also be the voice of the consumers, who are the customer's customers. To listen to these voices properly, the project manager must understand the customer's business model and strategy for reaching out to consumers as well as the customer's value management program.

Project VM must begin with a clear understanding of the customer's definition of value. Value mismatches generally lead to bad results. However, it should be understood that on long-term projects, the customer's definition of value may change.

Agreeing with everyone's definition of value works only if the project staffing function provides resources with the capability to equal or exceed the desired project value. Figure 5.40 shows that the best resources should be assigned to projects that provide high value

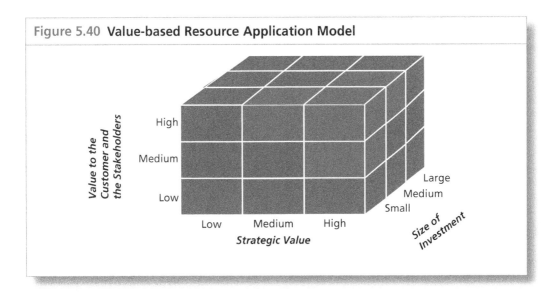

Figure 5.40 Value-based Resource Application Model

to customers and consumers, provide high strategic value to the project manager's company, and require a small investment by the project manager's company.

The Value Management Baseline

Configuration management is the process of managing changes in the deliverables, hardware, software, documentation, and even measurements. The value baseline supports configuration management processes. Traditionally, configuration management included input from multiple baselines but not from a value baseline. Therefore, at change control board meetings, it was somewhat difficult to quantitatively define exactly what added value would be achieved. Today, the answers to the critical questions to be addressed at the change control board meetings come from four baselines:

1. **Cost baseline:** Cost of the change
2. **Schedule baseline:** Impact on the schedule
3. **Risk baseline:** Risk effects
4. **Value baseline:** Value-added opportunities

All baselines are reference points against which a comparison is made between planned and actual progress. All baselines are established at a fixed point in time. However, some value baselines must evolve over time. Unlike the traditional baselines for time, cost, and scope, value baselines are highly dependent on when the measurements can be made, the measurement intervals, and the fact that value baselines are often displayed as step functions rather than linear functions.

Value baselines have several characteristics:

- The value baseline can be composed of attributes from other baselines.
- The value baseline should be shown to the customer. Other baselines may or may not be shown to the customers.
- The value baseline may or may not be a contractual obligation.
- Unlike other baselines, the value baseline can change from life cycle phase to phase.
- Stakeholders must understand the differences between actual and planned value, whether favorable or unfavorable.
- Baseline changes may require modifying or reworking the project plan or even result in project cancellation.
- The value baseline can change without any changes occurring in other baselines.
- On some projects, monitoring the value baseline in the early life cycle phases may provide no meaningful data.
- It is important to determine how often the value baseline must be updated.

- Value baselines may need to be continuously explained to the viewer. This may not be the case with other baselines.
- An increase of, say, 5% in the value baseline does not necessarily mean that the actual value has increased by 5%.
- Performance measurements for value may need to be customized rather than off-the-shelf techniques. This may need to be agreed on up front on the project.

Getting agreement on project value at the beginning of a project may be difficult. Even with an agreement, each stakeholder can still have their own subjective agenda on what value means to them:

- **Project manager:** Lowest cost
- **Project design team:** Functionality
- **Client:** Market share
- **Consumers:** Best buy

Stakeholder involvement is essential when establishing the value baseline. Stakeholders must clearly understand what is meant by:

- Normal value performance
- An increase in value
- A decrease in value

Factors that can cause the value baseline to deteriorate include:

- Faulty initial expectations of value
- Unrealistic expectations of value
- Unattainable expectations of value
- Poor alignment with business objectives
- Recessions
- Changing customer/consumer requirements and habits
- Unexpected crises

Selecting the Value Baseline Attributes

The value baseline is composed of a collection of attributes taken from other baselines. Each project's value baseline can be different. Figure 5.41 shows the six generic categories for the selection of value attributes. The attributes in the figure are represented as gears because changes in one attribute could result in changes to all of the other attributes. Each of the value attributes can have a different meaning to each of the stakeholders. This is shown in Table 5.19.

Sometimes metrics such as time and cost are not treated as value attributes but are still important to the client. As an example, consider a project where the life cycle cost of the deliverable is more important to the client than the original contract cost. Life cycle costing is the total cost to the consumer for the acquisition, ownership,

Figure 5.41 Value Metric Attributes

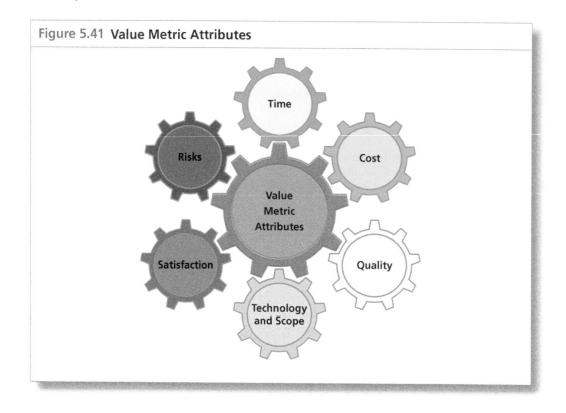

operations, support, and disposal of the deliverable. What if the project's cost increases by $100,000 but life cycle costing indicates a tenfold savings in costs associated with safety, reliability, operability, maintenance, and environmental factors? In this example, the cost metric is still important during project execution but is not regarded as critical enough to be used as a value attribute. The same could occur with the time metric, where getting to market with a quality product is more important than simply time to market.

TABLE 5.19   Interpretation of Attributes

GENERIC VALUE ATTRIBUTE	PROJECT MANAGER'S VALUE ATTRIBUTE	CUSTOMER'S VALUE ATTRIBUTE	CONSUMER'S VALUE ATTRIBUTE
Time	Project duration	Time to market	Delivery date
Cost	Project cost	Selling price	Purchase price
Quality	Performance	Functionality	Usability
Technology and scope	Meeting specifications	Strategic alignment	Safe buy and reliable
Satisfaction	Customer satisfaction	Consumer satisfaction	Esteem in ownership
Risks	No future business from this customer	Loss of profits and market share	Need for support and risk of obsolescence

Life cycle costing decisions can add significant value to customers and consumers. Project managers should seek out project value opportunities, but not necessarily at a significant expense to the project. Stakeholders may need to approve project value opportunity initiatives. Project value initiatives can lead to overachievement actions by the project team:

Overachievement Trends

- Exceeding specifications rather than meeting them
- Unnecessarily giving clients more than they ask for or need
- Identifying opportunities for the future at the expense of the project

Risks of Overachievement

- Increased complexities
- Greater overall uncertainty and risk
- Possible conflicting priorities
- Inability to meet time compression goals

We can now summarize the skills needed for effective project value management:

- Innovation skills
- Brainstorming skills
- Problem-solving skills
- Process skills
- Life cycle costing skills
- Risk management skills

6 DASHBOARDS

CHAPTER OVERVIEW

Dashboards are an attempt to go to paperless project management and yet convey the most critical information to the stakeholders the fastest way. Dashboards are communication tools for providing data to viewers. If the goal is to communicate the big picture quickly, perhaps in summary format, dashboards are the way to go. Other project management tools, such as written reports, may be necessary to provide supporting information and details.

Designing the dashboard is not always easy, however. Multiple dashboards may be required on the same project. There are rules that can be followed to make the design effort easier. Included in this chapter are white papers from companies that assist clients in dashboard design efforts.

CHAPTER OBJECTIVES

- To understand the characteristics of a dashboard
- To understand the differences between dashboards and scorecards
- To understand the different types of dashboards
- To understand the benefits of using dashboards
- To understand and be able to apply the dashboard rules

KEY WORDS

- Business intelligence (BI)
- Dashboards
- Rules for dashboard design
- Scorecards
- Traffic light reporting

6.0 INTRODUCTION

The idea behind digital dashboards was an outgrowth of decision support systems in the 1970s. With the surge of the web in the late 1990s, business-related digital dashboards began to appear. Some dashboards were laid out to track the flows inherent in business processes while others were used to track how well the business strategy was being executed. Dashboards were constructed to represent financial measures that even executives could understand. Figure 6.1 shows what a typical dashboard might look like.

Project Management Metrics, KPIs, and Dashboards: A Guide to Measuring and Monitoring Project Performance, Fourth Edition. Harold Kerzner.
© 2023 John Wiley & Sons, Inc. Published 2023 by John Wiley & Sons, Inc.
Companion Website: www.wiley.com/go/kerznermetrics4e

Figure 6.1 The Framework for a Typical Dashboard

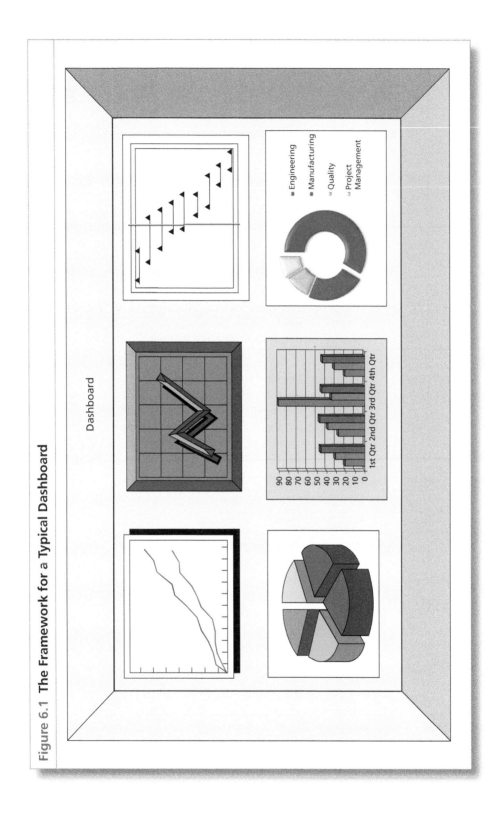

Dashboard

Perhaps the single most important event affecting dashboards was the introduction of the importance of key performance indicators (KPIs) as part of the Balanced Scorecard approach published by Robert S. Kaplan and David P. Norton in the mid-1990s.[1]

Later, in an article by Mark Leon, 135 companies were surveyed, and more than half were using dashboards.[2]

Even though dashboards are quite common in certain industries, the uses for the dashboards can vary significantly. As an example, here are some specific types of dashboards:

- Hospital workflows and bed management dashboard
- Museum dashboard
- Best places to work dashboard
- Casino management dashboard
- Dentist dashboard
- Energy dashboard
- Federal government dashboards to improve performance and cut waste
- Flex dashboard at the Federal Reserve Bank
- Food quality control dashboard
- Investor risk dashboard
- Sales compensation dashboard

The website www.dashboardspy.com has archives that describe a multitude of dashboards for almost every industry.

In today's business environment, the ability to make dashboard presentations is almost as indispensable as writing skills. People tend to take graphics for granted but do not realize what is wrong with various types of information graphics because this is not traditionally taught in schools. There are specialized seminars and webinars on dashboard design to fill this gap.

Many dashboards fail to provide value because of design issues, not technology. Effective dashboards are not bells and whistles, glitter, and bright lights. Dashboard design is effective communication. Most people fail to understand that information visualization is a science, not an art.

According to Stephen Few,

> The fundamental challenge of dashboard design is to display all the required information on a single screen, clearly and without distraction, in a manner that can be assimilated quickly.

---

1 Robert S. Kaplan and David P. Norton, *The Balanced Scorecard: Translating Strategy into Action* (Cambridge, MA: Harvard Business School Press, 1996).
2 Mark Leon, "Dashboard Democracy," *Computerworld*, June 16, 2003, 40-41. Available at www.computerworld.com/article/2570516/business-intelligence/dashboard-democracy.html.

If this objective is hard to meet in practice, it is because dashboards often require a dense display of information. You must pack a lot of information into a very limited space, and the entire display must fit on a single screen, without clutter. This is a tall order that requires a specific set of design principles.[3]

Few's comments can be modified to provide a definition of a dashboard as it relates to project management:

A project management dashboard is a visual display of a small number of critical metrics or key performance indicators such that stakeholders and all project personnel can see the necessary information at a glance in order to make an informed decision. Raw data is converted into meaningful information. All of the information should be clearly visible on one computer screen.

There are some simple facts related to dashboards:
- Dashboards are communication tools.
- Dashboards provide the viewers with situational awareness of what the information means now and what it might mean in the future.
- Properly designed dashboards provide business intelligence (BI) information.
- Dashboards are not detailed reports.
- Some dashboards simply may not work.
- Some dashboards may be inappropriate for a particular application and should not be forced on the stakeholders.
- More than one dashboard may be required to convey the necessary information; they should provide the right data and without overwhelming people with information.
- It is important that the information displayed focus on the future. Otherwise, viewers will get bogged down analyzing the past rather than thinking about what's ahead.
- Effective dashboards are a means for reducing potential stakeholder and executive meddling.
- During competitive bidding, dashboard displays may be the difference between winning and losing a contract.
- Infographics should be based on information design and data visualization best practices.
- With the growth in the use of KPIs, it is extremely important that stakeholders and other dashboard viewers have a good understanding of what is being measured. Deciding what to track is critical. Many projects fail because the dashboard designers insert

3 Stephen Few, "Dashboard Design: Beyond Meters, Gauges and Traffic Lights," *Business Intelligence Journal* Winter 2005 (January 1), p. 20. Available at www.bi-bestpractices.com/view-articles/4666. Few has also written excellent books on dashboards: *Show Me the Numbers* (Oakland, CA: Analytics Press 2004); *Information Dashboard Design* (Sebastopol, CA: O'Reilly, 2006); and *Now You See It* (Oakland, CA: Analytics Press, 2009).

too many bells and whistles, which can become distractions. Also, simply because an indicator on a dashboard is not flashing does not mean that things are going well.

- Dashboards are usually laid out in landscape rather than portrait view, especially if the dashboards are to be printed.
- Do not expect to hit a "home run" with the first dashboard design.

TIP The project manager must explain to the stakeholders how to identify when things are going well and when things are going poorly using the dashboard KPIs.

There are also traps that dashboard designers must understand. Two such traps involve dashboard security and the use of branding. Dashboard security refers to the process of restricting the information presented in the dashboard only to those who have the right to that information. The security system may be quite complex since each of the stakeholders may receive a different dashboard.

Also, because space on dashboards is limited, care must be taken to avoid heavy usage of company or project logos and other branding information. Company branding is always nice to have, but screen real estate is limited and expensive. Cluttering up a dashboard with too much information can lead to information overload.

Dashboard designers assigned to a project team must understand:

- The basics of project management
- The dashboard audience
- The purpose of the dashboard
- The end user's needs
- How the dashboard will be used
- How the measurements will be made
- How often the measurements will be made
- How the dashboard will be updated, and how often
- How to maintain uniformity in design, if possible
- Security surrounding the display of the data

Unlike business dashboards, which can be updated quarterly, project management dashboards focus on month-to-date and cumulative-to-date comparisons or the proximity to the target. Project management dashboards might also include real-time reporting. When designing a dashboard, the designer must know how frequently the dashboards must be updated.

Traditional business dashboards are designed for a broad audience. Project management dashboards are targeted dashboards containing viewer-specific metrics. There are generally two purposes for the dashboard data:

- The viewer sees the data and draws his/her own conclusions. These dashboards contain decision-making data and information necessary for problem identification.

■ The viewer sees the conclusions that the project manager wants him/her to see. This dashboard is common for internal reporting and status updating.

The data on project management dashboards are used for both purposes but are heavily biased toward viewers' needs and the metrics/KPIs selected. The data must be audience or viewer appropriate.

6.1 DOES EVERYONE KNOW WHAT A DASHBOARD REALLY IS?

There are many different types of information displays people need for exploring, analyzing, monitoring, and acting on data. In Dundas BI—our dashboard, reporting, and analytics platform—you're able to create different types of "views" (which we use for displaying data using data visualizations connected to metric sets), but by far the most common and most important view, is the **dashboard**.

What can you use dashboard views for in Dundas BI? Hang on—the answer is surprisingly controversial, but also demonstrates Dundas BI's capabilities and flexibility.

Dashboards!

You don't say! But what *exactly* is a dashboard? You'll find that answers vary, but over time, many thoughtful definitions are converging toward those of Stephen Few, who has spent much time trying to give the word a clear and specific meaning within the context of data visualization, while advocating for its best practice design. His definition is as follows:

> A dashboard is a visual display of the most important information needed to achieve one or more objectives that has been consolidated on a single computer screen so it can be monitored at a glance.

This section was written by Jamie Cherwonka, the R&D Director for Data Visualization at Dundas Data Visualization. Material in Sections 6.1, 6.7, 6.9, 6.10, 6.12, and 7.1 has been provided by Dundas Data Visualization, Inc. © 2021 Dundas Data Visualization. All rights reserved. Dundas Data Visualization is a leading, global provider of Business Intelligence (BI) and Data Visualization solutions. Dundas provides organizations with the most customizable, innovative, and scalable BI platform, enabling users access to all their data for better decisions and faster insights. Dundas is dedicated to helping their users make sense of their data to solve real business problems. For almost 30 years, Dundas has been helping organizations discover deeper insights faster, make better decisions, and achieve greater success. Dundas Data Visualization, Inc., www.dundas.com, Info@dundas.com, 500-250 Ferrand Drive, Toronto, ON, Canada, 1-800-463-1492, 1-416-467-5100.

A second definition later added:

...to rapidly monitor current conditions that require a timely response to fulfill a specific role.

True to its car analogy, your dashboard (in the context of business intelligence) is meant to include all of and only the information you need to monitor while you do your actual job, to know whether you need to take timely action on something. (Unfortunately, some car dashboards may be heading in the wrong direction with this—but that's a conversation for another time.) Determining what information that is, and for each person or role, can be a difficult task.

Dashboards should fit on your screen all at once without needing to interact with it to know that something requires your attention (sometimes that can make screen real estate scarce), and it may call your attention with visual alerts (or states). Once the dashboard has your attention, some interactions may be desired to support acting on the information, such as hover-over tooltips, drilling down a summary of data to more details, or navigating to a different screen. If the latest data is needed throughout the day or even every few seconds, the dashboard may automatically refresh itself.

Stephen Few is extremely precise in his choice of words and specifically states that a dashboard is *not* for data exploration, analysis, or a lookup report. It can't have scrollbars, or be organized into multiple screens, tabs, or require other clicks to access any data to be monitored—otherwise, you effectively have multiple dashboards. If it contains any information that would not trigger you immediately to take some action if it changed, it should be removed so that it doesn't steal space and attention from the other data.

If some of your dashboards don't quite fit this definition, you're not alone.

Dashboard Design

Whoever designs this dashboard, however, will likely need comprehensive analytics capabilities to filter, group by, and sort data, take subsets like top 10, apply formulas, and much more. This is why all of the analysis capabilities you can find in Dundas BI's full-screen metric set editor are also available while editing a dashboard. If you want to, you can drag your data onto the dashboard editor to instantly create a metric set and visualization, and use the same Data Analysis Panel as in our metric set editor. Switch between editing and viewing instantly to interact with the data to filter, sort, drill down, and more (you can choose to disable these interactions for your dashboard viewers when designing one).

Keep in mind, not all new dashboards necessarily have to involve as much work or require the user to start from scratch. All user types in Dundas BI are able to create new dashboards and drag in whatever they need from existing metric sets or visualizations from other pre-existing dashboards, and then add filters and interactions. While this may not necessarily be as good a result as a dashboard purposefully designed by an expert from the beginning, we want to offer our customers the choice to give all their users access to this "self-serve" ability, and they may indeed produce some helpful dashboards.

What should we call supplementary screens you can click to access from the actual dashboard for more details to support taking some action? They're not meant to be called dashboards, but the process for designing one is the same.

Guided Analysis

What if you have a display of multiple charts or other visualizations meant to be interacted with to perform some type of analysis? Perhaps there are specific tasks it's intended for, so additional analysis functions shouldn't be included. Is this still a dashboard?

No, but what is it called?

More on that later. For now, you can start with a dashboard, design it to support your needed analysis, and disable any unwanted functions if needed.

Data Exploration

Remember all those analysis tools I mentioned earlier that you have access to when editing a dashboard? They also provide you with a great environment for performing exploratory analysis. If you need to go beyond a single visualization at a time, the dashboard editor in Dundas BI can act as a blank canvas for rapidly changing, revisualizing, copying, and rearranging data and their visualizations. You can even apply a formula to data from multiple visualizations and display the combined result in a new visualization, called a "formula visualization." When one dataset is being analyzed and visualized differently on the same screen ("multi-faceted" analysis), this is aided by automatic data brushing and filters you can connect to all selected visualizations with one click.

This is why choosing some analytics options from the toolbar such as Histogram or Correlation Matrix in the full-screen metric set editor automatically brings your work onto a new blank dashboard: because the result should be in a new visualization, which you can then work with alongside the original.

You can also find the option "Edit Copy For Analysis" in the toolbar or context menu of a finished dashboard, even if someone else created it so that you can quickly access these analysis functions if needed based on the original dashboard's data.

More

While there's a particular intended meaning for the word "dashboard" for those who follow best practice design in data visualization, it's also a term with established meanings, and not all of those have anything to do with cars or aircraft. The following definitions come from the *Oxford Dictionary of English*:

- *"a graphical summary of various pieces of important information, typically used to give an overview of a business"*
- *"a home page on a website giving access to different elements of the site's functionality"*

Let's take that second definition for example. Using Dundas BI, you can easily create a dashboard to serve as your own customized portal *for* Dundas BI: in your user profile, set Default View to automatically display a particular dashboard whenever you log in.

Keep in mind, whenever you are designing something meant to fit within a single screen—optionally resizing itself for different screen sizes or else for a particular target size—a dashboard is the most appropriate view type to choose in Dundas BI. For example, you may want to use our extensive visualization library and feature set to create a customized infographic designed down to the pixel. (Why else would we have included a Word Cloud as an option for a dashboard—located under the "Fun" category?)

Not a Dashboard

There are some things everyone can agree are quite different than a dashboard, both technically and in practice. One is often called a "report," including in Dundas BI, which produces a multi-page report generally expected *not* to fit on one screen, and often designed with page headers and footers for formatting for PDFs and printing.

Small multiples, while sometimes able to fit themselves to your screen, don't do this necessarily, and they are often worked with on their own for analysis. Alternatively, both small multiples and our visual scorecards are often added to a dashboard alongside other content.

Conclusion

Even when knowing dashboard best practices and aspiring to create the one dashboard we can rely on to monitor for our role, we may fall short and/or create any number of other useful views of data. There are also certainly other types of displays we need from our BI software outside of dashboards. Unfortunately, the best practice definition for "dashboard" does not provide us with a technical distinction from other types of visualization displays designed to fit on a single screen, and the workflows of users working with these different displays overlap substantially. We also don't seem to have good, established alternative terms to use, and some of the suggestions offered so far are unlikely to catch on. This may be why we aren't the first business intelligence vendor to rely so much on this word (dashboard), especially from a technical point of view.

The reality is there are multiple meanings of words like "dashboard" and "report" out there, even if we all one day agreed how to solve this within the area of business intelligence. To ensure everyone always understands, maybe it's best to simply say you have a dashboard for monitoring purposes when those are your requirements, and the distinction is important. Or, feel free to double-click the name of your dashboard right away and type in your preferred alternative.

6.2 HOW WE PROCESS DASHBOARD INFORMATION

Hoping to design the "perfect" dashboard is wishful thinking. Even the best metrics and dashboards cannot compensate for the ways that people process information. Effective dashboard design requires an understanding of how people learn. Perhaps the biggest risk is that infographic designers do not understand the information they are working with and how people will perceive the information.

We generally accomplish more through our visual sense that other senses. Dashboards require us to think visually. The results of visualization can be:

- Short-term memory retention
- Long-term memory retention
- Visual working memory retention, which lies in between short- and long-term memory

Retention may be increased if we appeal to more than one sense. For example, combining a brief narrative with each metric in the dashboard may be the best approach. However, this may limit the space available on the dashboard or require scrolling down to other screens. Another approach might be to provide some of the narrative information in a dynamic rather than static format. This can be accomplished by inserting animation or motion graphics to the dashboard.

6.3 DASHBOARD CORE ATTRIBUTES

When designing a project management dashboard, the core attributes, as shown in Figure 6.2, are:

- **Aesthetic appearance:** The viewers want to use it.
- **Easily understood:** The material is easily comprehended.
- **Retention:** The material will be remembered.

In project management, all three attributes would most likely have equal importance. Perfection in dashboard design for project

Figure 6.2 **Dashboard Core Attributes**

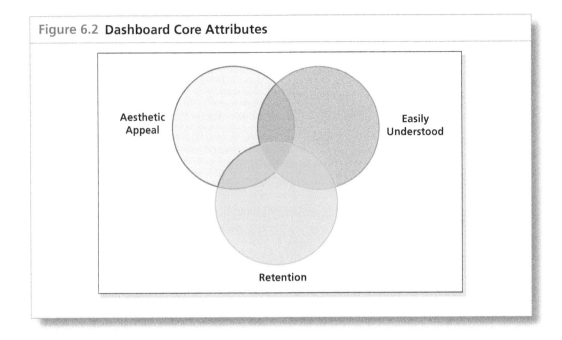

management applications would occur if all three circles came together. It is not always possible to design a perfect dashboard because each person has a different way of learning and retaining information. The more the circles overlap, the better the dashboard.

6.4 THE MEANING OF INFORMATION

Collecting dashboard information just for information's sake alone can be time-consuming and expensive. Not all data provides information that can be used. Meaningful information must satisfy the following criteria:

- The information must show a pattern over time (i.e., a trend). This cannot be accomplished with a discrete data point.
- The viewer must be able to understand what the data says. Several metrics viewed together may be required to get a clear picture of the project's health.
- The viewer must be able to understand how the measurements were made and the fact that there may be some "noise" in the data because of irregularities in measurement. The viewer must filter out the noise if possible. Vagueness is the enemy of metric measurement and data interpretation. Misrepresentation of data can lead to serious communication failure.
- Dashboard viewers must be able to quickly identify those items that need immediate attention. Some dashboard designers focus on glitter that prevents clear communications and focuses on perhaps the wrong objectives.

- The data must provide enough information such that the viewer can make an informed decision.
- The viewer must have enough information to fix a problem.
- If the dashboard viewer must scroll around the dashboard or change screens, then the overall purpose of having a dashboard may not have been achieved.
- When designing a dashboard, focus first on the data needed for informed decision making rather than the means of display.
- Dashboard success is achieved when the status of the project is quite apparent from the dashboard metrics. Dashboards are now part of decision support systems.

Data does not necessarily tell the viewer what actions to take but does function as an early warning indicator of potential problems. This can be seen from Table 6.1.

The last metric in Table 6.1 shows that there could be three or more causes for this early warning sign to exist. Therefore, more than one metric may be required to fully understand the health of the project.

In an ideal situation, all viewers would see the same dashboards. This would certainly reduce the time and effort for the creation of dashboards. However, in reality, this is impractical because the level of confidentiality of some information may preclude viewing by certain viewers. Dashboards are customized for individual viewers. Each viewer may have different needs based on the decisions they must make.

Some executives will not take ownership for dashboard reporting systems for fear of looking bad in the eyes of their colleagues if the system is not accepted by the users or workers, or it fails to provide meaningful results.

TABLE 6.1 Metrics and Early Warning Signs

METRIC	MEASUREMENT	EARLY WARNING SIGN
Number of scope changes	Very high	May have trouble meeting baselines and constraints
Amount of overtime	Greater than usual	Project is understaffed or major problems exist
Employee turnover	Highly skilled people removed from project to put out fires elsewhere	Possible schedule slippage
Amount of rework	Very high	We are using unskilled labor
Number of deliverables	We have a shortage of deliverables	Technical complexity exists, the project is understaffed, or workers lack necessary skills

6.5 TRAFFIC LIGHT DASHBOARD REPORTING

In the attempt to go to paperless project management, emphasis is being given to visual displays such as dashboard and scorecards. Executives and customers desire a visual display of the most critical project performance information in the least amount of space. Simple dashboard techniques, such as traffic light reporting, shown in Figure 6.3, can convey critical performance information.

The following are examples of the meaning of the indicators in Figure 6.3:

- **Red traffic light:** A problem exists that may affect time, cost, quality, or scope. Sponsorship or stakeholder involvement may be necessary.
- **Yellow or amber light:** This is a caution. A potential problem may exist, perhaps in the future if the situation is not monitored. The sponsors/stakeholders are informed, but no action is necessary at this time. If some action has been taken, the problem has not been resolved as yet.
- **Green light:** Work is progressing as planned. No involvement by the sponsors or stakeholders is necessary.

Figure 6.3 **Traffic Light Dashboard Indicators**

The colors red, yellow, and green may be interpreted differently by each viewer. Therefore, an explanation of the colors may be necessary. For example:

When discussing project risks:

- **Red:** Some risk events exist, and there are no workable mitigation strategies.
- **Yellow:** Some risk events have been identified, and mitigation strategies are being developed.
- **Green:** There are no risks.

When discussing project staffing:

- **Red:** Either sufficient resources are lacking or the assigned resources have questionable credentials.
- **Yellow:** Resource staffing issues exist, but staffing is ramping up.
- **Green:** Sufficient and qualified resources are available.

Traffic light reporting using the colors red, green, and yellow or orange is used often. However, most companies do not establish criteria for when to change colors. Also, to be decided is who has the authority to change the colors. As an example, a project manager discovers that one of the work packages performed in a functional area is two weeks behind schedule. The project manager's first inclination is to change the color of this work package from green to red. However, the functional manager asserts that his people will be working overtime and the planned schedule date should be met. Now the decision as to what color the light should be is a little more complicated.

Although a traffic light dashboard with just three colors is most common, some companies use many more colors. The information technology (IT) group of a retailer had an eight-color dashboard for IT projects. An amber color meant that the targeted end date had passed, and the project was still not complete. A purple color meant that this work package was undergoing a scope change that could have an impact on the triple constraints.

Although dashboards for project management applications are just in their infancy, companies have been using traffic light reporting for some time. It is common for project stakeholders to be briefed by the project manager without paperwork exchanging hands. The project manager displays the status of a project on a screen, using a computer and an LCD projector. Beside all of the work packages in the work breakdown structure is a traffic light. Senior management then takes keen interest in all of the work packages indicated in red. One Detroit-based company believes that in the first year of using this technique and going to paperless meetings, it saved $1 million, and savings are expected to increase each year.

6.6 DASHBOARDS AND SCORECARDS

Some people confuse dashboards with scorecards, but there is a difference between them. According to Wayne W. Eckerson,

> Dashboards are visual display mechanisms used in an **operationally** oriented performance measurement system that measure performance against targets and thresholds using right-time data.
>
> Scorecards are visual displays used in a **strategically** oriented performance measurement system that chart progress towards achieving strategic goals and objectives by comparing performance against targets and thresholds.[4]

Both dashboards and scorecards are visual display mechanisms within a performance measurement system that convey critical information. The primary difference between dashboards and scorecards is that dashboards monitor operational processes such as those used in project management, whereas scorecards chart the progress of tactical goals.

Table 6.2 and the description following it show how Eckerson compares the features of dashboards to scorecards.

TABLE 6.2   Comparing Features

FEATURE	DASHBOARD	SCORECARD
Purpose	Measures performance	Charts progress
Users	Supervisors, specialists	Executives, managers, and staff
Updates	Right-time feeds	Periodic snapshots
Data	Events	Summaries
Display	Visual graphs, raw data	Visual graphs, comments

Source: Wayne W. Eckerson, *Performance Dashboards: Measuring, Monitoring and Managing Your Business* (Hoboken, NJ: John Wiley & Sons, 2006), p. 13.

Material in Section 6.6 is from Wayne W. Eckerson, *Performance Dashboards: Measuring, Monitoring, and Managing Your Business* (Hoboken, NJ: John Wiley & Sons, 2006), especially Chapter 1, "What Are Performance Dashboards," pp. 3–25. Chapter 12 of Eckerson also provides an excellent approach to designing dashboard screens.

4 Wayne W. Eckerson, *Performance Dashboards: Measuring, Monitoring and Managing Your Business* (Hoboken, NJ: John Wiley & Sons, 2006), pp. 293, 295.

Dashboards

Dashboards are more like automobile dashboards. They let operational specialists and their supervisors monitor events generated by key business processes. Unlike automobiles, however, most business dashboards do not display events in "real time" as they occur; they display them in "right time" as users need to view them. This could be every second, minute, hour, day, week, or month, depending on the business process, its volatility, and how critical it is to the business. However, most elements on a dashboard are updated on an intraday basis, with latency measured either in minutes or in hours.

Dashboards often display performance visually, using charts or simple graphs, such as gauges and meters. However, dashboard graphs are often updated in place, causing the graph to "flicker" or change dynamically. Ironically, people who monitor operational processes often find the visual glitz distracting and prefer to view the data in its original form, as numbers or text, perhaps accompanied by visual graphs.

Scorecards

Scorecards, on the other hand, look more like performance charts used to track progress toward achieving goals. Scorecards usually display monthly snapshots of summarized data for business executives who track strategic and long-term objectives or daily and weekly snapshots of data for managers who need to chart the progress of their group of projects toward achieving goals. In both cases, the data are summarized, so users can view their performance status at a glance.

Like dashboards, scorecards also make use of charts and visual graphs to indicate performance state, trends, and variance from goals. The higher up the users are in the organization, the more they prefer to see performance encoded visually. However, most scorecards also contain (or should contain) a great deal of textual commentary that interprets performance results, describes action taken, and forecasts future results.

Summary

In the end, it does not really matter whether you use the term "dashboard" or "scorecard" as long as the tool helps to focus users and organizations on what really matters. Both dashboards and scorecards need to display critical performance information on a single screen so that users can monitor results at a glance.

Dashboards appear to be more appropriate for project management than scorecards. Table 6.3 shows some of the factors relating to this.

Although the terms are used interchangeably, most project managers prefer to use dashboards and/or dashboard reporting rather than scorecards. Eckerson defines three types of dashboards as shown in Table 6.4 and the description that follows:

TABLE 6.3 Comparison of Dashboards and Scorecards

FACTOR	DASHBOARDS	SCORECARDS
Performance	Operational issues	Strategic issues
Work breakdown structure level for measurement	Work package level	Summary level
Frequency of update	Real-time data	Periodic data
Target audience	Working levels	Executive levels

TABLE 6.4 Three Types of Performance Dashboards

	OPERATIONAL	TACTICAL	STRATEGIC
Purpose	Monitor operations	Measure progress	Execute strategy
Users	Supervisors, specialists	Managers, analysts	Executives, managers, staff
Scope	Operational	Departmental	Enterprise
Information	Detailed	Detailed/summary	Detailed/summary
Updates	Intraday	Daily/weekly	Monthly/quarterly
Emphasis	Monitoring	Analysis	Management

Source: Wayne W. Eckerson, *Performance Dashboards: Measuring, Monitoring and Managing Your Business* (Hoboken, NJ: John Wiley & Sons, 2006), p. 18.

Operational dashboards monitor core operational processes and are used primarily by frontline workers and their supervisors who deal directly with customers or manage the creation or delivery of organizational products and services. Operational dashboards primarily deliver detailed information that is only lightly summarized. For example, an online web merchant may track transactions at the product level rather than the customer level. In addition, most metrics in an operational dashboard are updated on an intraday basis, ranging from minutes to hours, depending on the application. As a result, operational dashboards emphasize monitoring more than analysis and management.

Tactical dashboards track departmental processes and projects that are of interest to a segment of the organization or a limited group of people.

Managers and business analysts use tactical dashboards to compare performance of their areas or projects, to budget plans, forecasts, or last period's results. For example, a project to reduce the number of errors in a customer database might use a tactical dashboard to display, monitor, and analyze progress during the previous 12 months toward achieving 99.9% defect-free customer data by 2007.

Strategic dashboards monitor the execution of strategic objectives and are frequently implemented using a Balanced Scorecard approach, although Total Quality Management, Six Sigma, and other methodologies are used as well. The goal of a strategic dashboard is to align the organization around strategic objectives and get every group marching in the same direction. To do this, organizations roll out customized scorecards to every group in the organization and sometimes to every individual as well. These "cascading" scorecards, which are usually updated weekly or monthly, give executives a powerful tool to communicate strategy, gain visibility into operations, and identify the key drivers of performance and business value. Strategic dashboards emphasize management more than monitoring and analysis.

Three critical steps must be considered when using dashboards: (1) the target audience for the dashboard, (2) the type of dashboard to be used, and (3) the frequency in which the data will be updated. Some project dashboards focus on the KPIs that are part of earned value measurement. These dashboards may need to be updated daily or weekly. Dashboards related to the financial health of the company may be updated monthly or quarterly.

6.7 CREATING A DASHBOARD IS A LOT LIKE ONLINE DATING

Here are some of the similarities between dashboards and online dating.

Finding Out the Needs of the Stakeholders

In the case of online dating, you are the stakeholder. You need to understand what it is that you are looking for. What do you expect out of a relationship? One rule you can borrow from sales 101 is to ask yourself "why" until you can no longer provide an answer to it. By examining and uncovering what it is that you truly need, the process of finding someone to fulfill that becomes much easier.

This process also holds true when building a dashboard, whether it is for your personal use or for an entire department. You need to discover what people are trying to solve or change with dashboards and what they want to get out of a dashboard solution. You can use the "why" exercise for this as well. Once you find out the root factors that are driving the dashboard initiative, the process of building it becomes that much easier.

Making a Connection

Online dating, or dating in general, is all about making those connections, right? You need to filter through a number of people, try to reach out to people in creative ways, and at the same time, not be creepy (you know that guy/girl, don't be that guy/girl). You eventually establish some communication with a handful of people, or maybe even just one, and build a connection.

The same is true for dashboards. You need to ensure that you can connect to your data sources before building out a dashboard. In some cases the data may not be that "clean," meaning it has not been stored in a way that allows you to easily pull the information. A simple example of this would be an Excel spread sheet that has no data in certain fields, or where columns/rows change the type of information they contain. Like online dating, you might have to sort through a bunch of crap before you are able to build that connection.

Choosing Your Key Performance Indicators

Yes, this is a dashboard term, and it might seem laughable, but it is needed.

In order to determine whether or not you can have a successful relationship with someone, you need to have that "checklist," right? When going on a date, or even when talking to someone online, you should have a set of measures based on what you determine your needs are. If those needs involve marriage, kids, and wilderness excursions, then perhaps the career-focused, technology-dependent person whose longest relationship is a couple of weeks isn't going to the be the best match. If your needs involve getting attention more than giving it, you might want to be wary of how many times the person talks about themselves versus asking about you.

KPIs are a must in dashboards. If you know what your stakeholders need from a dashboard, it is easier to choose the KPIs you wish to visualize, but it doesn't make it instantaneous. Based on the needs that you have access to, you can determine what measures will best help your stakeholder(s) achieve their goals from the dashboard. For example, if the need is to save money, you might want to use KPIs that would allow the stakeholder(s) to spot operational efficiencies, redundancies, and monitor savings.

Selecting Your Visuals

No, this doesn't have to do with visual appearance. You should have been able to address that with the first point. This has to do with selecting the right person, or people, who can help you realize your goal or needs. Selecting the right person is critical to the success of any relationship.

Selecting the right visuals is also critical to the success of any dashboard.

You need to be able to display the information in the most effective way possible to the viewer. The purpose of a dashboard is to provide information at a glance that enables the users to take action. If the information is not presented in the most concise and efficient way possible, you increase the amount of time it takes to glean actionable insight from the dashboard and detract from meeting the needs of your stakeholders.

Building on the Momentum

As you browse through hundreds of profiles and the person on the other side does the same, it's crucial that you follow up on a good discussion and quickly move onto the next stage to see if this can be a good fit. While most discussions start the same way with your usual spiel that works well, it is key for your success to be able to keep on impressing and find interest in each other past that first contact. Of course, you want to ensure that you engage in an authentic way without sharing false information.

For dashboards, you also have to start with a great first impression—a picture is worth 1000 words. But if all you have is a good first impression with no substance behind it, stakeholders will quickly lose interest. This is where a well-designed dashboard allows users to further interact with the data and uncover more of the value it gives as they further explore. Being able to improvise and customize to stakeholder desires can also help ramping up the relationship.

Maintenance

For online dating, this should be self-explanatory. Every relationship requires work, some more than others, in order to be successful. You need to communicate and check in to ensure that you're on the same page, that you're still on course, and that you're meeting each other's needs.

Just as every relationship requires maintenance, so too does every dashboard.

The needs of the stakeholder(s) can change over time, and you need to update your dashboard accordingly. This is also true of online dating. Ideally one's needs won't change, but in reality, they do, and you need to adjust accordingly.

6.8 BENEFITS OF DASHBOARDS

Digital dashboards allow viewers to gauge exactly how well the project is performing overall and to capture specific data. The benefits of using dashboards include:

- Visual representation of performance measures
- Ability to identify and correct negative trends
- Ability to measure efficiencies/inefficiencies
- Ability to generate detailed reports showing new trends
- Ability to make more informed decisions based upon collected intelligence
- Align strategies and overall goals
- Save time over running multiple reports
- Gain total visibility of all systems instantly[5]

5 Wikipedia, "Dashboard (Business)," available at https://en.wikipedia.org/wiki/Dashboard_(business)#cite_note-Briggs-2, retrieved August 2, 2017.

In order for a project to continuously improve, four steps are required:

1. Measure performance and turn it into data.
2. Turn data into knowledge.
3. Turn knowledge into action.
4. Turn action into improvements.

Dashboards are tools that allow this to happen.

6.9 IS YOUR BI TOOL FLEXIBLE ENOUGH?

A Flexible BI Tool—What Does It Mean and Why Does It Matter?

So, you spent the last few weeks/days/hours (use the appropriate metric based on how fast your BI tool is) to create a good dashboard or analyze data you want to present to your team in your next meeting. You are excited about it. You even added a nice touch with a new visualization to show a business forecast. But when you present it to your colleagues, you're asked:

- "Why is that line green and not purple? Shouldn't it be purple like our brand colors?"
- "What is that line chart showing?"
- "Can you change the scale? I just can't read it like that."

You then have to explain that it's the BI tool's defaults and you're not sure those can be changed, and if they can, it will take time.

Unfortunately, in this case and many others, "small" design and layout concerns can take users' attention away from the actual content. This is just one, simple example of how the flexibility (in this case visual) of your BI tool can determine your BI solution success and the adoption by the user communities within your organization.

Many BI tools these days seem to be similar on the surface—providing similar capabilities (connections to common data sources, visualizations and dashboards, analytical options, mobile consumption options, etc.). The promise of these BI tools is often similar as well—easy to use, scalable, good performance. However, once you dive slightly deeper into a real-life project or task, you often find that your BI tool (even though once meeting your initial requirements) doesn't always provide the necessary options, resulting in users having to compromise or underdeliver. Why is that the case?

Section 6.9 was written by Dundas Data Visualization; © 2021 by Dundas Data Visualization, Inc. All rights reserved.

Typical BI solutions suffer from a lack of flexibility. While they come standard with prebuilt solution types, they can't adapt to changing business needs and requirements.

Why Is Flexibility So Important?

The ability to meet a specific design is key to get the necessary buy-in from a large audience. As BI tools are more and more visual, the ability to display the data exactly the way the users want to consume it ("pixel perfect") is often an important business requirement. Often, the right styling in place enables a professional look and feel and impacts a potential user's emotional factor, thus increasing the chances of improved adoption of the tool across an organization. This is even more important when the design is for an external group of users or when the solution is embedded into another application with existing styling.

For a BI tool to allow you to realize a PowerPoint wireframe or a Photoshop mockup exactly as designed, a great deal of flexibility is required around the way you can lay out and layer the information as well as control the visualizations styling via built-in configurations or your own preference (i.e., via Cascading Style Sheets). For example, looking at the bar chart in Figure 6.4, there are many elements one may wish to control beyond just changing the bar colors.

Figure 6.4 *Source: ©2021 by Dundas Data Visualization, Inc. All rights reserved*

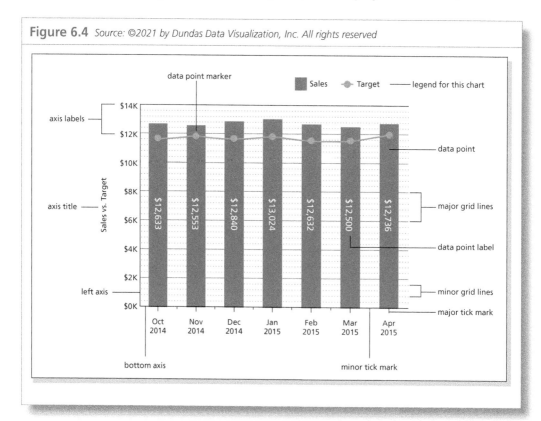

In addition, web-based views (such as dashboards) require the support not only of data-bound visualizations but also other design elements, such frames, labels, buttons, sliders, and others.

Stay up to Speed with Your Changing Business Needs

As your business evolves and new challenges come into play, your BI and analytics needs will change as well. That is why it is good practice to have an iterative process for your dashboard and always consider it a "live project." Being able to quickly modify and change elements on your existing views without breaking the view, or the way it is accessed by others, without having to re-create it, is critical to support changing needs in a timely fashion.

Be Independent (with Fewer Tools and Users Involved to Get Your Job Done)

Many "business-friendly" tools can be intuitive yet limited in scope, requiring users to rely on multiple products to meet their needs. Some may be limited around the way they can prepare data and join it across different sources, others may be limited in the way they can analyze the data or display it (in dashboards/reports/scorecards/specific visuals). Whatever the case may be, getting the job done via a combination of tools, that don't always play nicely together, or by depending on another person to change another system is not an optimal approach. Flexible BI tools can serve as a one-stop shop, allowing users to get all of their common BI needs in a single place, ensuring great efficiency and improved effectiveness.

Adapt to Each and Every User

A one-size-fits-all-users approach can send self-service business intelligence efforts off track. So can a lack of oversight and up-front development. As different users have different skills and needs, it is important to be able to cater to those users in a dedicated way to support the way they can and want to work. For example, for some users, too many options such as the ability to revisualize (change visualizations types) can be overwhelming and confusing. Being able to control and limit the options per user, or group, enables users to experience the system as they expect and according to their skills. Smart systems take this idea further and offer users tailored options automatically via different user roles, as well as offer automatic updates based on user behavior.

Be Ready for the Unknown

You can do a lot of planning prior to purchasing your BI tool or starting your project. You may have a long list of functionality requirements or you may have even conducted a small proof of concept, but once your project starts, new system requirements often arise. Flexible systems shine when it comes to scenarios that were unknown at the time the systems were designed. Make sure your tool gives you the necessary internal hooks to adapt to new challenges and to further extend the solution functionality and integration with other systems via its open application program interfaces.

Flexible systems are those that can be used for more than one purpose or mission, and especially for purposes that are developed after the systems are deployed.

Is your BI tool flexible enough to meet your future needs?

6.10 FOUR EASY STEPS TO IMPLEMENTING A SUCCESSFUL BUSINESS INTELLIGENCE SOLUTION

Venturing into a new Business Intelligence (BI) project can be as exciting as it is daunting. For many people, BI is something they know they need, but they don't quite know where to get started. One of the things I see time and time again are enterprises and software vendors that jump right into developing their first BI solutions without the necessary planning in place. For example, these companies neglect to:

- Establish a clear picture of what the problem they are trying to resolve is.
- Understand who their actual target audience is or what they want.
- Outline a clear set of deliverables to target completion, etc.

All of these issues lead companies down a spiraling path of endless development, redefining of requirements, and wasted effort that rarely results in successful solutions.

Throughout my time working with various companies both new and experienced in BI, I have developed a basic guide to ensure that any BI project is a successful one. This approach aims to minimize endless documentation and instead focuses on gathering a clear understanding of the problems that need solving along with the end users' expectation. So, without further ado, here is my **four-step guide to implementing a successful Business Intelligence solution.**

Step 1: Understand the Business Needs

Outcome of this Step: Determine a list of business requirements

Before you start getting into "*how*" you are going to implement your BI solution, you need to understand "*why*." The purpose of every

---

Section 6.10 was written by Dany Nehme, a Senior BI Consultant at Dundas Data Visualization, Inc. © 2021 by Dundas Data Visualization. All rights reserved.

solution is to solve a problem, and Business Intelligence is no different. Without an understanding of what the problem you are trying to solve is, there is nothing to guide your implementation. By understanding the problem, you can begin to outline the requirements of your solution.

How to get there: Understand your audience

At the end of the day, you are building a solution for your end-users and stakeholders. If the solution does not meet their expectations, then it *won't be considered a success*. That's why it's important to understand who your audience is, what problems they currently have, and what they are hoping to gain from a BI solution. I recommend meeting with members of each of your user groups (face-to-face whenever possible) and having these discussions to better understand their needs. During these meetings, you'll want to ask questions such as:

What problems and pain points are you hoping to alleviate with a BI solution? As mentioned before, without knowledge of the problems, there is no way to develop a solution. Who better to tell you their problems than the people you are building the solution for?

Which current business areas are lacking visibility? Sometimes it goes beyond knowing what problems you are currently facing, and instead recognizing there are problems that you may not even be aware of. Business Intelligence opens the floodgates when it comes to possibility, and that often means being able to get clarity on avenues that were previously unavailable.

What kind of experience are they looking to have with a BI solution? You can develop what you believe to be the greatest dashboard(s) of all time, perfect for being viewed on large, mounted monitors. After days of work, the deadline comes, and you present your baby to the stakeholders. Within a couple of seconds, all your hard work is torn to shreds as they utter, "Can you make this work on my mobile devices?". I have seen this specific scenario and many others like it happen countless times. I can't stress enough how easy and important it is to avoid such wasted efforts by simply talking to your audience, understanding the type of experience they are looking to have, and making sure your solution matches their expectations.

Step 2: Keep It SMART

Outcome of this step: Determine a list of SMART objectives

One of the best ways to ensure the success of a project is to determine what its objectives are. Objectives provide you with a clear set of goals that you can guide your solution toward. By meeting those objectives, you can say that *your solution achieved exactly what it is supposed to.*

With that being said, *any* list of objectives just won't cut it. For example, let's say you proceed with the objective of *making more sales.*

How do you actually determine if an objective like this has been achieved? Are you trying to make more sales compared to the previous year? Are you trying to make more sales during the month following the implementation of the solution compared to the previous month? If you make one more sale than in the previous period, does that count? This is where SMART objectives come in. SMART stands for:

- Specific
- Measurable
- Achievable
- Realistic
- Tangible

With SMART objectives, we can eliminate any ambiguity, progress becomes much easier to track, and unrealistic objectives are avoided.

Now, let's revisit the example objective I had, but this time I'll make sure it meets the SMART objective criteria: Our objective is to **increase the total year-end sales dollar amount by at least 10% compared to the previous year**.

As you can see, this is a much more specific objective. I have a timeline to achieve it by (end of the year), and a clear target I am trying to hit (110% of last year's total sales). Any objectives that do not follow the SMART guidelines should not be considered objectives of your project until they have been revamped to do so.

Step 3: Determine Your Deliverables

Outcome of this step: Determine a list of project deliverables and sub-list of KPI definitions for each deliverable

At this point, you should have a solid understanding of why you need a BI solution. Now, it's time to think about how you will get there. Before any development begins, you need to know what you are actually developing. Are you developing a dashboard, multiple dashboards, or an automated report? Are you preparing data for self-service analysis? A good BI tool will offer you limitless possibilities in terms of what you can create, so before you get lost in the development process, determine what the deliverables of the project will be. It's important to distinguish and separate your project objectives from your project deliverables. Remember, objectives focus on elements that are external to the project, while deliverables focus on elements internal to the project.

How to get there: Use what you know

You've spoken to your stakeholders and target audience and based on these discussions you should have an understanding of how many different user groups you are dealing with, what problems they are looking to resolve, and what type of experience each wants. Use this knowledge to determine what deliverables will best achieve the results your audience is looking for. Think about how much overlap you have between each of your user groups and their requirements. Can some of these be addressed within a single deliverable, or are they separate pieces to the overall solution?

For example, your target audience might include members of your Business Development and Finance departments, as well as upper management who oversee both those teams. The individual members of those two departments are looking for a way to monitor and track their specific and detailed day-to-day operations and performance, while upper management wants to receive scheduled updates on the overall performance of both departments. Upper management also wants the ability to see the detailed, day-to-day information, but only when there are any issues. In this case, your deliverables could be:

1. A detailed-level Business Development dashboard
2. A detailed-level Financial dashboard
3. A scheduled, high-level report on both departments

At this stage, you will also want to start thinking about which KPIs you'll include on each of your deliverables. As this is a Business Intelligence solution, the driving force of the solution's information will be the KPIs that it presents. Now that you know what deliverables you will be providing, think about which KPIs can best address the business requirements and SMART objectives associated with your solution.

Remember, real-estate on dashboards and reports is precious, so *every KPI should serve a purpose*. If you cannot tie a KPI back to one of the solution's business objectives, then it probably is not needed. Then, for each of your deliverables, come up with a sub-list of the KPIs it will include, along with the business objective each KPI will address.

To make the development of each KPI more seamless, putting the right amount of planning into defining each KPI is critical. Here are some things I recommend considering when defining your KPIs:

- What measures and dimensions should the KPI include?
- How can the KPI be filtered?
- Are there any targets or comparisons to be made?
- Are there any conditional states that signify if the results are good or bad?
- Where is this data coming from?

The idea is that these definitions should be clear enough that even users *without* a technical background should be able to easily understand them. By creating these definitions, the transition to developing your KPIs will be a smooth one, as it will simply be a matter of applying what has already been defined.

Step 4: To the Drawing Board

Outcome of this step: Create mockups for each deliverable

The fourth and final step to this guide is to create mockups for your deliverables. In this phase, you'll want to begin by laying out each deliverable's set of KPIs in a drawing. Yup, a physical drawing! These drawings can start out roughly (treat them as you would a brainstorming session) as you'll want to easily be able to make changes. I prefer working on a whiteboard at first.

To simplify this step, I like breaking it down into two parts:

Determining your data visualizations. At this point, you should have your KPIs well defined; you should know what they are going to include and what business objectives they are going to serve. This is where you will want to utilize your data visualization skills to determine the best visualization option to make understanding the data as easy as possible. A user should be able to tell if a KPI is meeting its objective or not within a simple glance, and the right visualization will allow for that.

Design the layout of your KPIs and user interface (UI). By now you should know who the target audience of each deliverable will be, so use this information to create a suitable design. Think about each group of users that will be interacting with specific dashboards or reports, what their skillsets are, and what their user stories would be. Use this information to create mockups that address each user group's desired user experience. You should also know how many KPIs your deliverable will have, as well as what types of filters you will want to include. Based on these details, you can determine how to divide that available space and whether you want to have a scrolling view or organize items onto different layers.

This stage is one that tends to be revisited and reworked quite frequently, so be sure to communicate with your stakeholders before making any final decisions. Any feedback that they provide can go right back into adjusting your mockups, which is probably much easier than trying to adjust a finished solution.

Closing Comments

The point of these steps is to guide you in the right direction when starting your next Business Intelligence project. What's important to remember, is that this guide provides you with a solid plan to base your development on, rather than going in blindly. Many organizations might already have steps in place to get the same results; however, these are ones that I have seen lead to successful BI solutions countless times.

Are there any steps in this guide that you have put into action before? Are there steps that you feel are missing that you can't start your BI projects without? We'd love to hear from you your success following these steps along with other approaches you've followed in the past.

6.11 **RULES FOR DASHBOARDS**

This chapter discusses certain rules for dashboards, such as rules for colors, for metaphor selection, and for positioning the metaphors. However, there are certain overall rules should be considered as well. Some of these include:

- Dashboards are communication tools to provide information at a glance.
- Be sure to understand aesthetics, especially in symmetry and proportions.
- Be sure to understand computer resolution considerations involving readability.
- Begin dashboard design with an understanding of the user's needs.
- Be sure to understand content selection and the accuracy of the information.
- Dashboard design can be done with simple displays.
- Dashboard design can be done with simple tools.
- Use the fewest metrics necessary.
- Determine the fewest metrics that can be retained in short term memory.
- Using too many colors or sophisticated, complex metaphors leads to distractions.
- Metrics should be limited to a single screen.
- Be sure to understand the available space; know the number of windows/frames available within the dashboard.
- Perfection in design can never be achieved.
- Asking for assistance with the design effort is not an embarrassment.
- Monitor the health and user friendliness of the dashboard.

6.12 **THE SEVEN DEADLY SINS OF DASHBOARD DESIGN AND WHY THEY SHOULD BE AVOIDED**

The basic purpose of a traditional dashboard is to make it easy for company management and staff to access crucial and current business data at little more than a glance. And, make no mistake about it, "easy" is the key word here—if the user has to work too hard to understand key information contained in a dashboard, then it has failed in its mission.

The fact is well-designed dashboards get heavily used—and badly designed ones don't. If the dashboard layout doesn't deliver data clearly and concisely, then there's no point in making the effort to create it in the first place. With that in mind let's review the "Seven Deadly Sins of Dashboard Design"—and why they should be avoided.

Deadly Sin #1: Off the Page, Out of Mind

As noted, the reason for creating a dashboard is to gather together relevant data in one place for instant analysis. When a dashboard design forces us to either scroll down or navigate to another screen entirely, that "instant" factor is lost—we lose track of what we're looking at and our short-term memory is challenged. Not only that, but the data contained on a second screen or in an area that requires scrolling is perceived as being less important—and easy to dismiss. And if it actually is easy to dismiss, it probably shouldn't be part of the dashboard in the first place!

Deadly Sin #2: And This Means ... What?

An effective dashboard puts the data into a relevant context, so we understand what the numbers mean and what response might be called for. Just supplying random figures—such as 30,000 orders have been placed—without comparing it to company targets or showing whether the category in question is improving or growing worse, means we might as well be punching buttons on a calculator at random to see what happens. The proper context allows for the proper interpretation of the information—and motivates the correct action.

Deadly Sin #3: Right Data, Wrong Chart

Have you ever been served a pie slice that's only about a millimeter thick? Probably not. Yet many dashboards insist on serving up pie charts with so many razor-thin slices that you can barely make out what they're supposed to signify. Conversely, simple bar graph info might be better represented through a pie chart (for some very specific cases). Using the wrong type of chart to represent data can prevent proper analysis. The right chart gives you at-a-glance clarity that visually communicates efficiently and effectively.

Similarly, visualizing quantitative data in the wrong way can cause severe misinterpretation. For example, a three-dimensional bar chart can cause bars at the back to appear smaller than they really are. Also, when you use a bar graph and don't start from zero, proportions can become extremely distorted. Don't judge the visualization by the formula used to create that visualization; judge it visually just as the user will, to make sure it makes proportional sense.

Deadly Sin #4: Not Making the Right Arrangements

If the dashboard data isn't arranged in the right way, it won't have the visual impact you're looking for. This becomes even more important when the dashboard is interactive and one visual can affect another or when you have filter controls that can affect all or some of the visualizations on the dashboard. Always group together related metrics/controls for easy analysis/comparison instead of placing them willy-nilly on the page. The design should be done in a logical fashion that facilitates the user's comprehension of all the data.

Deadly Sin #5: A Lack of Emphasis

Part of the job of a dashboard is to immediately draw your eyes to the most important information. If that data is draped in drab colors, you might check it last—or not at all. It's also a mistake to use bright prominent colors all across the dashboard—nothing stands out, even when it should. Again, you want to guide the user to read the dashboard the way it's meant to be read; that means properly highlighting the information that requires the most attention.

Dashboard designers can also get carried away and add a lot of useless graphics that just detract from the data. Don't get us wrong, it's great to use graphic elements that enhance the information on display but stay away from any kind of *Project Runaway* experimentation when it comes to business dashboards. That means junking the unnecessary pictures, logos, frames, and borders and sticking to a consistent display design or integrated theme.

Deadly Sin #6: Debilitating Detail

A large part of the dashboard's value comes from the fact that it provides "big-picture" measurements of trends and movements, without a company employee having to dig through pages of reports to come up with the same results. Why, then, should a dashboard bother to offer too much detail in its presentation? For instance, money amounts don't need to be itemized down to the penny: It's easier to deal with $43.4 thousand than $43,392.98. Similarly, overcomplicating such simple items as the date (April 23 is easier to read than 4/23/12) doesn't do anyone any favors. Again, the dashboard's prime attribute should be its simplicity, and that should be reflected in every aspect of the design.

Deadly Sin #7: Not Crunching the Numbers

A proper dashboard should clearly communicate the meaning of its data rather than forcing the user to figure it out. That means utilizing the right metric to send the right message. For example, if you want company executives to monitor actual revenue versus the target revenue for the month, it's not enough to simply show the two numbers through two separate lines that are side by side on a chart. It's much better to graph the actual revenue as a percentage of the target number—or even graph the variance (e.g., +7%, –6%) around the 0 axis. Nobody keeps up with their algebra for a reason, so don't make the dashboard users do the math, make the dashboard do it for them.

By avoiding these Seven Deadly Sins, you can end up with a heavenly dashboard design, one that will communicate quickly and efficiently the data points you want communicated, so that those involved can do their jobs better. When form and function are both taken into account, the dashboard will achieve its prime objectives.

6.13 BRIGHTPOINT CONSULTING, INC.: DESIGNING EXECUTIVE DASHBOARDS

Introduction

Corporate dashboards are becoming the "must-have" business intelligence technology for executives and business users across corporate America. Dashboard solutions have been around for over a decade but have recently seen resurgence in popularity because of the advance of enabling business intelligence and integration technologies.

Designing an effective business dashboard is more challenging than it might appear because you are compressing large amounts of business information into a small visual area. Every dashboard component must effectively balance its share of screen real estate with the importance of the information it is imparting to the viewer.

This article will discuss how to create an effective operational dashboard and some of the associated design best practices.

Dashboard Design Goals

Dashboards can take many formats, from glorified reports to highly strategic business scorecards. This article refers to operational or tactical dashboards employed by business users in performing their

Material in Section 6.13 has been taken from BrightPoint Consulting white paper "Designing Executive Dashboards," by Tom Gonzalez, managing director, BrightPoint Consulting, Inc., © 2005 by BrightPoint Consulting, Inc. Reproduced by permission. All rights reserved. Mr. Gonzalez is the founder and managing director of BrightPoint Consulting, Inc., serving as a consultant to both Fortune 500 companies and small–medium businesses alike. With over 20 years of experience in developing business software applications, Mr. Gonzalez is a recognized expert in the fields of business intelligence and enterprise application integration within the Microsoft technology stack. BrightPoint Consulting, Inc. is a leading technology services firm that delivers corporate dashboard and business intelligence solutions to organizations across the world. BrightPoint Consulting leverages best of breed technologies in data visualization, business intelligence, and application integration to deliver powerful dashboard and business performance solutions that allow executives and managers to monitor and manage their business with precision and agility. For further company information, visit BrightPoint's web site at www.brightpointinc.com. To contact Mr. Gonzalez, e-mail him at tgonzalez@brightpointinc.com.

daily work; these dashboards may directly support higher-level strategic objectives or be tied to a very specific business function. The goal of an operational dashboard is to provide business users with *relevant* and *actionable* information that empowers them to make effective decisions in a more efficient manner than they could without a dashboard. In this context, "relevant" means information that is directly tied to the user's role and level within the organization. For instance, it would be inappropriate to provide the CFO with detailed metrics about website traffic but appropriate to present usage costs as they relate to bandwidth consumption. "Actionable" information refers to data that will alert the user as to when and what type of action needs to be taken in order to meet operational or strategic targets. Effective dashboards require an extremely efficient design that takes into account the role a user plays within the organization and the specific tasks and responsibilities that user performs on a daily/weekly basis.

Defining Key Performance Indicators

The first step in designing a dashboard is to understand what key performance indicators (KPI) users are responsible for and which KPIs they wish to manage through their dashboard solution. A KPI can be defined as a measure (real or abstract) that indicates relative performance in relationship to a target goal. For instance, we might have a KPI that measures a specific number, such as daily Internet sales with a target goal of $10,000. In another instance we might have a more abstract KPI that measures "financial health" as a composite of several other KPIs, such as outstanding receivables, available credit and earnings before tax and depreciation. Within this scenario the higher-level "financial" KPI would be a composite of three disparate measures and their relative performance to specific targets. Defining the correct KPIs specific to the intended user is one of the most important design steps, as it sets the foundation and context for the information that will be subsequently visualized within the dashboard.

Defining Supporting Analytics

In addition to defining your KPIs, it is helpful to identify the information a user will want to see in order to diagnose the condition of a given KPI. We refer to this non-KPI information as "supporting analytics" because it provides context and diagnostic information for end users in helping to understand why a KPI is in a given state. Often these supporting analytics take the form of more traditional data visualization representations such as charts, graphs, tables, and, with more advanced data visualization packages, animated what-if or predictive analysis scenarios.

For each KPI on a given dashboard, you should decide if you want to provide supporting analytics and, if so, what type of information would be needed to support analysis of that KPI. For instance, in the case of a KPI reporting on aging receivables, you might want to provide the user a list of accounts due with balances past 90 days. In this case when a user sees that the aging KPI is trending in the wrong direction, he/she could click on a supporting analytics icon to bring up a table of accounts due sorted by balance outstanding. This information would then support the user in his/her ability to decide what, if any, action needed to be taken in relationship to the condition of the KPI.

Choosing the Correct KPI Visualization Components

Dashboard visualization components fall into two main categories: KPIs and supporting analytics. In either case, it is important to choose the visualization that best meets the end users' need in relationship to the information they are monitoring or analyzing.

For KPIs there are five common visualizations used in most dashboard solutions. The following list describes each component's relative merits and common usage scenario.

1. **Alert icons:** The simplest visualization is perhaps an alert icon, which can be a geometric shape that is either color-coded or shaded various patterns based on its state. Potentially, the most recognizable alert icon is a green, yellow, or red circle, whereby the color represents a more or less desirable condition for the KPI.

   When to use: These types of visualizations are best used when they are placed in the context of other supporting information, or when you need a dense cluster of indicators that are clearly labeled. Traditional business scorecard dashboards that are laid out in table-like format can benefit from this visualization in which other adjacent columns of information can be analyzed, depending on the state of the alert icon. These types of icons are also useful in reporting on system state, such as whether a machine or application is online or not. Be cautious of using icons that depend exclusively on color to differentiate state, as 10% of the male population and 1% of the female population is color-blind; consider using shapes in conjunction with color to differentiate state.

2. **Traffic light icons:** The traffic light is a simple extension of the alert icon, and has little advantage over the alert icon in terms of data visualization. Like the alert icon, this component only offers one dimension of information, but it requires 100% of the screen real estate. The one advantage of the traffic light icon is that it is a more widely recognized symbol of communicating a "good state," "warning state," or "bad state."

   When to use: In most cases a simple alert icon is a more efficient visualization, but in situations where your dashboard is being used by a wide audience on a less frequent basis, a traffic light component will allow users to more quickly assimilate the alert information because of their familiarity with the traffic light symbol from real-world experience.

3. **Trend icons:** A trend icon represents how a key performance indicator or metric is behaving over a period of time. It can be in one of three states: moving toward a target, [moving] away from a target, or static. Various symbols may be used to represent these states, including arrows or numbers. Trend icons can be combined with alert icons to display two dimensions of information within the same visual space. This can be accomplished by placing the trend icon within a color- or shape-coded alert icon.

    When to use: Trend icons can be used by themselves in the same situation you would use an alert icon, or to supplement another more complex KPI visualization when you want to provide a reference to the KPI's movement over time.

4. **Progress bars:** A progress bar represents more than one dimension of information about a KPI via its scale, color, and limits. At its most basic level, a progress bar can provide a visual representation of progress along a one-dimensional axis. With the addition of color and alert levels, you can also indicate when you have crossed specific target thresholds as well as how close you are to a specific limit.

    When to use: Progress bars are primarily used to represent relative progress toward a positive quantity of a real number. They do not work well when the measure you want to represent can have negative values: The use of shading within a "bar" to represent a negative value can be confusing to the viewer because any shading is seen to represent some value above zero, regardless of the label on the axis. Progress bars also work well when you have KPIs or metrics that share a common measure along an axis (similar to a bar chart), and you want to see relative performance across those KPIs/metrics.

5. **Gauges:** A gauge is an excellent mechanism by which to quickly assess both positive and negative values along a relative scale. Gauges lend themselves to dynamic data that can change over time in relationship to underlying variables. Additionally, the use of embedded alert levels allows you to quickly see how close or far away you are from a specific threshold.

    When to use: Gauges should be reserved for the highest level and most critical metrics or KPIs on a dashboard because of their visual density and tendency to focus user attention. Most of these critical operational metrics/KPIs will be more dynamic values that change on a frequent basis throughout the day. One of the most important considerations in using gauges is their size: too small, and it is difficult for the viewer to discern relative values because of the density of the "ink" used to represent the various gauge components; too large, and you end up wasting valuable screen space.

With more sophisticated data visualization packages, gauges also serve as excellent context-sensitive navigation elements because of their visual predominance within the dashboard.

Supporting Analytics

Supporting analytics are additional data visualizations that a user can view to help diagnose the condition of a given KPI or set of KPIs. In most business cases, these supporting analytics take the form of traditional charts and tables or lists. While the scope of this article is not intended to cover the myriad of best practices in designing traditional charting visualizations, we will discuss some of the basics as they relate to dashboard design.

When creating supporting analytics, it is paramount that you take into account the typical end user who will be viewing the dashboard. The more specialized and specific the dashboard will be, the more complexity and detail you can have in your supporting analytics. Conversely, if you have a very high-level dashboard, your supporting analytics will generally represent higher-level summary information with less complex detail.

In the following list, we will discuss some of the most common visualizations used for designing supporting analytics.

1. **Pie charts:** Pie charts are generally considered a poor form of data visualization for any data set with more than half a dozen elements. The problem with pie charts is that it is very difficult to discern proportional differences with a radially divided circle, except in the case of a small data set that has large value differences within it. Pie charts also pose a problem for labeling, because they are either dependent on a color or pattern to describe the different data elements, or the labels need to be arranged around the perimeter of the pie, creating a visual distraction.

   When to use: Pie charts should be used to represent very small data sets that are geared to high-level relationships between data elements. Usually pie charts can work for summary-level relationships but should not be used for detailed analysis.

2. **Bar charts:** Bar charts are an ideal visualization for showing the relationship of data elements within a series or multiple series. Bar charts allow for easy comparison of values because of the fact that the "bars" of data share a common measure and can be easily visually compared to one another.

   When to use: Bar charts are best suited for categorical analysis but can also be used for small time series analysis (e.g., the months of a year). An example of categorical analysis is examining sales of products broken down by product or product group, with sales in dollars being the measure and product or product group being the category. Be careful in using bar charts if you have a data set that can have one element with a large outlier value; this will render the visualization for the remaining data elements unusable. This is because of the fact that the chart scale is linear and will not clearly represent the relationships between the remaining data elements. An example is shown in Figure 6.5. Notice that, because widget 2 has sales of $1.2 mm, you cannot easily discern that widget 3 has twice as many sales ($46,000) as widget 1 ($23,000).

3. **Line charts:** Line charts are ideal for time series analysis where you want to see the progress of one or more measures over time. Line charts also allow for comparative trend analysis because you can stack multiple series of data into one chart.

   When to use: Use line charts when you would like to see trends over time in a measure versus a side-by-side detailed comparison of data points. Time series line charts are most commonly used with the time dimension along the X-axis and the data being measured along the Y-axis.

4. **Area charts:** Area charts can be considered a subset of the line chart, where the area under or above the line is shaded or colored.

   When to use: Area charts are good for simple comparisons with multiple series of data. By setting contrasting color hues, you can easily compare the trends over time between two or more series.

5. **Tables and lists:** Tables and lists are best used for information that either contains large lists of nonnumeric data, or data that has relationships not easily visualized.

Figure 6.5 Typical Bar Chart *Source: BrightPointConsulting white paper, "Designing Executive Dashboards" by Tom Gonzalez, managing director, BrightPoint Consulting, Inc., © 2005 by BrightPoint Consulting, Inc. Reproduced by permission. All rights reserved.*

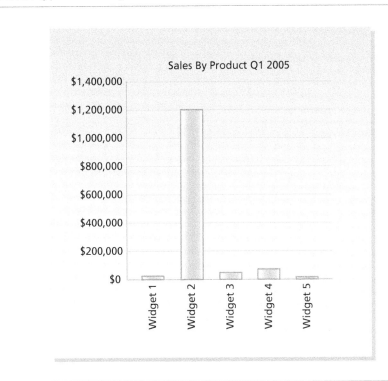

When to use: You will want to use tables or lists when the information you need to present does not lend itself to easy numeric analysis. One example is a financial KPI that measures a company's current liquidity ratio. In this case, there can be a complex interrelationship of line items within the company's balance sheet, where a simple table of balance sheet line items would provide a more comprehensive supporting analytic than a series of detailed charts and graphs.

A Word about Labeling Your Charts and Graphs

Chart labels are used to give the users context for the data they are looking at, in terms of both scale and content. The challenge with labeling is that the more labels you use and the more distinctive you make them, the more it will distract the user's attention from the actual data being represented within the chart.

When using labels, there are some important considerations to take into account. Foremost among these is how often your user will be viewing these charts. For charts being viewed on a more frequent basis, the user will form a memory of relevant labels and context. In these scenarios, you can be more conservative in your labeling by using smaller fonts and less color contrast. Conversely, if a user will only be seeing the chart occasionally, you will want to make sure everything is labeled clearly so that the user does not have to decipher the meaning of the chart.

Putting It All Together: Using Size, Contrast, and Position

The goal in laying out an effective dashboard is to have the most important business information be the first thing to grab your user's visual attention. In your earlier design stages, you already determined the important KPI's and supporting analytics, so you can use this as your layout design guide. Size, contrast, and position all play a direct role in determining which visual elements will grab the user's eye first.

Size: In most situations, the size of a visual element will play the largest role in how quickly the user will focus their attention upon it. In laying out your dashboard, figure out which element or group of elements will be the most important to the user and make their size proportionally larger than the rest of the elements on the dashboard.

This principle holds true for single element or groups of common elements that have equal importance.

Contrast: After size, the color or shade contrast of a given element in relationship to its background will help determine the order in which the user focuses attention on that element. In some situations, contrast alone will become the primary factor, even more so than size, for where the user's eye will gravitate. Contrast can be achieved by using different colors or saturation levels to distinguish a visual element from its background. A simple example of this can be seen in the Figure 6.6.

As you can see, the black circle instantly grabs the user's attention because of the sharp contrast against the white background. In this example, the contrast even overrides the size of the larger circle in its ability to focus the user's visual awareness.

Position: Visual position also plays a role in where a user will focus his or her attention. All other factors being equal, the top-right side of a rectangular area will be the user's first focal point, as seen in Figure 6.7. The next area a user will focus is the top-left side, followed

Figure 6.6 Contrasting Colors *Source: BrightPoint Consulting white paper "Designing Executive Dashboards" by Tom Gonzalez, managing director, BrightPoint Consulting, Inc., © 2005 by BrightPoint Consulting, Inc. Reproduced by permission. All rights reserved*

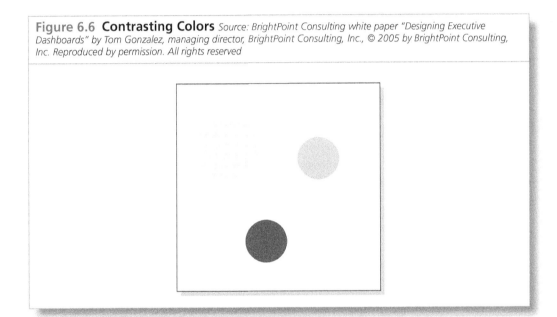

Figure 6.7 Positioning of Icons *Source: BrightPoint Consulting white paper "Designing Executive Dashboards" by | Tom Gonzalez, managing director, BrightPoint Consulting, Inc., © 2005 by BrightPoint Consulting, Inc. Reproduced by permission. All rights reserved*

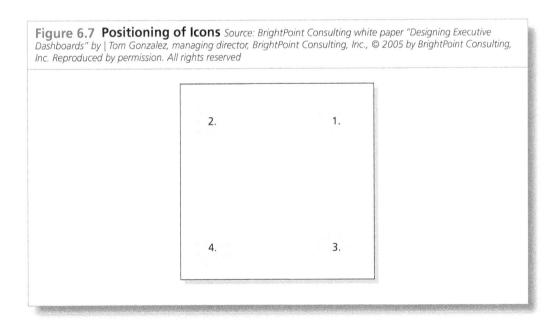

by the bottom right, and finally the bottom left. Therefore, if you need to put an element on the dashboard that you don't want the user to have to hunt around for, the top-right quadrant is generally the best place for it.

Position is also important when you want to create an association between visual elements. By placing elements in visual proximity to each other, and grouping them by color or lines, you can create an implied context and relationship between those elements. This is important in instances when you want to associate a given supporting analytic with a KPI or group together related supporting analytics.

Validating Your Design

You will want to make sure that your incorporation of the preceding design techniques achieves the desired effect of focusing the user's attention on the most important business information and in the proper order. One way to see if you have achieved this successfully is to view your dashboard with an out-of-focus perspective. This can be done by stepping back from your dashboard and relaxing your focus until the dashboard becomes blurry and you can no longer read words or distinguish finer details. Your visual cortex will still recognize the overall visual patterns, and you will easily see the most attention-grabbing elements of your design. You want to validate that the elements attracting the most visual attention correspond with the KPIs and supporting analytics that you had previously identified as being most critical to the business purpose of your dashboard.

Please bear in mind, the design guidelines presented in this article should be used as general rules of thumb, but it is important to note that these are not hard-and-fast rules that must be followed in every instance. Every dashboard has its own unique requirements, and in certain cases you will want to deviate from these guidelines and even contradict them to accomplish a specific visual effect or purpose.

Additional Readings

Fitts, P. M. "The Information Capacity of the Human Motor System in Controlling the Amplitude of Movement." *Journal of Experimental Psychology*, 47 (1954): 381–391.

Tractinsky, Noam. "Aesthetics and Apparent Usability: Empirically Assessing Cultural and Methodological Issues." In *CHI '97: Proceedings of the ACM SIGCHI Conference on Human Factors in Computing Systems* (Atlanta, GA) (March 22–27), pp. 115–122.

Tufte, Edward, *Envisioning Information*. Cheshire, CT: Graphics Press, 1990.

6.14 **ALL THAT GLITTERS IS NOT GOLD**

In the previous sections, we discussed the use of various metaphors and icons to display information in dashboards. Designing a perfect dashboard may be impossible. An image that works well for one dashboard may be inappropriate for another. Also, there are both advantages and disadvantages to all images and colors. My belief is that, if the image works and provides the necessary and correct information for the stakeholder and the stakeholders understands the image, continue using it.

To understand the complexity in selecting the right image, let us consider gauges. Gauges are instruments for ascertaining or regulating the level, state, dimensions, or forms of things. Gauges are also verbs that imply an act of measuring something. Like other images that are frequently used on dashboards, gauges have advantages and disadvantages.

Advantages

- Can be used to display both quantitative and qualitative information
- Aesthetically pleasing image
- Can display data using colors, digital indicators, and pointers

Disadvantages

- Occupies more dashboard space than traditional images
- May be more difficult to read than other images
- May be more difficult to update
- Cannot be used to show trends or whether performance is increasing or decreasing with regard to a target
- Can reflect metric information but not KPI information
- Cannot show the speed at which a change has occurred
- Cannot identify how much attention is needed to correct an unfavorable situation

Several rules can be used for dashboard designs. In-depth explanations of each of these can be found on the Internet. As an example: **Rules for selecting the right artwork:** Selecting the right image is critical. As an example, gauges cannot show trends. Options for images include:

- Gauges
- Thermometers
- Traffic lights
- Area charts
- Bar charts
- Stacked charts
- Bubble charts
- Clustered charts
- Performance trends
- Performance variances

- Histograms
- Pie charts
- Rectangles with quadrants
- Alert buttons
- Cylinders
- Composites
- **Rules for positioning the artwork (number of windows and frames):** There must be a speed of perception. Also, the upper left (or upper right, based on the designer's preference) is usually considered to be more important than the lower-right corner.
- **Rules for visualization (readability and symmetry and proportions):** The image and information should be easy to read and aesthetically pleasing to the eye.
- **Rules for accuracy of the information (context selection):** The image must provide reasonably accurate information for informed decision making without requiring an interpretation by the viewer. However, some stakeholders are more interested in trends than absolute performance.
- **Rules for color selection:** Factors that must be considered include:
  - Colors
  - Positioning of the colors
  - Brightness
  - Orientation
  - Saturation
  - Size
  - Texture
  - Shape

As mentioned before, perfection in dashboard design may not be possible. Even the simplest designs can have possible flaws for the viewer. As an example, consider the following area charts:

- **Traditional area chart (Figure 6.8):** This displays the trend over time or categories.
- **Stacked area chart (Figure 6.9):** This displays the trend of the contribution of each value over time or categories.
- **100% stacked area chart (Figure 6.10):** This displays the trend of the percentage each value contributes over time or categories.

These charts are good for looking at trends. Some stakeholders may be interested more in trends than hard numbers. However, to get a precise value at a specific time period would require detailed measurements from the charts, and this would introduce the opportunity for error.

Another commonly used image is the bar chart. As an example, consider the following bar charts:

- **Clustered bar chart (Figure 6.11):** This compares values across categories.
- **Stacked bar chart (Figure 6.12):** This compares the contribution of each value to a total across categories.

Figure 6.8 Area Chart *Source: Nils Rasmussen et al., Business Dashboards (Hoboken, NJ: John Wiley & Sons, 2009), L p. 40.*

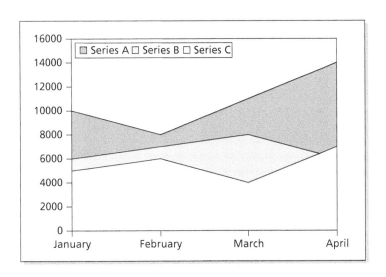

Figure 6.9 Area Chart, Stacked *Source: Nils Rasmussen et al, Business Dashboards (Hoboken, NJ: John Wiley & Sons, 2009), p. 94*

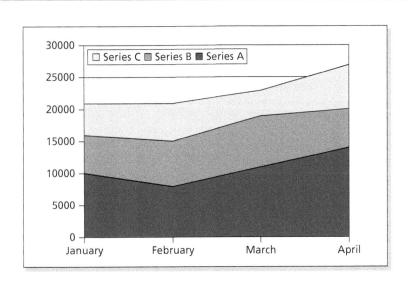

Figure 6.10 Area Chart, 100% Stacked *Source: Nils Rasmussen et al., Business Dashboards (Hoboken, NJ: John Wiley & Sons, 2009), p. 41. I*

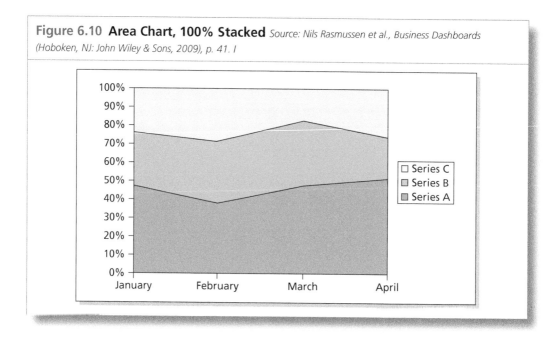

Figure 6.11 Bar Chart, Clustered *Source: Nils Rasmussen et al., Business Dashboards (Hoboken, NJ: John Wiley & Sons, 2009), p. 95.*

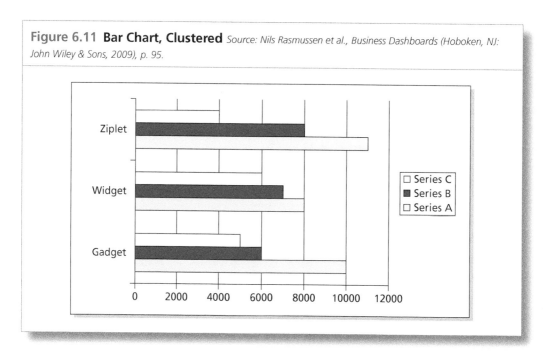

- **100% stacked bar chart** (Figure 6.13): This compares the percentage each value contributes to a value across categories.

   In these figures, viewers can be distracted if part of the bar appears in the same color as the background of the image. Also, in the stacked

Figure 6.12 Bar Chart, Stacked *Source: Nils Rasmussen et al., Business Dashboards (Hoboken, NJ: John Wiley & Sons, 2009), p. 41. I*

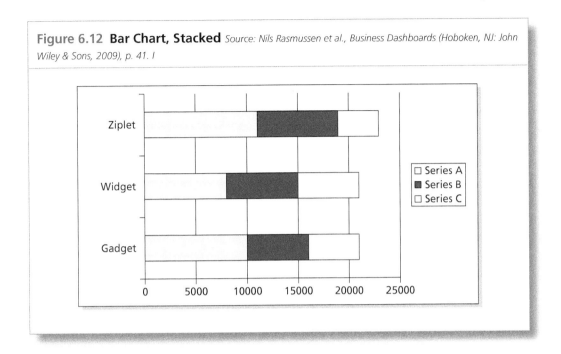

Figure 6.13 Bar Chart, 100% Stacked *Source: Nils Rasmussen et al., Business Dashboards (Hoboken, NJ: John Wiley & Sons, 2009), p. 96. I*

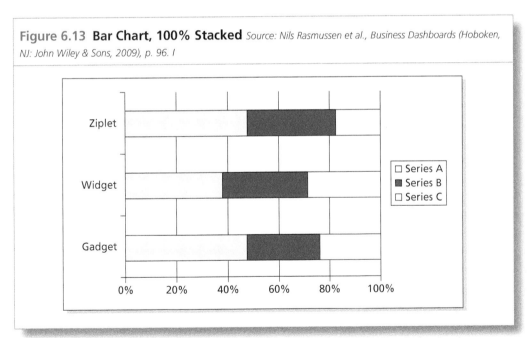

bar charts, getting the exact value of Series B and Series C may require measurement that can lead to error.

Bubble charts, as shown in Figure 6.14, are more appropriate to business dashboards than project dashboards. The chart compares three sets of values, as line charts do, but with a third value displayed as the size of the bubble marker.

Column charts are similar to bar charts. As an example, consider the following three column charts:

- **Clustered column chart** (Figure 6.15): This compares values across categories.
- **Stacked column chart** (Figure 6.16): This compares the contribution of each value to a total across categories.
- **100% stacked column chart** (Figure 6.17): This chart compares the percentage each value contributes to a value across categories.

Some form of column chart appears in almost all dashboards. However, care must be taken in the selection of the colors. In Figures 6.15 and 6.16, the shades of the colors on the columns may create a visual problem. The shades in Figure 6.17 are easier to read, provided that the individual is not color blind.

Gauges are used to show a single value. Typically, gauges, such as those shown in Figure 6.18, will also use colors to indicate whether the value that is displayed is "good," "acceptable," or "bad."

Icons can be found in a variety of shapes. Most popular are traffic lights (oval circles) or arrows used in conjunction with dashboards or scorecards to visualize and highlight variances. This is shown in Figure 6.19. Colors like green, yellow, and red are used to indicate values as "good," "acceptable," and "bad." The color green can have more than one meaning. For example, in some icons, green may indicate that a change is need rather than no change is necessary.

Figure 6.14 Bubble Chart *Source: Nils Rasmussen et al., Business Dashboards (Hoboken, NJ: John Wiley & Sons, 2009), p. 96.*

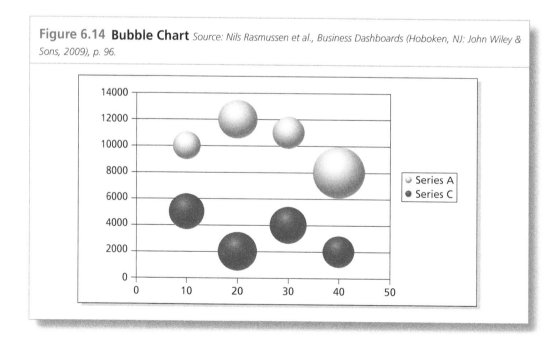

Figure 6.15 Column Chart, Clustered *Source: Nils Rasmussen et al., Business Dashboards (Hoboken, NJ: John Wiley & Sons, 2009), p. 97.*

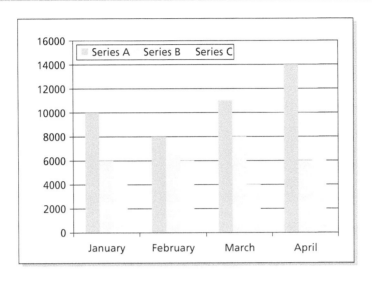

Figure 6.16 Column Chart, Stacked *Source: Nils Rasmussen et al., Business Dashboards (Hoboken, NJ: John | Wiley & Sons, 2009), p. 97.*

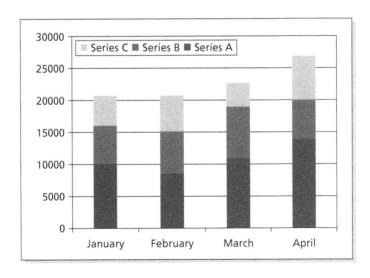

Figure 6.17 Column Chart, 100% Stacked *Source: Nils Rasmussen et al., Business Dashboards (Hoboken, NJ: John Wiley & Sons, 2009), p. 98.* I

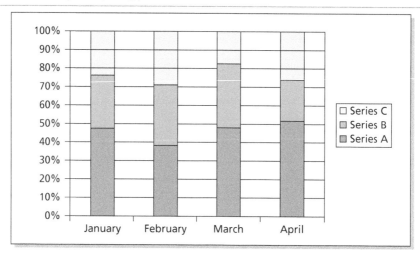

Figure 6.18 Gauges *Source: Nils Rasmussen et al., Business Dashboards (Hoboken, NJ: John Wiley & Sons, 2009), p. 98.*

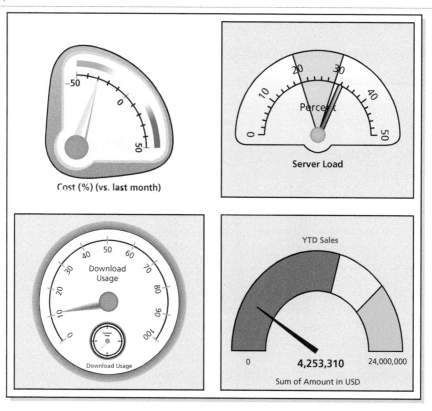

Line charts are also images that can be used to show trends. However, no more than three or four lines should appear on a chart. Examples of line charts are:

- **Traditional line chart** (Figure 6.20): This displays trends over time or categories.
- **Stacked line chart** (Figure 6.21): This displays the trend of the contribution of each value over time or categories.
- **100% stacked line chart** (Figure 6.22): This displays the trend of the percentage each value contributes over time or categories.

Perhaps the most important word in dashboard design is *simplicity*. Colorful graphics, intricate designs, and three-dimensional (3-D) artwork can distract the viewer from the more critical information. Figure 6.23 shows primary and secondary stakeholders. When you first look at the figure, you are intrigued by the 3-D effect, which adds nothing to the information you want to convey. Putting the information in a table or line chart would have achieved the same effect and might have been easier to understand. Also, there are no numbers in the figure, so the viewer may not be sure exactly how many stakeholders are in each category.

Figure 6.24 is similar to Figure 6.23 but more complex. When you first look at the figure, your eye focuses on the 3-D effect, then you must read the words over and over again to understand what you are looking at even though numbers are provided. Finally, the milestones that were completed within time and cost could have been all

Figure 6.19 Icons *Source: Nils Rasmussen et al., Business Dashboards (Hoboken, NJ: John Wiley & Sons, 2009), p. 99.*

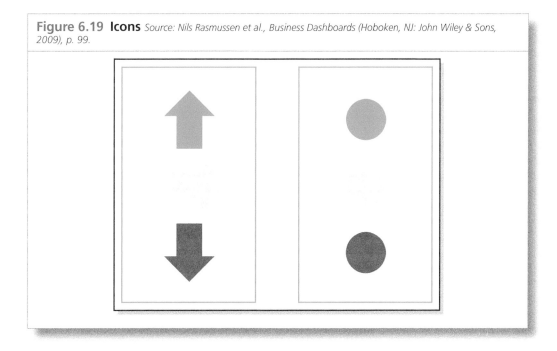

Figure 6.20 Line Chart *Source: Nils Rasmussen et al., Business Dashboards (Hoboken, NJ: John Wiley & Sons, 2009), L p. 99.*

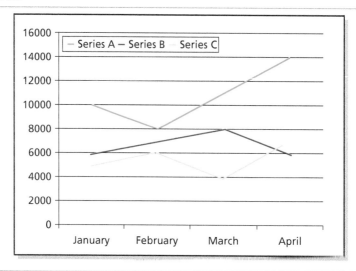

Figure 6.21 Line Chart, Stacked *Source: Nils Rasmussen et al., Business Dashboards (Hoboken, NJ: John Wiley & Sons, 2009), p. 100.*

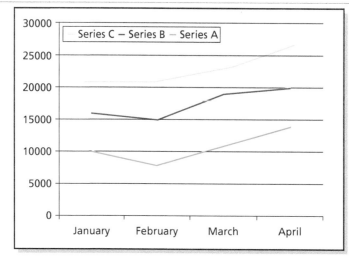

Figure 6.22 Line Chart, 100% Stacked *Source: Nils Rasmussen et al., Business Dashboards (Hoboken, NJ: John Wiley & Sons, 2009), p. 100.*

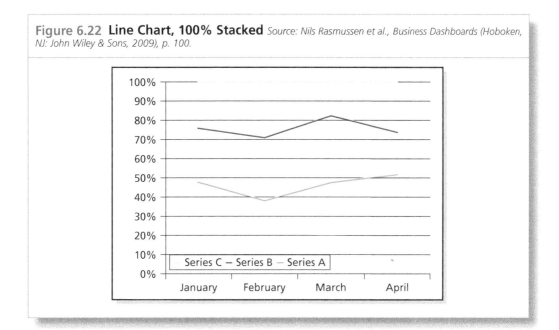

Figure 6.23 Tiered Stakeholder Identification in 3-D

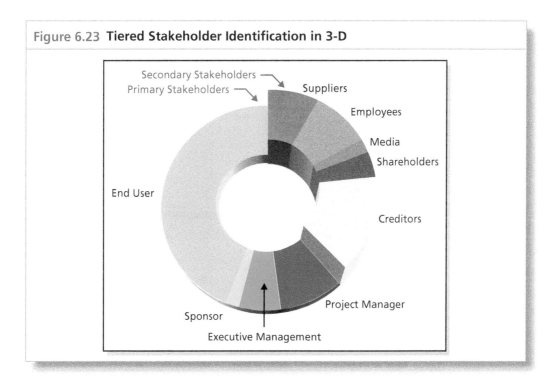

of the work packages that did not have a major impact on the project's success, whereas other milestones may have a significant impact. This problem might be overcome by allowing the viewer to drill down to more depth.

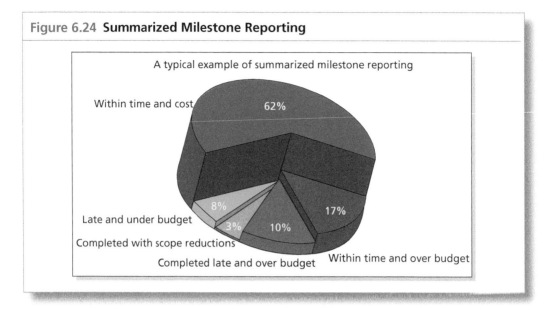

Figure 6.24 Summarized Milestone Reporting

A typical example of summarized milestone reporting

Within time and cost — 62%

17%

8%

Late and under budget — 3% 10%

Completed with scope reductions

Completed late and over budget   Within time and over budget

Figure 6.25 illustrates the current breakdown of labor hours on a project. The figure lacks numerical values for each slice of the pie and would be easier to read as a column chart.

Figure 6.26 represents a 3-D pie chart that would be part of a dashboard for the PMO. The chart illustrates the most common reasons why projects have failed in the past. Once again, even though the image looks impressive in 3-D, the information could be presented more clearly in a line chart and with numbers included. In its current format, all of the slices of the pie look like they are the same

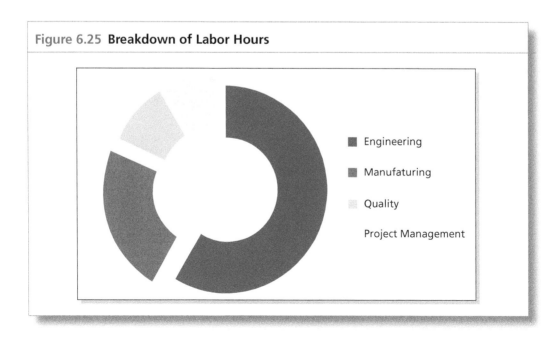

Figure 6.25 Breakdown of Labor Hours

- Engineering
- Manufaturing
- Quality

Project Management

Figure 6.26 **Causes of Failure**

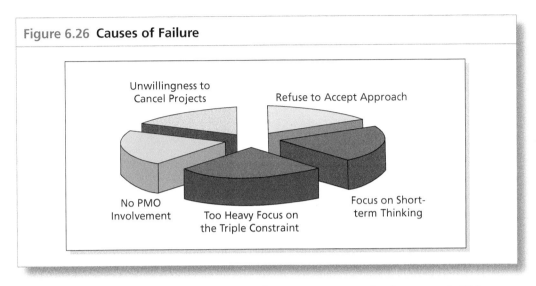

size. This may not be the case. As a general rule, any embellishments that are not relevant to the data have no place in the chart.

Square pie charts may very well replace the traditional pie chart. Figures 6.27 and 6.28 show the square pie chart in two different rotations. With the square pie chart, the colored-in cells can be added up to get the percentages. Even with the 3-D effect, distortion does not affect the readability of the chart. Certain elements can be

Figure 6.27 **A Square Pie Chart**

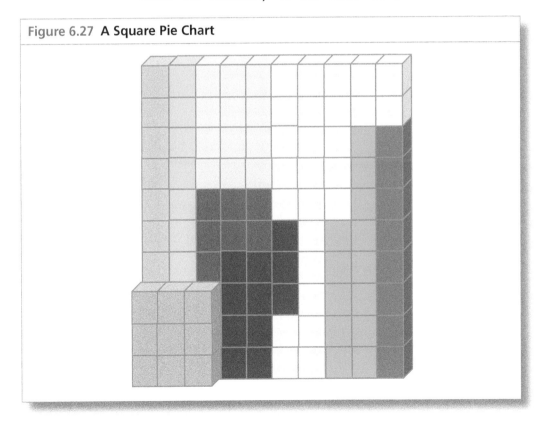

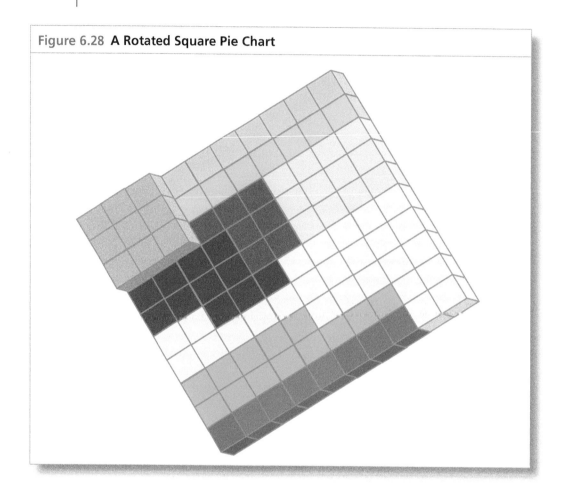

Figure 6.28 A Rotated Square Pie Chart

highlighted by elevating them above the other cells. However, there may be some limitations to the use of square pie charts:

- Each cell generally represents whole number rather than decimals.
- The image may not impress people if one or more of the percentages are very large.
- Cell color selection and placement must be done carefully, and more time may be necessary to create and interpret this type of chart than with the traditional pie chart.
- The image may require more real estate on the dashboard screen than the traditional pie chart, thus limiting the number of metrics.
- The 3-D rotation of the chart must be done without sacrificing the aesthetic value of the image and its readability for accuracy.

Figure 6.29 shows the total cost breakdown for four work packages. Although the chart looks impressive, there is no background grid with which the viewer can make assessments. Also, the use of red or shades of red might lead the viewer to believe that the labor dollars are excessive or a problem area.

Figure 6.30 shows the cost overrun data for four work packages. In this case, there is a grid, but it is difficult to determine the overrun magnitude of labor and procurement. Also, for Work Package 4, should the front or back side of the 3-D bar be used? If the front side of the bar is used, the cost overrun is 11%, whereas the back side of the bar illustrates a 12% overrun.

Figure 6.31 shows the cumulative month-end cost performance index (CPI) and schedule performance index (SPI) data. On the grid,

Figure 6.29 Total Cost Breakdown Per Work Package

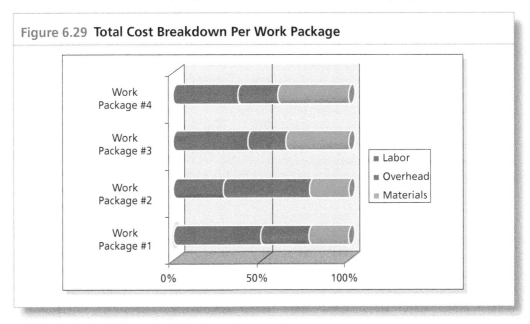

Figure 6.30 Cost Overrun Data

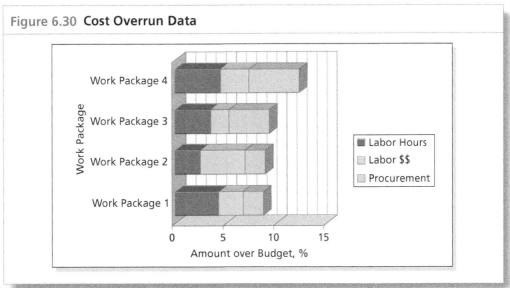

the parity line at 1.0 should probably be highlighted to show the nearness to the targeted value. Also there should be more grid lines so that meaning numbers can be determined.

There are advantages to using 3-D column charts. However, inserting too much into the charts can make them difficult to use. Figure 6.32 illustrates the complexity in making exact value determinations for Series 1 and Series 2. Also, it might be better to use neutral or standard

Figure 6.31 Cumulative Month End CPI and SPI Data

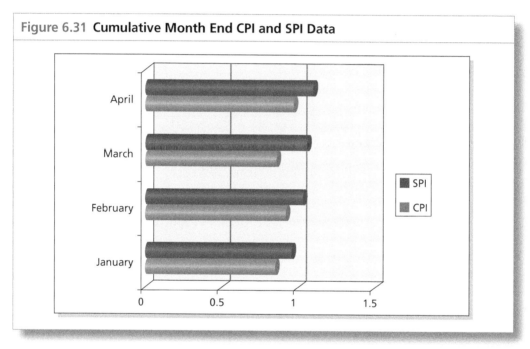

Figure 6.32 3-D Column Chart

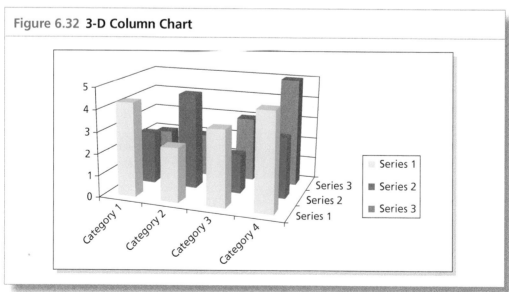

colors rather than colors designed to emphasize a special situation. Figure 6.33 shows typical neutral colors.

Another common mistake is in the use of textures and gradients, as shown in Figure 6.34. While there are benefits to this in conducting presentations, they may not be appropriate for dashboards.

Figure 6.35 shows a column chart with bright colors. The purpose of bright colors is to emphasize a good or bad situation. If all of the colors are bright, as in this figure, the viewer may not know what is or is not important.

Figure 6.33 Possible Colors

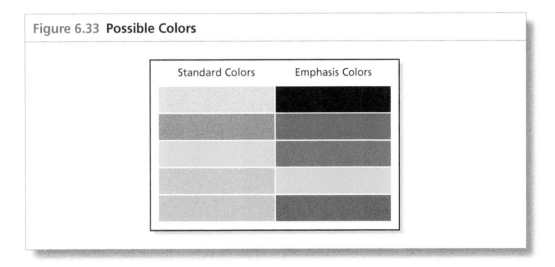

Figure 6.34 Column Chart with Gradients

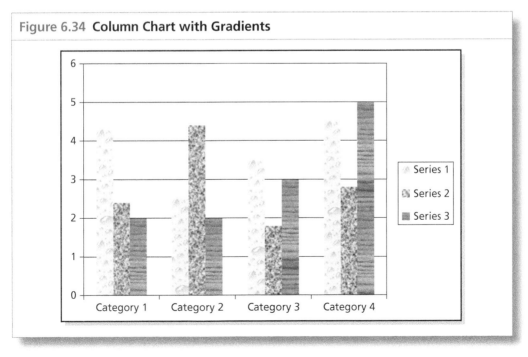

When using a column chart, standard colors should be used and the shading should go from lightest to darkest for easy comparison, as shown in Figure 6.36. Also, creating shadows or exotic colors behind the columns can be distracting and should be avoided because the shadows contain no information or data.

Figure 6.35 Column Chart Using Bright Colors

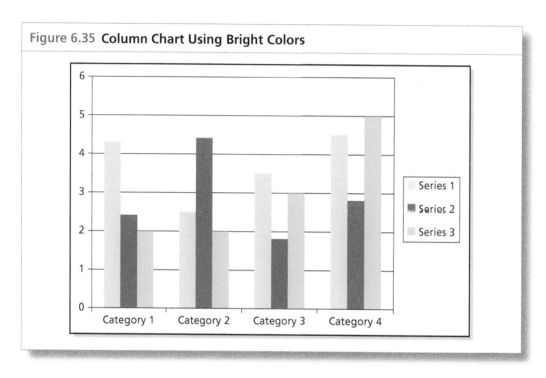

Figure 6.36 Column Chart Using Shading

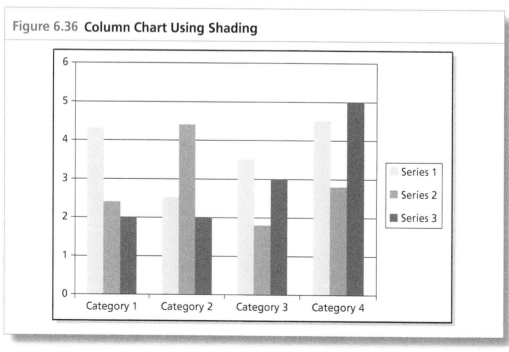

Background colors or shading can play tricks on the eye. For example, in Figure 6.37, the inner squares are all the same size, yet some people perceive the inner square that is on the right side to be larger than the other squares.

Another example appears in Figure 6.38. The outer circles represent the total cost of a work package, in dollars, and the inner circle represents the dollar value of the labor hours that are part of the total cost. Again, the eye may be deceived because all of the inner circles are the same size. Because some inner circles consume a larger percentage of the outer circles, some inner circles appear larger.

Figure 6.37 Background Colors with Shading

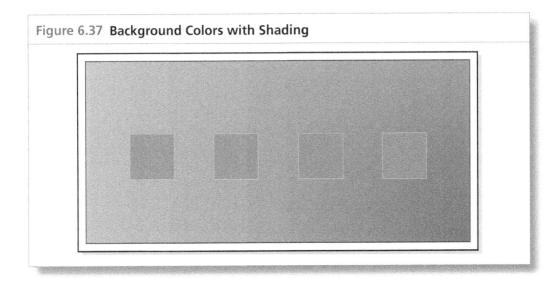

Figure 6.38 Concentric Circle Charts

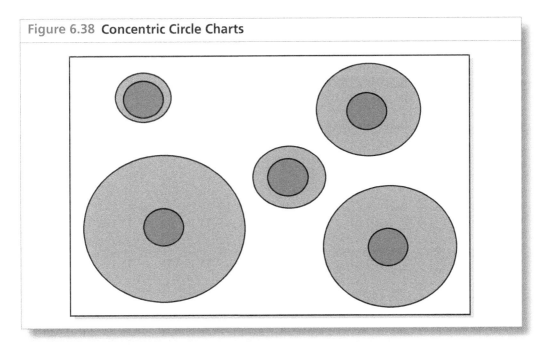

Radar charts, as seen in Figure 6.39, are usually avoided because they are often hard to read, even for people that use them frequently. The information in a radar chart can be displayed in a column or bar chart. However, there are situations where radar charts can be quite effective.

This chapter has emphasized that the least amount of metrics that can be used for informed decision making should be placed on the dashboard. Unfortunately, more information may be necessary. Sometimes the viewer must have the option to drill down to additional levels for clarification. For example, in Figure 6.40, the column on the left represents buttons. When the button is illuminated in red, the metrics on the screen are for Work Package #1 only. The viewer has the option of depressing any of the buttons.

Based on the amount of depth in the information needed by the stakeholders, some dashboards must be designed for in-depth levels of detail. This becomes a costly effort if each stakeholder requires a different level of detail.

Figure 6.41 is an attempt to show the cost and schedule variances as the project progresses. The chart is good if used just to see the trends in the variances. If actual numbers are required for decision making, however, then the data should be represented in a table.

Some charts are more appropriate when illustrated as a log-log plot or semi-log plot. Figure 6.42 shows a typical learning curve that would be used as part of a manufacturing project. While most project managers may be familiar with this type of chart, it should not be used in dashboards to be presented to stakeholders.

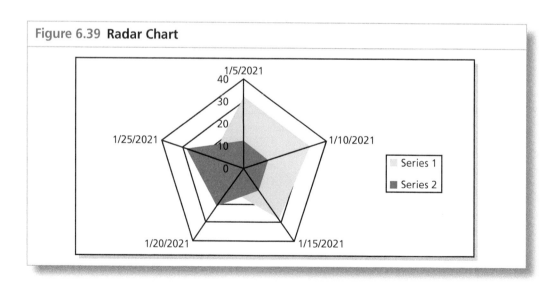

Figure 6.39 Radar Chart

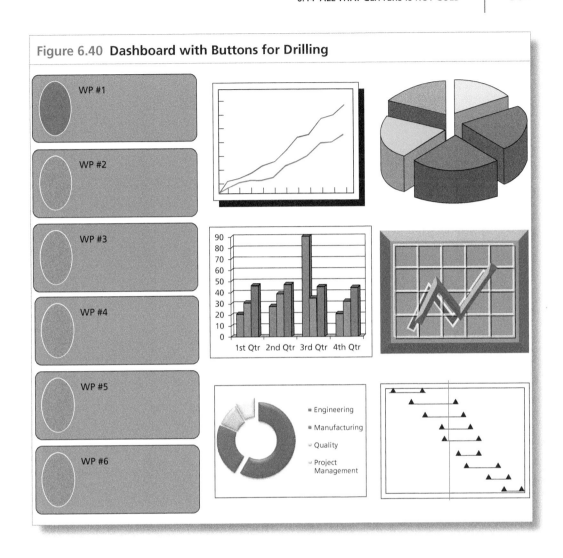

Figure 6.40 Dashboard with Buttons for Drilling

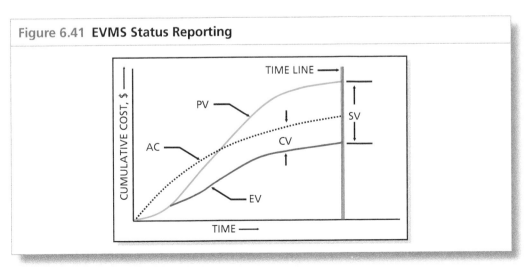

Figure 6.41 EVMS Status Reporting

Pointers are indicators that point to a value on a dial, gauge, or scale. An example of a pointer on a sliding scale is shown in Figure 6.43. Care must be taken not to overload an image with pointers because the viewer of the image must continuously refer to the legend to determine the meaning of each pointer.

Figure 6.42 Learning Curve on a Log-Log Plot

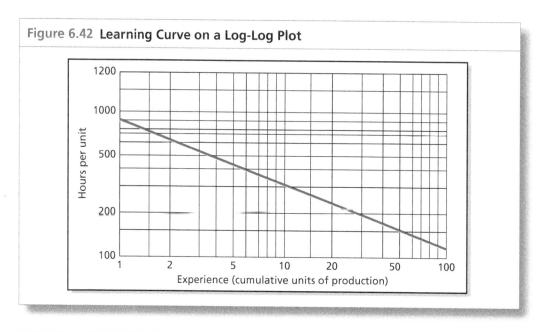

Figure 6.43 Pointers on a Vertical Sliding Scale

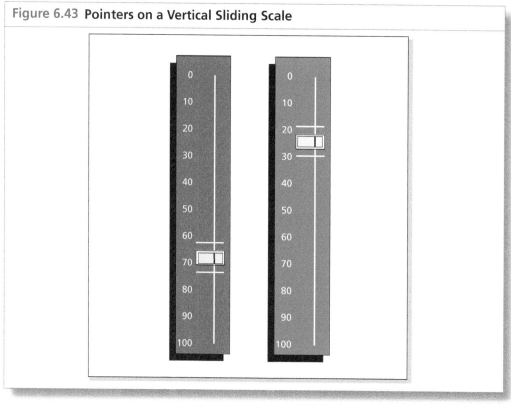

Not all metrics displayed on a dashboard are easily understood. An example might be a company that has cyclical revenue streams as shown in Figure 6.44. Some companies prefer to identify all quarters on a display and use dotted lines to identify which quarters should be compared. This is shown in this figure. Others may prefer just to show the comparison of Q1 and Q2 for each year on one image and Q3 and Q4 comparisons on another image. Viewers must understand what they are looking at when viewing cyclical data.

A heat map is a graphical representation of data where the individual values are represented using colors. Meteorologists often use heat maps when discussing the weather. Many different color schemes can be used to illustrate a heat map, with perceptual advantages and disadvantages for each. Rainbow color maps are often used, as humans can perceive more shades of color than they can of gray, and this would purportedly increase the amount of detail perceivable in the image. Figure 6.45 shows a heat map for the importance of certain constraints. The darker the color, the more important the constraint.

Figure 6.44 **Cyclical Data**

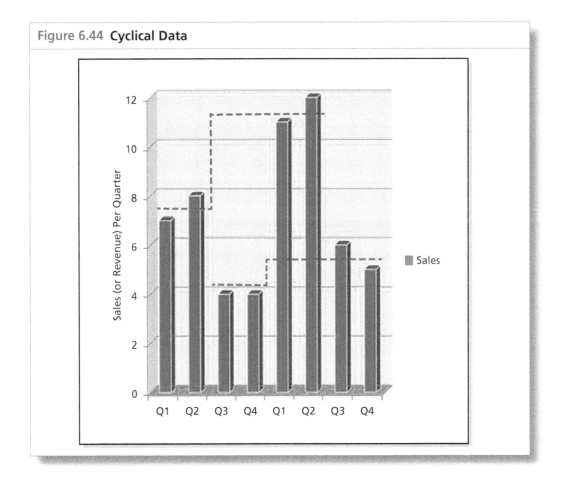

Figure 6.45 Heat Map

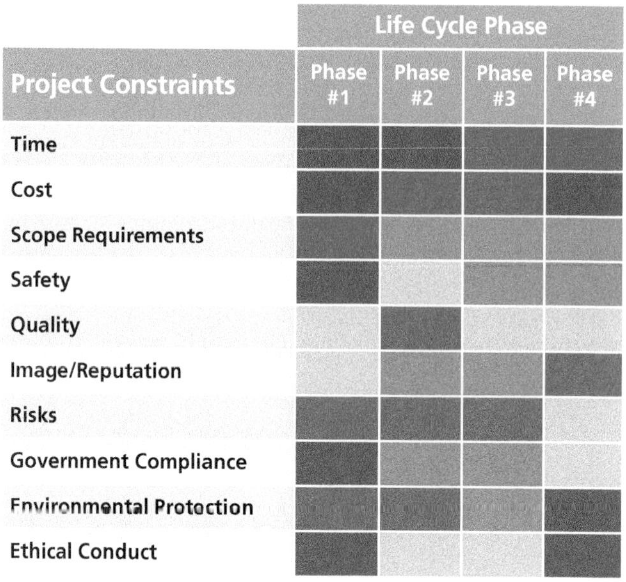

Project Constraints	Life Cycle Phase			
	Phase #1	Phase #2	Phase #3	Phase #4
Time				
Cost				
Scope Requirements				
Safety				
Quality				
Image/Reputation				
Risks				
Government Compliance				
Environmental Protection				
Ethical Conduct				

6.15 USING EMOTICONS

Sometimes symbols are used to make the dashboard look elegant. Emoticons can be used in dashboards, but their meaning can be misinterpreted. For example, in Figure 6.46:

- Does a happy face mean good or very good?
- Does a sad face mean bad or very bad?
- The color red is normally used to reflect something bad. What does a red smile indicate?
- The color green normally indicates something favorable. What does a green frown indicate?

Information flag buttons can be used for drill-down capability to get clarification on the exact meaning of the emoticon.

Figure 6.47 provides additional examples of other emoticons that can be used. Once again, they can be misinterpreted. For example, does the police officer emoticon mean the project has stopped or that there is just a small delay? Does the devil emoticon mean that the project is in serious trouble or that the project manager is trying to resolve a problem?

Figure 6.46 Using Emoticons

Cost Variances	
Work Package #3: Cost Variance	
Work Package #4: Cost Variance	
Work Package #7: Cost Variance	
Work Package #8: Cost Variance	
Work Package #9: Cost Variance	

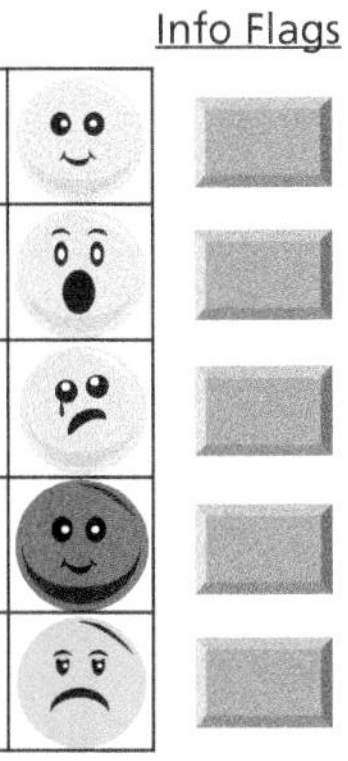

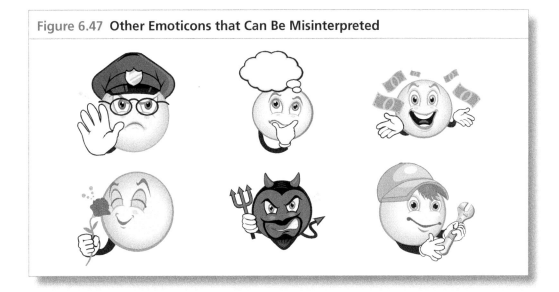

Figure 6.47 Other Emoticons that Can Be Misinterpreted

6.16 MISLEADING INDICATORS

Misleading indicators are displays that can generate confusion as a result of the way that the image was represented. Misleading indicators often contain chart junk, which are unnecessary lines, labels, images, colors, and data that can detract from the image's integrity and value. Sometime chart junk is needed for aesthetic appeal. However, most often the result can be decision making that is not in the best interest of the project.

Perhaps the most common cause of a misleading indicator is from a failure to consider the sensitivity of the measurement system that was used. The result of using a misleading indicator can be:

■ Not being able to validate the data
■ Showing an image that can be manipulated
■ Showing images susceptible to more than one interpretation
■ Showing images that can be misunderstood
■ Showing images that cannot be used to assess performance risks

The laws of dashboard design related to accuracy and aesthetics are related in the way that the images are visualized by the viewer. Bad visualization can destroy the credibility of the dashboard. Data is often visualized by size and height. Some shapes are easy to size correctly while others can lead to a faulty interpretation. As an example, look at the column chart in Figure 6.48.

When first visualizing this chart, the eye focuses on the two columns where it is quite apparent that the column for March is twice the size as the column for February. Viewers may therefore erroneously conclude that the favorable variance for March is twice the favorable variance for February. This would be true using the Y-axis on the left side of the figure. But if we were to use Y-axis on the right-hand side of Figure 6.48, then the favorable variance increase in March is only 50% of February's variance rather than a 100% increase.

Another potential problem is when we use the area of a circle to represent data as shown in Figure 6.49. For simplicity's sake, let us assume the radius in Circle A is one inch and we are trying to show

Figure 6.48 Column Chart Showing Favorable Variances

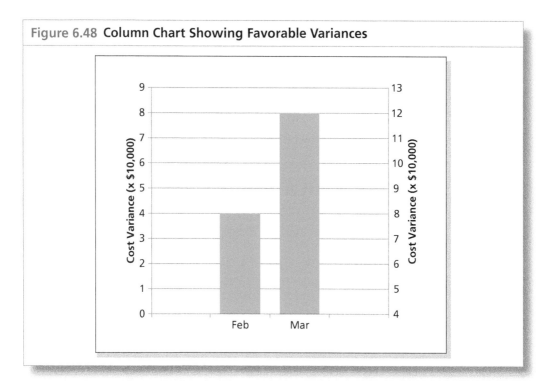

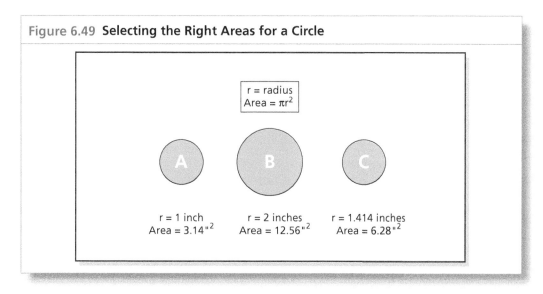

Figure 6.49 Selecting the Right Areas for a Circle

using Circle B that there has been a significant improvement of 100% by showing that Circle B is twice the area of Circle A. If we simply try to double the radius from one inch to two inches to create Circle B, then the area of Circle B is four times the area of Circle A rather than the target of twice the area. This could be misleading to the viewer. To show that a new circle has twice the area of Circle A, the new circle should have a radius of 1.414 inches, as shown in Circle C.

6.17 AGILE AND SCRUM METRICS[6]

Introduction: Agile Overview

Let us start by defining Agile. It is not a Project Management methodology or a framework. Agile *is* a philosophy that is flexible and responsive. It involves continuous learning and improvement and is an approach that requires a cultural shift where silos are removed, small teams are formed, and are empowered.

---

6 Material in this section has been provided by Keith E. Wilson, MBA, B.Comm., PMP, CSM, CSPO, KMP, SPC 5.0, DALSM, CDAI, CDC, MCP, MCT. © 2021 by Keith Wilson. All rights reserved. Reproduced with permission. Keith is a Senior Consultant and Trainer with over 25 years of successful coaching, training, management, and consulting experience. He has extensive Agile Training/Coaching experience and makes Agile work for his clients. He has earned the following certifications: SAFe Program Consultant (SPC 5), Certified Scrum Master (CSM) and Certified Scrum Product Owner (CSPO), Kanban Management Professional (KMP), PMI Disciplined Agile- Senior Scrum Master (DASSM), Coach (CDAC), and Certified DA Instructor (CDAI), Project Management Professional (PMP), Microsoft Certified Trainer and Professional (MCP, MCT). He is a welcomed trainer at numerous Fortune 500 corporations, universities, and associations worldwide and successfully consulted, developed courses, coached, and trained thousands of people virtually, in person, and through on-demand recordings in all areas of Agile Scrum, Kanban, SAFe, Disciplined Agile, Project, Program and Portfolio Management, Business Analysis, Microsoft Project, Project Online/Server, SharePoint, Microsoft Azure DevOps, Leadership and Interpersonal Skills.

Agile is a collection of principles focused on enabling teams to deliver work in small, workable increments, thus delivering value to their customers with ease. Evaluation of the requirements, plans, and results take place continuously, which helps the team to respond to changes.

In 2001 a group of software developers got together in Utah and shared the problems associated with software development projects that were rarely on time, in budget, the requirements not met, and just not bug-free. So, they developed the Agile Manifesto, which states that:

We are uncovering better ways of developing products by doing it and helping others do it. Through this work we have come to value:

Individuals and interactions *over processes and tools*

Working solutions *over comprehensive documentation*

Customer collaboration *over contract negotiation*

Responding to change *over following a plan*

That is, while there is value in the items on the right, we value the items on the left more.

I have substituted solutions for software as over the last 20 years Agile has branched out from software and just information technology to organizations working to reaching business agility. See Figure 6.50. Although Scrum is the dominant form of Agile, it is not the only method and practice as Illustrated in the 14th state of Agile report.

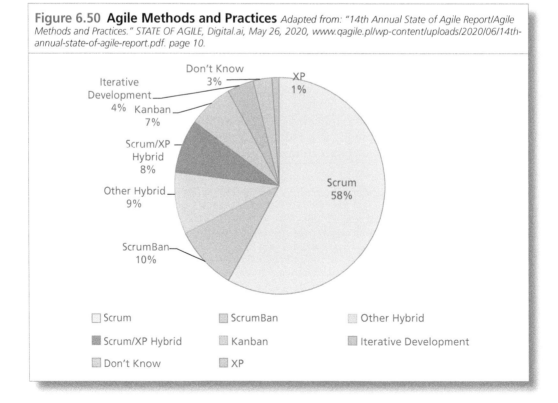

Figure 6.50 Agile Methods and Practices *Adapted from: "14th Annual State of Agile Report/Agile Methods and Practices." STATE OF AGILE, Digital.ai, May 26, 2020, www.qagile.pl/wp-content/uploads/2020/06/14th-annual-state-of-agile-report.pdf. page 10.*

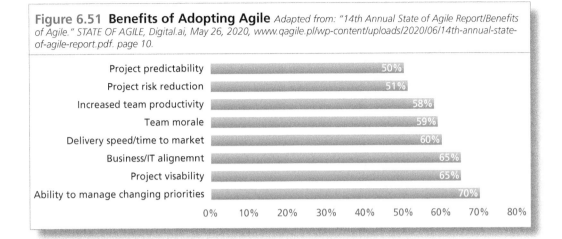

Figure 6.51 Benefits of Adopting Agile *Adapted from: "14th Annual State of Agile Report/Benefits of Agile." STATE OF AGILE, Digital.ai, May 26, 2020, www.qagile.pl/wp-content/uploads/2020/06/14th-annual-state-of-agile-report.pdf. page 10.*

There are no shortages of studies that highlight the benefits of adopting Agile as shown in Figure 6.51. As a result, since 2013 we have seen a hockey stick effect concerning the number of organizations moving to Agile and its expansion beyond IT.

Organizational flexibility is important when undergoing adopting Agile which means openness to change and flexibility to respond to new information while aligning to Principle 1 from the Agile Manifesto.

#1 Our highest priority is to satisfy the customer through early and continuous delivery of valuable products.

Agile Metrics

Agile metrics are the quantifiable measures regarding the quality of work, productivity, progress, and level of success of the project, and the performance and well-being of the Agile team. Metrics help key stakeholders to identify the areas of strength and weakness and resolve possible bottlenecks. Metrics can also lead to higher predictability in regards to the timelines while emphasizing the quality of the product developed.

The Agile Manifesto principle ten states

Simplicity – the art of maximizing the amount of work not done – is essential.

Therefore, Agile Teams need to select the key metrics to track and utilize.

The metrics are primarily meant for the Agile team. They should be the ones collecting, using, and sharing the metrics. In addition, the metrics should be used within the conversations among the team members. It needs to act as a starting point of discussions during sprint planning.

https://projectresources.cdt.ca.gov/agile/agile-readiness-assessment

General Agile Metrics

Value Delivered

Implementing features with high value should be the top priority of all Agile Teams. See Figure 6.52. An upward trend in this metric shows that things are on track, and a downward trend means the implementation of lower-value features occurring.

Net Promoter Score

Customer loyalty is an important factor to determine the success of an organization and the Net Promoter Score is a good gauge of loyalty. The Net Promoter Score shown in Figure 6.53 measures how much the customers are willing to recommend the product or service to others. It is an index that ranges from −100 to 100.

Epic Progress Chart

Agile teams can be aligned with a product vision and in turn, the vision can be broken down into Epics—Feature, Stories, and Tasks as shown in Figure 6.54.

Epics are the container that captures and manages the largest initiatives that occur within a portfolio. https://www.scaledagile.com

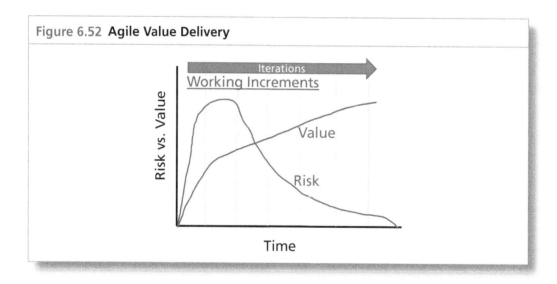

Figure 6.52 **Agile Value Delivery**

Figure 6.53 **Net Promoter Score**

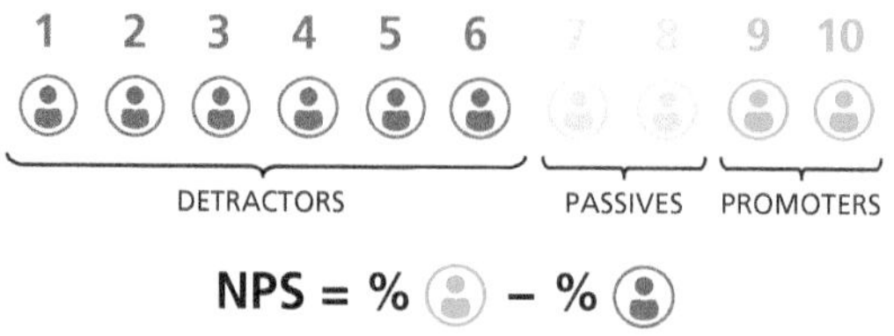

Figure 6.54 **Epic Progress Chart**

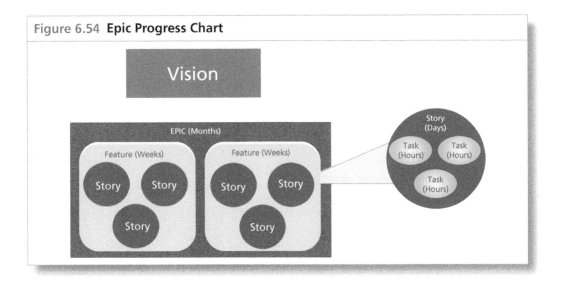

Daniel Pink in his book *Drive: The Surprising Truth about What Motivates Us* indicated that meaningful work is one of the top three motivators. By ensuring the team is aware of how their work ties into a Product Vision and its Epics coupled with visibility into the Epic and release burndown make that possible.

Drive: The Surprising Truth about What Motivates Us by Daniel H. Pink

Epic Progress Report Tracking
Epic Progress Report Tracking: epic progress is somewhat similar to traditional project metrics. In Figure 6.55, we see the baseline of the first estimate for the job size (red line) in comparison to the current estimate and we see how much of the epic has been finished compared to the original estimate.

Features
Features, as shown in Figure 6.56, are visible "units" of functionality that the customer recognizes, and it's at this level of detail that the customer can prioritize their needs. https://www.scaledagile.com

Feature Kanban Board
A feature Kanban board, as shown in Figure 6.57, will show the flow of work associated with developing the features with the goal to optimize the flow.

Work Item Aging
The purpose of the work item aging chart is to detect the timeline for unfinished tasks. Figure 6.58 shows the aging work in progress and indicates the time that passes between the start and completion of the current task.

Impediment Aging

This chart, Figure 6.59, shows the number of days open impediments are open during a sprint.

Control Chart

The control chart shown in Figure 6.60 is an objective measurement tool to determine if the process under inspection can reliably

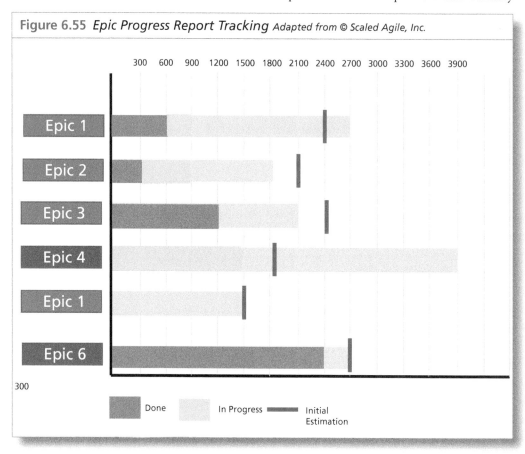

Figure 6.55 *Epic Progress Report Tracking* Adapted from © Scaled Agile, Inc.

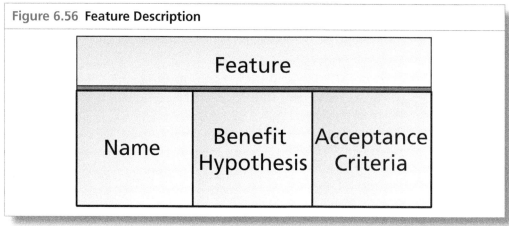

Figure 6.56 **Feature Description**

Figure 6.57 Feature Kanban Board

Discover	Define	Ready for Development	Doing	Done
Feature 5 Feature 6	Feature 4	Feature 3	Feature 2	Feature 1

Figure 6.58 Work Item Aging

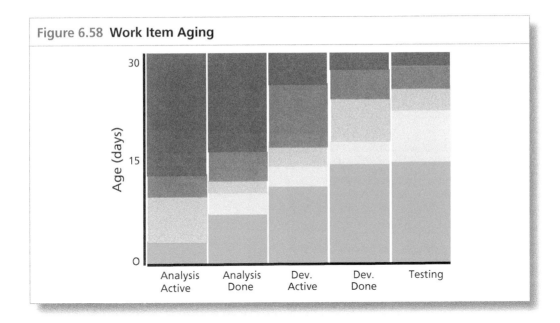

produce what is expected of it in the future. By examining the patterns of the data on the control chart we can understand if the process is "in control" or "out of control." This graph shows the team's velocity over time.

Escaped Defects

An escaped defect metric as shown in Figure 6.61 shows how many bugs were experienced by users in production. Ideally, a Scrum Team should fully test stories and completely avoid escaped defects.

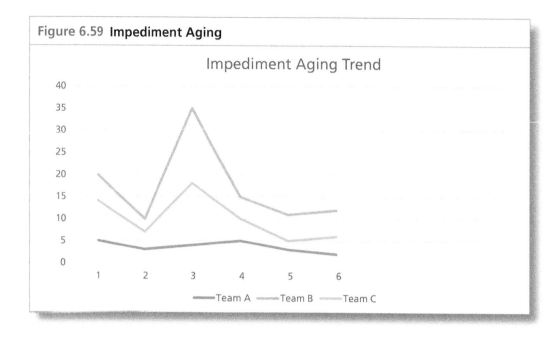

Figure 6.59 **Impediment Aging**

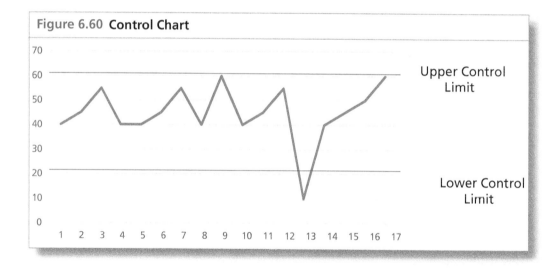

Figure 6.60 **Control Chart**

Scrum Metrics

Scrum is a lightweight framework that helps people, teams, and organizations generate value through adaptive solutions for complex problems.

In a nutshell, Scrum fosters an environment where:

1. A product owner orders the work for a complex problem into a product backlog.

2. The self-organized, cross-functional Scrum Team turns a selection of the work into an increment of value during a Sprint.
3. The Scrum Team and its stakeholders inspect the results and adjust for the next Sprint.

https://scrumguides.org/scrum-guide.html

Scrum is relatively simple as it only has 3 roles, 3 artifacts, and 5 events as illustrated in Figure 6.62:

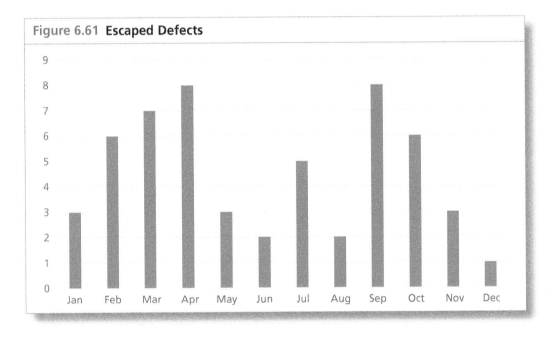

Figure 6.61 **Escaped Defects**

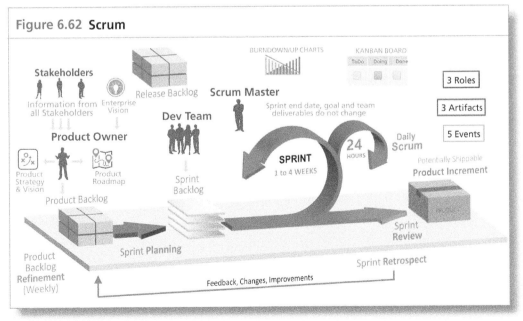

Figure 6.62 **Scrum**

Sprint Burndown Chart

The burndown chart in Figure 6.63 is a graphical representation of the estimated Scrum story points or tasks as compared to the actual story points or tasks completed. Before commencing a sprint, the Scrum master will determine the team's capacity which is a measure of the Team's availability during the upcoming Sprint. If needed the historical velocity which is the measured results of a team average story point per sprint over time will be adjusted.

Sprint Burnup Chart

The Burnup Chart in Figure 6.64 provides a visual representation of a sprint's completed work compared with its total scope. It offers insights into the team's progress throughout the sprint.

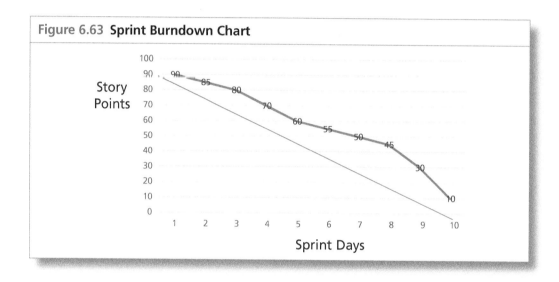

Figure 6.63 **Sprint Burndown Chart**

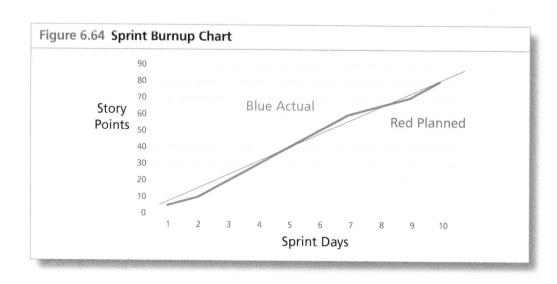

Figure 6.64 **Sprint Burnup Chart**

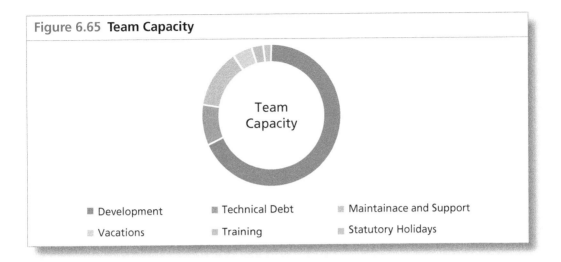

Figure 6.65 **Team Capacity**

- Development
- Vacations
- Technical Debt
- Training
- Maintainace and Support
- Statutory Holidays

Team Capacity
Team capacity, as shown in Figure 6.65, may fluctuate over time due to vacations or sick leave, statutory holidays, work for other teams, or their non-teamwork. This team capacity chart shows the available time for development and considers the time spent on technical debt clean up, maintenance and support, training, vacation statutory holidays.

Team Velocity
Velocity, as shown in Figure 6.66, is a measure of the team's rate of progress calculated by summing the size of the user stories that the team can complete during an iteration. Or simply the team's velocity = the number of story points (completed during an iteration). Velocity charts compare the planned versus completed velocity.

Impediment Tracking Chart
Types of impediments include:

- Blockers (for a user story)
- People issues
- Technical issues
- Operational issues
- Managerial Issues
- Organizational issues
- Process issues
- External things
- Business or customer issues
- Culture and waterfall attitudes

The impediment tracking graph in Figure 6.67 illustrates the number of blocked items per sprint.

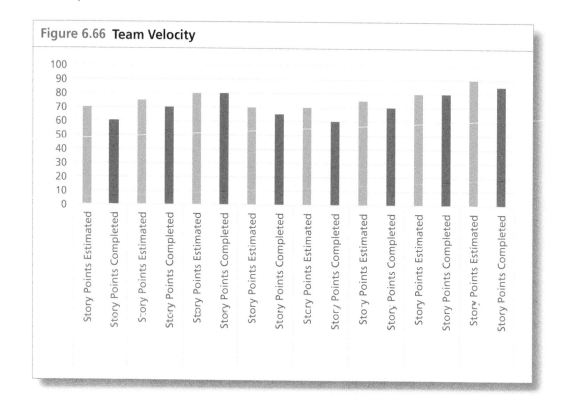

Figure 6.66 **Team Velocity**

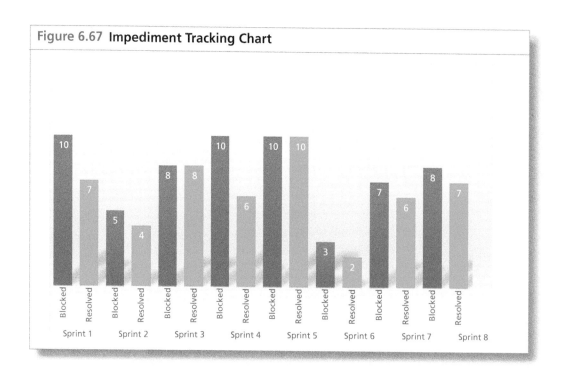

Figure 6.67 **Impediment Tracking Chart**

Other Sprint Charts

Task by Status

This graph, shown in Figure 6.68, communicates the status of each task associated with the stories the team is actioning during a sprint.

Sprint Bugs by Priority

The sprint bug chart depicted in Figure 6.69 communicates the bugs by stage of development and the bug criticality.

Figure 6.68 Task by Status

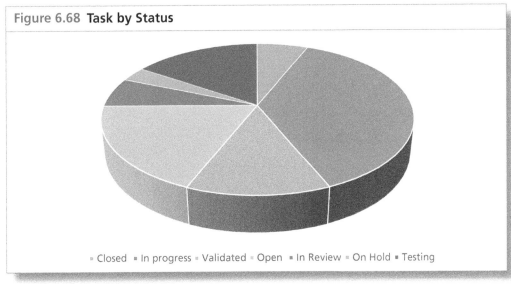

≡ Closed ■ In progress ≡ Validated ≡ Open ■ In Review ≡ On Hold ■ Testing

Figure 6.69 Sprint Bugs by Priority

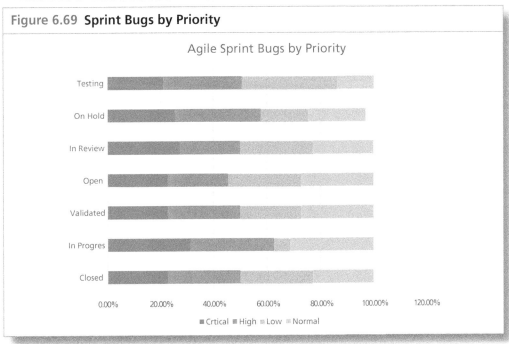

Sprint Tracking Chart

This simple chart shown in Figure 6.70 illustrates the status of all product backlog items in a sprint.

Commitment Efficiency

Commitment Efficiency shown in Figure 6.71 compares the user stories completed at the end of a sprint to the user stories committed at the start of the sprint.

Throughput

The throughput chart identified in Figure 6.72 calculates items processed under a specified period to aid in assessing the quantity of work across a period.

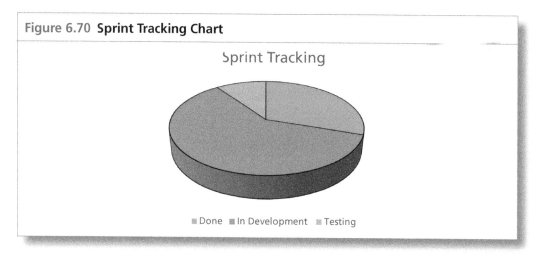

Figure 6.70 Sprint Tracking Chart

Figure 6.71 Commitment Efficiency

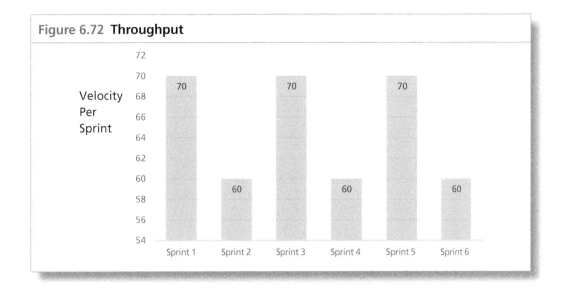

Figure 6.72 **Throughput**

Iteration Metrics

SAFe® for Lean Enterprises suggests the following additional Agile metrics shown in Figure 6.73.

Scaled Agile Metrics

SAFe® for Lean Enterprises is a knowledge base of proven, integrated principles, practices, and competencies for Lean, Agile, and DevOps. The Scaled Agile Framework encompasses a set of principles, processes, and best practices that helps larger organizations adopt

Figure 6.73 **Iteration Metrics**

Functionality	Sprint 1	Sprint 2	Quality and test automatic	Sprint 1	Sprint 2
# Stories (loaded at beginning of iteration)			% SC with test available/test automated		
# Accepted Stories (defined,built,tested and accepted)			Defect count at start of iteration		
% Accepted			# New test cases		
# Not accepted (not archived within iteration)			# New test cases automated		
# Not pushed to next(rescheduled in next iteration)			# New manual test cases		
# Not accepted: deferred to later date			Total automated tests		
# Not accepted: deleted from Backlog			Total manual tests		
# Added (during iteration; should typically be 0)			% Tests automated		
			Unit test coverage percentage		

agile methodologies, such as Lean and Scrum, to develop and deliver high-quality products and services faster.

SAFe® synchronizes alignment, collaboration, and delivery for large numbers of teams. It is particularly well-suited for complex projects that involve multiple large teams at the project, program, and portfolio levels.

Value Stream Key Performance Indicators

Value Stream Key are the set of steps the organization undertakes to deliver that specific product or service to the customer. Performance Indicators (KPIs), as shown in Figure 6.74 are the quantifiable measures used to evaluate how a value stream is performing against its forecasted business outcome.

Program (Value) Predictability Measure

The Program (Value) Predictability measure Objectives Business Value Planned vs. Actual and is shown in Figure 6.75. It requires the sponsoring executives to assign a "business value score" during planning meetings to each team's objectives for the program increment. At the end of the Program increment business owners collaborate with each Agile team to score the actual business value achieved for each of their teams' objectives.

Program Board: Feature Delivery, Dependencies, and Milestones

The SAFe Agile dependency board in Figure 6.76 shows features the teams are working on and the associated dependencies.

SAFe Program Predictability

The SAFe Program Predictability Chart in Figure 6.77 is a control chart that shows each team's objectives completed per program increment and if the team's result was in an acceptable range.

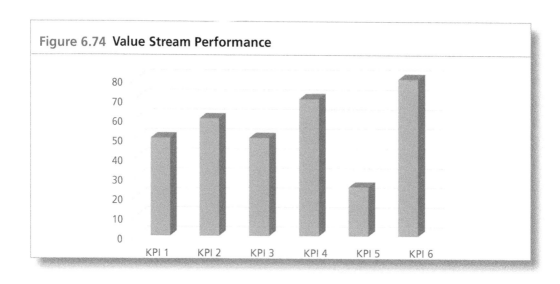

Figure 6.74 **Value Stream Performance**

Figure 6.75 **Program (Value) Predictability Measure**

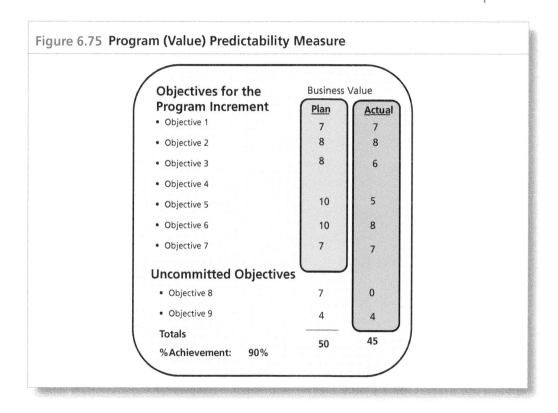

Figure 6.76 **Program Board**

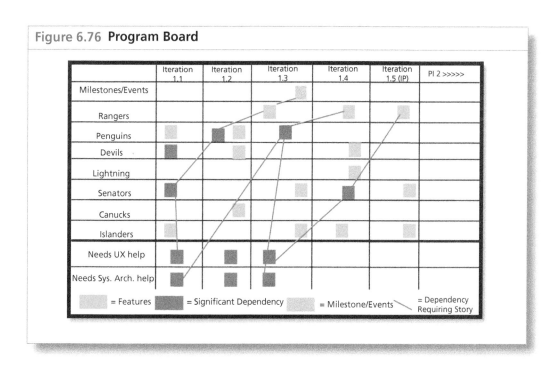

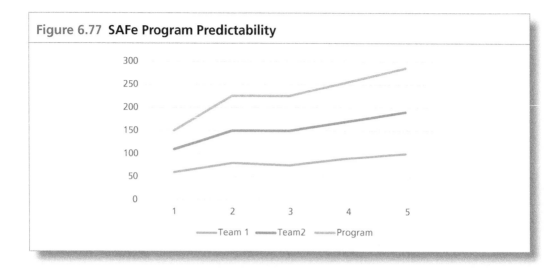

Figure 6.77 **SAFe Program Predictability**

Lean Kanban Metrics

Kanban Definition
Kanban is a visual system for managing work as it moves through a process. Kanban visualizes both the process (the workflow) and the actual work passing through that process. The goal of Kanban is to identify potential bottlenecks and fix them so work can flow through them cost-effectively at an optimal speed.

Kanban Principles

- Start with what you do now!
- Agree to pursue evolutionary change.
- Initially, respect current processes, roles, responsibilities, and job titles.
- Encourage acts of leadership at all levels.

Kanban Practices
Kanban practices are shown in Figure 6.78.

Kanban Board
A Kanban Board is shown in Figure 6.79.

Queues
Instead of limiting the time during active work, reduce the waiting time in queues by identifying where the task is stuck and take measures to cut down the work in queues.

Cumulative Flow Diagram
A "cumulative flow diagram" (CFD) as illustrated in Figure 6.80 shows work items moving through states, which provides visibility

Figure 6.78 Kanban Practices

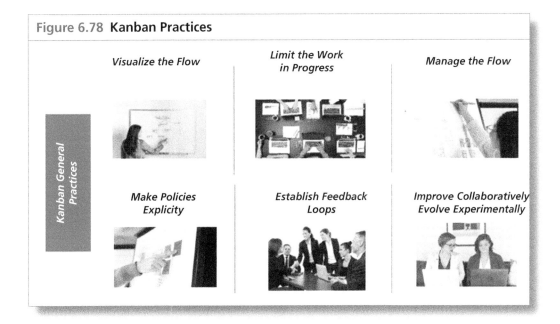

Figure 6.79 Kanban Board

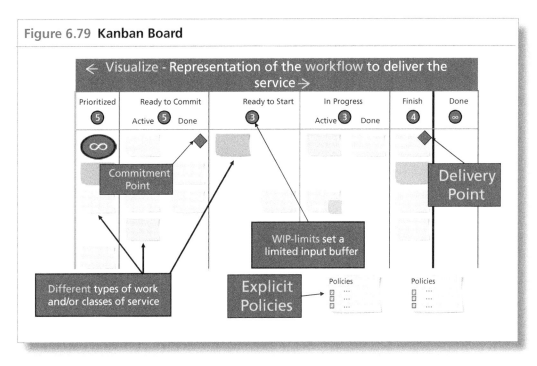

on cycle time, work-in-progress, and an estimated time required to complete. The cumulative flow diagram provides good feedback on flow and lack of flow, which is caused by impediments.

An ideal CFD represents a steady growth of completed work and a continuous delivery of value and tracks.

Figure 6.80 Cumulative Flow Diagram

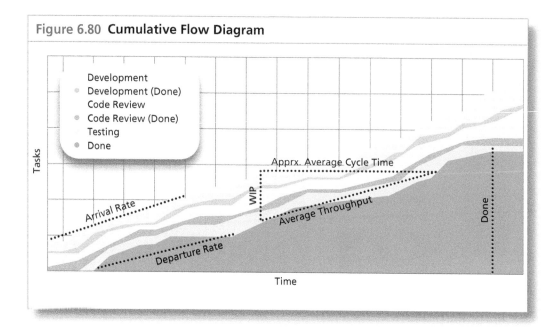

Figure 6.81 CFD Analysis.

PATTERN	MEANING	SUGGESTION FOR IMPROVEMENT
Differences in gradient	Mismatched arrivals and departures	Enforce WP limits
Flatline	Periods of activity	Track process bottlenecks
The S-curve	Points with little progress	Determine the cause workflow (queue) states
Bulging bands	Increasing WP	Split process steps
Stair steps	Work is delivered in batches	Consider implementing a continuous delivery model
Disappearing bands	Process state being skipped	Rethink the state's purpose

Cumulative Value of delivered work

Cycle Time—The average time between the start and end of production of one unit

Lead Time—The latency between the start and end of the entire production

Reducing Lead Time is the hallmark of LEAN, identifying deviations quickly will help the team track and optimize the flows and resolve bottlenecks as they occur.

Suggestions on how to interrupt the pattern in the CFD is shown in Figure 6.81.

The ultimate guide to reading cumulative flow diagrams www.getnave.com

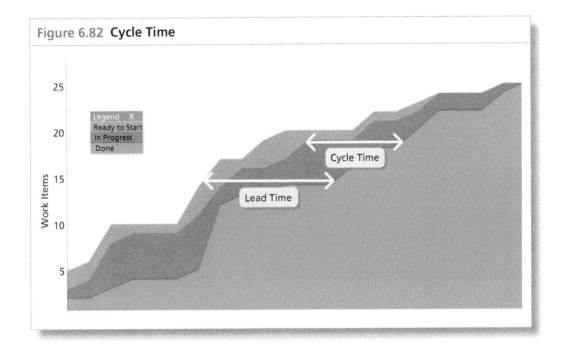

Figure 6.82 **Cycle Time**

Little's Law

$$Lead\ Time = \frac{Number\ of\ Things\ in\ Process}{Average\ Completion\ Rate}$$

Little's law was developed by John Little and states that the lead time, shown in Figure 6.82, can be calculated by dividing the number of items in Process by the average completion rate.

Understand Little's Law:

- Faster processing time decreases wait
- Control wait times by controlling queue lengths.

Summary

Despite Agile being simple to understand there are challenges with the successful implementation as shown in Figure 6.83.

Selecting the right metrics will help to inspect adapt and improvise. Leveraging principle 10 from the Agile Manifesto *"Simplicity – the art of maximizing the amount of work not done – is essential"* and chose the best fit-for-purpose metric for your team.

6.18 **DATA WAREHOUSES**

The expansion and use of project management into more business applications and nontraditional projects, such as strategic projects, have forced project managers to learn new skills and to have access to

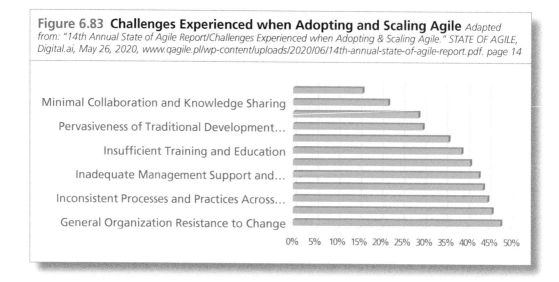

Figure 6.83 Challenges Experienced when Adopting and Scaling Agile *Adapted from: "14th Annual State of Agile Report/Challenges Experienced when Adopting & Scaling Agile." STATE OF AGILE, Digital.ai, May 26, 2020, www.qagile.pl/wp-content/uploads/2020/06/14th-annual-state-of-agile-report.pdf. page 14*

more information to navigate in these unchartered waters. Project managers are now expected to make both project and business decisions. There are now new challenges related to problem solving and decision making. All of this requires more information than is currently available in the earned value management system.

As project management applications grew, emphasis was initially placed upon creating a project management information system (PMIS) based almost entirely upon earned value management efforts using just time, cost, and scope metrics. Additional information, if necessary, came from personal contacts. Most decisions were based upon information related to the triple constraints and the impact it might have on the well-defined requirements listed in the statement of work. Today, project managers are expected to have access to the necessary information, tools, and processes to support complex problem analysis and decision making. Advances in technologies, the need for competing constraints, the expansion of project management into strategic projects, and the growth of information warehouses are driving companies toward business intelligence (BI) systems.

Throughout the life of a project, especially strategic projects, there is a significant amount of data that must be collected including information related to the project business case, project benefits realization plan, project charter, project master plan, customer interfacing, and market analyses. The knowledge contained in information warehouses, as well as the amount of information and speed of access, provides companies with a source of competitive advantage.

If the intent of a strategic project such as one requiring innovation is to create a commercially successful product, then the team members must understand the knowledge needed in the commercialization life cycle phases even though many of the team members may not be active participants during commercialization. Making the wrong decisions in the early project stages can create havoc during commercialization efforts. Team members should review commercialization data records of previously introduced products so that they understand the downstream impact of their decisions. This information should be included in information warehouses.

It can be extremely difficult to get people to use a knowledge management system correctly unless they can recognize the value in its use. With the amount of information contained in knowledge management systems, care must be taken in how much information to extract, and the format in which it appears as metrics and KPIs. Companies invest millions of dollars in developing information warehouses and knowledge management systems. There is a tremendous amount of rich but often complicated data about customers, their likes, and dislikes, and buying habits. This knowledge is treated as both tangible and intangible assets. But the hard part is trying to convert the information into useful knowledge. Simply stated, we must show that the investment in a knowledge management system contributes to a future competitive advantage.

The Growth of Business Intelligence Systems

Simply having knowledge repositories or information warehouses may not be sufficient to support future projects in an effective manner. BI systems are often considered as the next step after knowledge repositories or information warehouses and combine business information with technologies in a manner that allow project managers to make timely strategic and/or operational business decisions related to their projects.

The components of a BI system are data gathering, data storage, and knowledge management. Metrics information is a critical component of BI systems. The information contained in the BI systems can be historical, current, and predictive. The information can come from several sources including strategic and operational project management benchmarking studies conducted by PMOs.

BI technologies are designed to handle large amounts of "big data," whether structured, semi-structured, or unstructured, and present the data on meaningful dashboards such that project teams can make better business decisions and take advantage of business opportunities, especially when managing strategic projects. The technologies used in BI systems allow companies to look at external data (i.e., information from the markets in which the company operates) and internal data (i.e., financial and operational data) together and create business intelligence information to support strategic, tactical, and operational projects. BI systems facilitate corporate decision support systems by transforming raw data into meaningful and competitive business intelligence. However, there are still companies that believe that BI systems are merely the growth of business reporting systems.

Project managers will need to learn new decision-making tools including digitalized economics, artificial intelligence, and the Internet of things (IOT). With large amounts of data, teams may have to rely upon analytical statistics which includes:

- Descriptive data analytics: analysis of historical data including past successes and failures
- Predictive data analytics: analysis of the data to make predictions of what might happen
- Prescriptive data analytics: look at the reasons why things may happen, options for risk mitigation of future work, and options to take advantage of opportunities

Big Data

The growth of big data will most likely impact most companies worldwide. For effective analysis of the data, project teams will need workers that possess data science capability. The skills will include statistical methods, computational intelligence, and optimization techniques.

There are numerous mathematical models that currently exist to support project decision-making efforts using big data. A list includes:

- Financial models (ROI, IRR, NPV, payback period, benefit-to-cost ratio, breakeven analysis)
- Time (scheduling models)
- Money (cash flow models)
- Resources (competency models)
- Materials (procurement models)
- Work hours (estimating models)
- Environmental changes models
- Consumer tastes and demand models
- Inflation effects models
- Unemployment effects models
- Changes in technology models
- Simulation and games models
- Mental models

The expected benefits from using big data effectively include:

- Detection of patterns and trends related to time, cost, and scope
- Comparison to other projects as well
- Identification of the root causes of problems
- Better use of "what if" scenarios
- Better tradeoffs on competing constraints
- Better tracking of assumptions and constraints
- Better tracking of VUCA and the enterprise environmental factors
- Better response to out-of-tolerance situations
- Better capacity planning decisions involving resource utilization

- Ability to make strategic rather than just operational decisions
- Ability to make change management decisions
- Decision-making can be pushed down the organizational hierarchy, but there will be "rules for decision-making" established
- Emphasis on long-term perspectives rather than just short-term
- A reduction in the risk of making the wrong decision because of a lack of information

Project teams seem to focus on the knowledge management portion of the BI system. This includes:

- How performance metrics are created and reported
- How benchmarking information can be extracted
- Statistical and predictive analytics
- Data visualization techniques and dashboard design
- Business and project reporting for executives and stakeholders

PMOs and project teams are now participating in designing and managing BI systems projects. For some companies, this will bring additional challenges related to understanding BI systems and data visualization practices, including new ways of designing BI dashboards and interfacing with stakeholders.

Creating real-time dashboards that can change over the life of the project, and where information will be extracted from a variety of information warehouses, brings forth the problem with data quality and credibility of information:

> Data quality is a general challenge when automatically integrating data from autonomous sources. In an open environment the data aggregator has little to no influence on the data publisher. Data is often erroneous, and combining data often aggravates the problem. Especially when performing reasoning (automatically inferring new data from existing data), erroneous data has potentially devastating impact on the overall quality of the resulting dataset. Hence, a challenge is how data publishers can coordinate in order to fix problems in the data or blacklist sites which do not provide reliable data. Methods and techniques are needed to; check integrity, accuracy, highlight, identify and sanity check, corroborating evidence; assess the probability that a given statement is true, equate weight differences between market sectors or companies; act as clearing houses for raising and settling disputes between competing (and possibly conflicting) data providers and interact with messy erroneous web data of potentially dubious provenance and quality. In summary, errors in signage, amounts, labeling, and classification can seriously impede the utility of systems operating over such data.[7]

7 Wikipedia contributors, "Mashup (web application hybrid)," *Wikipedia, the Free Encyclopedia*, https://en.wikipedia.org/wiki/Mashup_(web_application_hybrid)#Data_ quality. Accessed June 15, 2017.

Another issue is that not all companies will need or require a data warehouse. Not all stakeholders will have data warehouses, and even if they do, the project manager must never assume that all of the information needed for the dashboards will be found in the information data warehouse.

6.19 DASHBOARD DESIGN TIPS

Here are some rules of thumb to follow when you design the layout of a dashboard.

Colors

You have a large number of colors to choose from, and although it is tempting to use a variety of different colors to highlight various areas of importance in a dashboard, most experts agree that too many colors and the "wrong" colors are worse than too few colors.

Note: Remember that some people are color blind, so if you use colors, try to use various shades. A good test is to print out a screenshot of the dashboard on a black-and-white printer and see whether you are able to distinguish what will now be shades of gray from each other.

Fonts and Font Size

Using the right or wrong fonts and font sizes is like the use of colors; it can make or break the entire look and feel that a user gets from looking at a dashboard. Here are some tips:

- Do not mix a number of different font types; try to stick with one. Use one of the popular business fonts, such as Arial.
- Do not mix a number of different font sizes, and do not use too small or too large fonts. Remember that dashboards likely will be used by many middle-aged or older employees who cannot easily read very small fonts. Ideally, you should use a font size of 12 or 14 points and apply boldface in headers (maybe with the exception of a main header that could be in some larger font). Text or numbers should be in fonts of 8 to 12 points, with 10 points being the most often used. When it comes to font sizes, the challenge for a designer is always the available areas on the screen (also referred to as screen real estate). With overly large fonts, titles, legends, descriptions, and so on may get cut off or bleed into another section, and fonts that are too small make the display hard to read.

Section 6.19 has been taken from Nils Rasmussen, Claire Y. Chen, and Manish Bansal, *Business Dashboards* (Hoboken, NJ: John Wiley & Sons, 2009).

Use Screen Real Estate

Per definition, most dashboards are designed to fit a single viewable area (e.g., the user's computer screen) so users can easily get a view of all their metrics by quickly glancing over the screen. In other words, the moment a "dashboard" is of a size that requires a lot of scrolling to the sides or up and down for users to find what they looking for, it is not really a dashboard anymore but a page with graphics. Almost always, users get excited about the possibilities with a dashboard, and they want more charts and tables than what will easily fit on a screen. When this happens, there are various options:

- Use components that can be expanded, collapsed, or stacked so that the default views after login still fit on a single screen, but users can click a button to expand certain areas where they need to see more.
- Use many dashboards. If there is simply too much information to fit on a single dashboard (e.g., move sales-related information to a "sales dashboard" and higher-level revenue and expense information to a "financial dashboard"). Many dashboard technologies have buttons or hyperlinks that let us link related dashboards together to make navigation easy and intuitive for the users.
- Use parameters to filter the data the user wants to see. For example, a time parameter can display a specific quarter in a dashboard instead of showing all four quarters in a year.

Component Placement

If you have two tables or grids or scorecards and four charts, how should they be organized on the screen? Here are a few tips:

- Talk with the key users to find out which information is most important so they can establish a priority. Based on this, place the components in order of importance. Most users read from left to right and start at the top, so that could be the order you place the components. The idea is that if the users have only a few seconds to glance at a dashboard, their eyes first catch what is most important to them.
- A second consideration for component placement is workflow. In other words, if users typically start by analyzing metrics in a scorecard component and then want to see a graphical trend for a certain metric they click on in a scorecard, that chart component should be placed adjacent to the scorecard to make it easy for users to transfer their view to the chart as they click on the scorecard.

6.20 TEAMQUEST CORPORATION

Several companies provide metric management and dashboard software solutions to their clients. In this are two white papers which serve as excellent summaries of the information for dashboard design efforts.

White Paper #1: Metric Dashboard Design

Designers of metrics management dashboards need to incorporate three areas of knowledge and expertise when building dashboards. They must understand the dashboard users' needs and expectations both for metrics and for the presentation of those metrics; they must understand where and how to get the data for these metrics; and they must apply uniform standards to the design of dashboards and dashboard suites in order to make them "intuitive" for the end users.

This paper outlines dashboard design best practices and design tips, and will help dashboard designers ensure that their projects meet with end user approval. It concludes with a checklist of design considerations for dashboard usability.

Increasing User Adoption of Metrics Dashboards

Users turn to metrics management solutions to find out what is going on with the business in order to make informed, reasoned decisions.

Good metrics management dashboards show key performance indicators (KPIs) in context so that they are meaningful and present them in a way that allows users to instantly understand the significance of the information. This presentation lets users quickly evaluate choices and make decisions with full confidence that these decisions are supported by facts.

The white papers in this section have been taken from two TeamQuest Corporation white papers, "Metrics Dashboard Design" and "Proactive Metrics Management." Both white papers are reproduced by permission. TeamQuest Corporation is a metrics management software vendor that develops proactive, web-based corporate performance monitoring and enterprise reporting applications. TeamQuest's proactive metrics management applications empower business users to see KPIs in real time and allow business managers to accurately gauge performance. With TeamQuest, organizations can harness corporate data into powerful visual metrics that: Automate the reporting and monitoring of KPIs, as well as data transformations; Enable discovery of new insights into the business, and to react quickly to those—rather than making after-the-fact corrections; and Trigger positive change by focusing on factors that directly impact corporate performance. TeamQuest's customers include Global 1000, Fortune 500, and mid-size organizations in ITSM, support, financial, insurance, retail, and other industry sectors. TeamQuest Headquarters: 80 Aberdeen St., Suite 400, Ottawa ON K1S 5R5. Phone: 1-613-236-1644, Toll Free: 1-866-636-6065. Email: info@teamquest.com; www.teamquest.com.

Dashboards are neither detailed reports nor exhaustive views of all data. Good metrics management solutions can offer users the option to "drill down" to as much detail as they require, or even link into reporting systems, but these are only ancillary functions. The primary function of metrics management dashboards is to support—even induce—proactive decision making.

Know the End Users
Users want dashboards that respond to their business requirements.

There is no substitute for understanding end users' needs and getting involved in dashboard development. Even more important than understanding product capabilities is understanding the people who will be using the dashboards, what they need to know to improve the business, and what sort of dashboard organization and displays will work best for them.

Use Context to Make Metrics Meaningful
Users need to understand what the metrics mean before they can make decisions. Data is meaningful only in context.

In order to easily understand metrics, users must see them in context—their context. In fact, context and presentation are integral to any metric; without them the metric is simply meaningless numbers.

Dashboard designers should take time to learn what contextual information users require in order for metrics to be meaningful for them and to facilitate decisions and actions.

Contextual information will differ depending on the specific area being managed. For example, dashboard users in finance may need to track actual expenditures against budget targets, while a support desk may need to track the number of trouble tickets exceeding mean resolution times by more than 15%. In an environment where a metrics management solution is being used to help improve processes, users may need to monitor trends, compared to performance during another given period. (See Figure 6.84.)

On the left side of Figure 6.84, the much-used pie chart effectively shows proportions but does not tell anything about performance or progress toward targets. In the right side of Figure 6.84, a simple bar chart effectively shows proportions of allocated budgets and monies spent, targets, and performance: progress toward targets (dotted lines).

Whatever the case, the dashboard designer should spend time understanding not only the data behind the indicator but also the data that creates the context for the indicator (such as revenue targets, time period coverage, and maximum capacities). This rule should guide the dashboard designer through every step of design, from data gathering to detailed composition of displays.

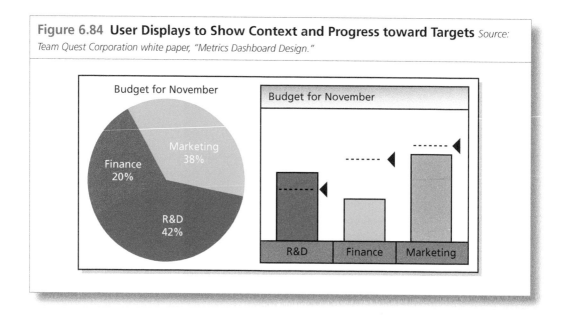

Figure 6.84 User Displays to Show Context and Progress toward Targets *Source: Team Quest Corporation white paper, "Metrics Dashboard Design."*

Data Retrieval

Users need to have confidence in the integrity of the metrics they use to manage the business. They need to know that they are acting on facts, not guesses.

There is no easy formula that will guarantee the value of data brought to dashboards. Nonetheless, it is essential to determine the sources, ownership, and quality of the data to be used before starting dashboard design.

The following guidelines will help dashboard designers deliver the metrics dashboard users need and will use.

Identify KRAs and KPIs

Users need to know their metrics so they can make informed decisions for their areas of responsibility.

A common—and effective—approach to understanding what users need to know is to work with end users to identify the key results areas (KRAs) for which they are responsible, then the key performance indicators (KPIs) they need to monitor and manage to improve performance in their areas of responsibility.

Get the Data Whatever Its Format or Location

Users want metrics derived from complete data, no matter where that data is stored, so they do not have to guess. Once the metrics users require have been identified, the data should be retrieved, *no matter where it is located*.

A well-designed metrics management dashboard provides both a current synthesis and details of key metrics. Often the data required to provide this synthesis and these details is spread across a variety of databases technologies and even spreadsheets at different physical locations. It is, nonetheless, essential that all required data be retrieved. A dashboard that provides metrics based on partial information is of little value when a global view is needed.

Refresh the Data According to Users' Needs

Users need metrics that are up to date so they can act on current and probable future situations.

The timeliness of a metric is as important as the metric data itself. Find out how current data must be for it to be valuable to users, and set the polling frequencies for the queries that retrieve the data accordingly.

Metrics used to monitor hourly call levels to a help desk are worse than useless if data is gathered every Sunday at 6:00 a.m. Similarly, if sales personnel report sales once a week on Thursdays, there is little point in polling the database every hour for updates to this data.

Usability Design Best Practices

Users do not want to be surprised by the dashboard design. They need to be able focus on metrics and decisions.

A metrics management dashboard should function for the user in the same way that an automobile dashboard or traffic signs function for an automobile driver.

Just as a driver knows to stop at a red light, a metrics dashboard user should understand without deliberation that, for example, a red thermometer means that corrective action is required. This requirement means that dashboard design must be consistent with common usage and practices as well as across all dashboards.

Offer Users a Choice of Views

Users need to be able to view metrics differently, so they can see the relationships between the metrics that affect their specific concerns. For example, financial metrics dashboards showing actuals versus budgets could offer views by department and by line item or profit center.

Whatever the view offered, dashboards should be consistent, and the current view should be clearly identified by titles and labels.

Use Commonly Accepted Symbols, Colors, and Organization

Users need to understand dashboards instantly and without specialized training so they can focus on their jobs, not on deciphering data.

Use symbols, colors and organization in ways that are commonly accepted and therefore easily understood. (See Figure 6.85.)

As an example, on the left side of Figure 6.85, the users might be confused by the use of color combined with the expressions on the faces. Red usually means that something is amiss, but the red face has a smile. Users might not know if action is required. On the right side of Figure 6.85, green means that everything is going well. The message is reinforced by the smiley face. Red usually means that action is required. This message is reinforced by the frown.

Figure 6.85 Using Color to Improve Communication of Key Information *Source: Team Quest Corporation white paper, "Metrics Dashboard Design."*

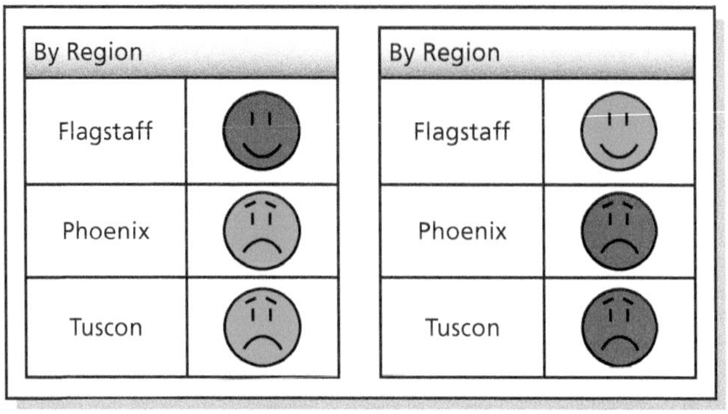

Red is the color most commonly used to identify something that requires attention. Therefore, unless there is a compelling reason to use another color to identify, for example, sales that are below targets, use red to communicate this information. Similarly, where appropriate, use commonly accepted symbols, such as stop signs or caution signs, rather than new symbols that users will have to learn.

Organize information using commonly accepted norms. Generally, this means that the most important information is placed at the top of a dashboard and secondary information and details are placed below.

Note, however, that symbols may differ between countries and cultures. Take these differences into account if the dashboards will be used in different parts of the world.

Establish Clear Dashboard Navigation and Hierarchies

Users need to be able to find information—more detail, less detail, or a different view—instantly.

Few dashboards are deployed individually. Typically, users require a suite of dashboards. Clearly establish an organization for the dashboard suite. Use an easily recognizable hierarchy of dashboards and consistent links for navigating between dashboards.

A good rule is to use the dashboard with top-level information for a user's area of responsibility as that user's entry point, then provide links from individual metrics to dashboards with more details about that area of the business. This is known as drill-down capability.

Maintain Consistency of Design

Users do not want to have to learn how to read each dashboard.

Establish and implement a limited set of templates with consistent use of color, symbols, and navigation, and use them throughout the dashboard suite.

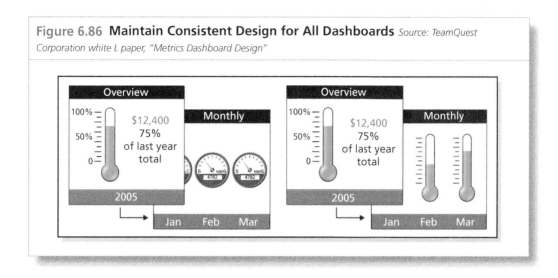

Figure 6.86 Maintain Consistent Design for All Dashboards *Source: TeamQuest Corporation white L paper, "Metrics Dashboard Design"*

Presentation of information on dashboards should be consistent. For example, if dashboards show trends, percentages, and absolute numbers, place each type of information in the same place on every dashboard and use the same display for these types of information across similar dashboards.

If thermometers are used to show progress of revenue against targets in the company financials roll-up dashboard, do not use gas gauges to show revenue against targets in the regional detail dashboards; use thermometers. In the left side of Figure 6.86, the different displays used to the overview (thermometer) and the monthly values (gas gauges) may confuse users who drill down to get more detail. On the right side of Figure 6.86, consistent use of thermometers (large for the overview and small for the details) ensures that the users will immediately and intuitively associate the two displays.

Similarly, if display threshold of red-yellow-green are set to 40, 60, and 80% for the company roll-up dashboard, use the same thresholds for the details unless there is a compelling reason to do otherwise.

Use Color Judiciously

Users expect color to provide important information they need, not distract them.

Generally, color can be used to create four effects on a dashboard. It can:

- Identify the status of key metrics and areas that require attention. For example, use red to identify expenditures that have increased more than 15% over the same period the previous year.
- Identify types of information. Color can be used to help users instantly identify the type of information they are looking at. For example, dark green can be used for monetary values, and dark blue for quantities of items.

Figure 6.87 Sample Dashboard with Grouped Metrics *Source: TeamQuest Corporation whitepaper, L "Metrics Dashboard Design"*

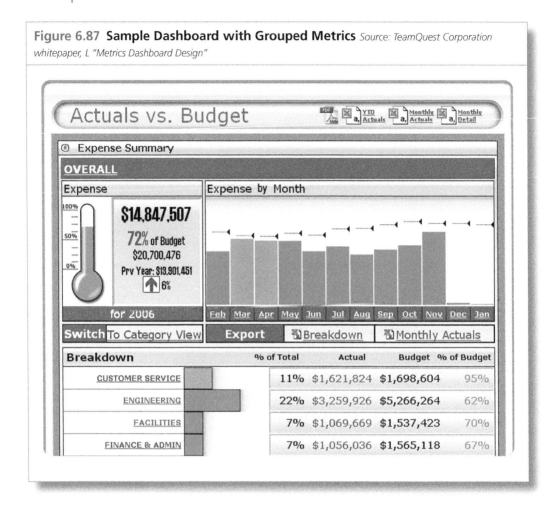

- De-emphasize areas or items. Border areas, backgrounds, and other supporting dashboard components (the dashboard skins) should use plain, unobtrusive colors that help define dashboard areas without distracting from the information displayed.
- Identify the dashboard type or its level. Different background colors or the color of dashboard titles can help users identify what they are looking at. For example, financial dashboards could use a green skin, while help desk dashboards could use beige as a reassuring color.

Use Dashboard Groups to Improve Organization

Users need to see metric groups and hierarchies so they can understand relationships between different areas of the business.

Group displays together by the type of metric displayed or by functional area. In Figure 6.87, TeamQuest's Profit Accelerator dashboards group related information together to improve dashboard "legibility."

For example, on a dashboard showing financial roll-ups, put top-level information into one group that shows progress against targets in three ways: absolute numbers against targets; performance compared to the same period in the previous year, or against average performance for the same period during the last five years; and the trend, based on performance over the last 60 days. Alternately, show absolute numbers against targets for each region or department.

Whatever the rationale used to group information on dashboards, be consistent. Use the same rationale when establishing groups on different dashboards.

Set dashboard groups to "open" or "closed" based on the hierarchy of information. Closed groups can be used to provide more detailed metrics or complementary metrics on the same dashboard without distracting the user's eye from the primary information.

Display Selection and Design

Users must be able to understand *what* they are being shown without stopping to analyze *how* it is being shown.

Select the display symbols that are most appropriate for displaying the information the dashboard users need.

Identify a limited set of symbols and use these on all dashboards. Be consistent with location, size, and color of supplementary information associated with the symbols. Consider using different sizes of the same display symbol for different levels of information.

Avoid displays that are overly complex, colorful, or animated. Such qualities are very effective when correctly used. However, the more complex the display, the more difficult it is to process. Overly complex dashboard displays distract users from the information they need and want. Consider offering views that separate the information into grouped displays or even different dashboards.

Do not hesitate to work with the dashboard vendor to design new display symbols that will improve presentation of information.

Actual Values, Percentages, and Trends

Different users need different information in order to make informed decisions and take appropriate action.

Decide with the dashboard users what metrics they need to monitor and manage. Different users may require different information from the same data.

For example, a CEO may want to know trends in budget expenditures, while a CFO and department managers may be more interested in actual numbers (actuals versus budgets). Use data to present the information users need, and make sure that the type of information is clearly identified. A thermometer showing revenue trends that is interpreted as showing actual numbers can lead to costly misinterpretations.

Timestamps

Users need to know when the metric was updated so they know the age of the data on which they base their actions.

Time is an essential part of a metric's context. Ensure that users can know when the data for the metrics was retrieved. In many cases this information is the key to understanding what the metric means. Ensure that this information is available but that it does not crowd the display. Consider putting the date and time in a mouse-over.

Titles and Labels

Users need to know instantly what they are looking at so they can focus on what it means for the business.

Give all dashboards meaningful, descriptive titles. Descriptive titles are generally more intuitive than cryptic or symbolic ones. Assign labels to dashboard groups and display symbols so that users can clearly identify information. Keep labels in the same location and use the same color standards throughout the dashboard suite.

Do not abuse labels. Overuse of labels can crowd out essential information.

Mouse-overs

Users often want more information to help them understand the significance of a metric.

Mouse-overs are an effective way to include detailed metric information without crowding the dashboard.

Information such as last and next run time metrics for another corresponding period, etc., can be included in mouse-overs to help users understand the significance of the primary information on the dashboard.

Parameter-based Views

Users want to see only the metrics that help them do their jobs. Different users give different weight to different information, as seen in Figure 6.88.

In Figure 6.88, parameters can be used to filter data so that users see only the information they need. In this example, data is filtered by region. All users see the same dashboard, but users in Region East see information for their region while users in Region West see information for their region.

Parameters are user-set variables that can be used to filter metric data delivered to the dashboards shown to different users. Consider designing dashboards with parameters, so that users get views of the dashboards based on their needs and permissions.

For example, the same dashboard suite might be used by a help-desk manager and individual employees in the department. The manager may need a consolidated view of all calls, by type, while each employee might need to see the calls for which he or she is responsible, a view created by filtering the information based on the user ID.

If the dashboards use parameters, display these prominently so that users will know what the metrics they are being shown represent.

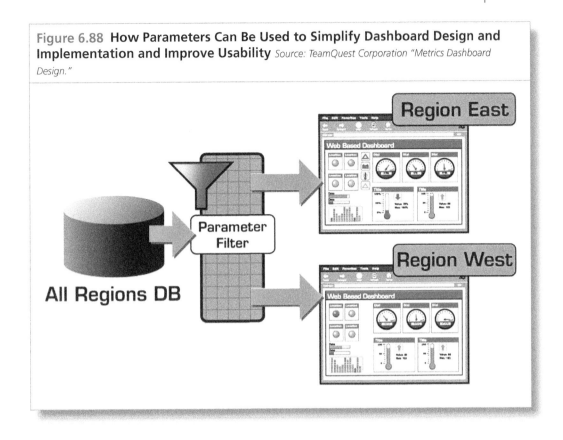

Figure 6.88 **How Parameters Can Be Used to Simplify Dashboard Design and Implementation and Improve Usability** *Source: TeamQuest Corporation "Metrics Dashboard Design."*

Use Thresholds and Threshold-triggered Actions

Users benefit m ost from dashboards that help them take corrective actions before problems occur.

Thresholds and threshold-triggered actions can be used to transform metrics monitoring into proactive management. Use thresholds to trigger actions that alert users to potential trouble areas, or even initiate corrective action by running scripts. In Figure 6.56, thresholds can be used with alerts to transform dashboards from passive monitoring devices into active vehicles inducing corrective action and, especially, prevent decisions and actions.

For example, set a threshold to send emails to the person responsible for managing a budget if spending reaches more than 70% of budget before the middle of the month. Or, in a help-desk environment, launch a script to change phone messages and reschedule nonessential activities if wait times increase beyond SLA requirements.

Roll-ups and Drill-downs

Users need to get both the big picture and the details.

A dashboard is most useful when used as part of a suite of complementary dashboards.

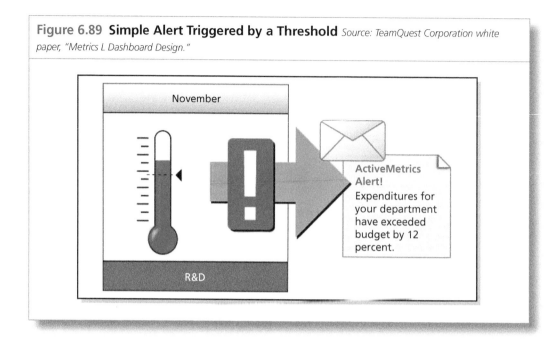

Figure 6.89 Simple Alert Triggered by a Threshold *Source: TeamQuest Corporation white paper, "Metrics L Dashboard Design."*

Group together on a dashboard the metrics users need to see together, and then use drill-downs and roll-ups to provide details and overviews.

This technique provides users the metrics they need without crowding too much information on any single dashboard.

Animation

Users appreciate a bit of fun but are annoyed by too many gimmicks.

Animation, such as blinking lights, moving figures, and other such displays, can add interest to dashboards. Use these sorts of features very judiciously, however, and consider providing nonanimated versions of dashboards or an "animation off" button.

Animation can be fun the first few times, but if not properly used it can become an annoyance. Consider using animation only for special projects, such as a month-end race to motivate the sales team, or to draw attention to new dashboard features for a limited period of time.

Visual Noise

Users turn to dashboards to be informed, not dazzled.

Avoid cluttering dashboards and dashboard displays with unnecessary paraphernalia, such as ornate frames, patterned backgrounds, or 3-D effects that add no value to the information displayed.

A minimalist approach is almost always the best approach.

Usability Checklist

Attention given to dashboard design can pay enormous dividends, both in user satisfaction with the dashboards and, especially, in improved business performance founded in proactive metrics management and reasoned, informed decision-making processes.

TABLE 6.5 User Questions and Design Solutions	
USER QUESTIONS	**DESIGN SOLUTIONS**
What am I looking at?	Use clear, descriptive titles and labels.
Does this mean "good" or "bad"?	Use standard, culturally accepted colors and symbols.
Are things getting better or worse?	Employ thresholds, and show meaningful comparisons and trends.
What is being measured, and what are the units of measure?	Clearly identify the units of measure, and provide actual values.
What is the target or norm?	Clearly show targets and norms, and design displays that show progress toward these.
How recent is the data?	Provide a date and timestamp for each metric.

TABLE 6.6 Common User Questions and Design Solutions	
USER QUESTIONS	**DESIGN SOLUTIONS**
How can I get more details?	Provide drill-down links to groups with detailed information.
How can I get a broader view?	Provide links to roll-up and overview dashboards.
What do I do with this information: what action should I take?	Always place data in context and, where possible, suggest advisable actions based on the metric.
When should I check for an update?	Provide the date and time when the metric will be updated. When business needs warrant, allow ad hoc updates.
How do I get metrics that are not on these dashboards?	Be ready to develop new dashboards. Users will want them!

Dashboard users want their questions about the metrics they are viewing on a dashboard answered even before they can formulate the questions. In fact, by the time a question about business performance is asked, it is often too late to take corrective action. Typical questions are shown in Table 6.5.

Valuable dashboards provide up-to-date information right at the user's fingertips so that problem areas can be addressed as they arise and opportunities can be taken advantage of immediately.

Ensuring that a dashboard and its components answer the questions in Table 6.6 should help dashboard designers create more effective dashboards.

White Paper #2: Proactive Metrics Management

Choosing a Metrics Management Solution

A business decision is only as good as the metrics that inform it. The quality of these metrics depends on three factors: accuracy, timeliness, and presentation. Understanding the importance of metrics, businesses are investing in defining and monitoring key performance indicators (KPIs). These indicators must not only accurately reflect the state of the business and current trends, but they must also be presented to users in a manner that induces effective, proactive decision making.

This paper reviews the capabilities of metrics management software and outlines the most important features and capabilities to look for in a solution.

Metrics Collection and Delivery

One of the key goals of businesses today is the collection and delivery of the metrics decision makers need to make timely, informed, and well-reasoned choices. Even in businesses where the key performance indicators (KPIs) have been defined and the data that drives them is available, the collection and delivery of the metrics can fail. Metrics that are timely, accurate, in context, and easy to understand are not immediately available to those who need them most: boards, executives, managers, and others at all levels of the organization.

No one is to blame. Until recently the tools needed for effective delivery of these metrics were simply not available. Reporting provides detailed profiles of the state of affairs as they were at a specific moment in the past. Business intelligence (BI) systems provide comprehensive data and deliver metrics, but these systems are prohibitively expensive and the metrics they provide are limited to those derived from the data inside the BI system. In today's business reality, metrics must be gathered from multiple, disparate systems and sources across the organization.

The Metrics Management Investment

Since the last generation of dashboards appeared on the market at the end of the 1990s, "dashboard solutions" have often been promoted as the answer to decision makers' need to know. Unfortunately, however, dashboards alone are not a solution. They are an essential part of the solution; they are effective for delivering metrics and for presenting them and making them accessible. But dashboards are only as good as the metrics they present; their value as decision-making tools is a function of both how well they present metrics and of the quality of the data behind these metrics.

What is needed, then, to transform decision making from educated guesswork based on partial information and hunches to a truly rational and informed process based on timely, accurate metrics is not simply dashboards. This transformation requires a proactive metrics management tool—one that can gather data from across multiple systems and deliver metrics to dashboards in context and with as much or as little detail as decision makers require.

Selecting a Solution

Due to the emergence of metrics management software and to the diversity of dashboard-type products on the market, the search for a solution can be confusing. Metrics management dashboards, as shown in Figure 6.90, should provide users with

Figure 6.90 Sample TeamQuest Metrics Management Dashboard *Source: TeamQuest Corporation white paper, "Pro-Active Metrics Management"*

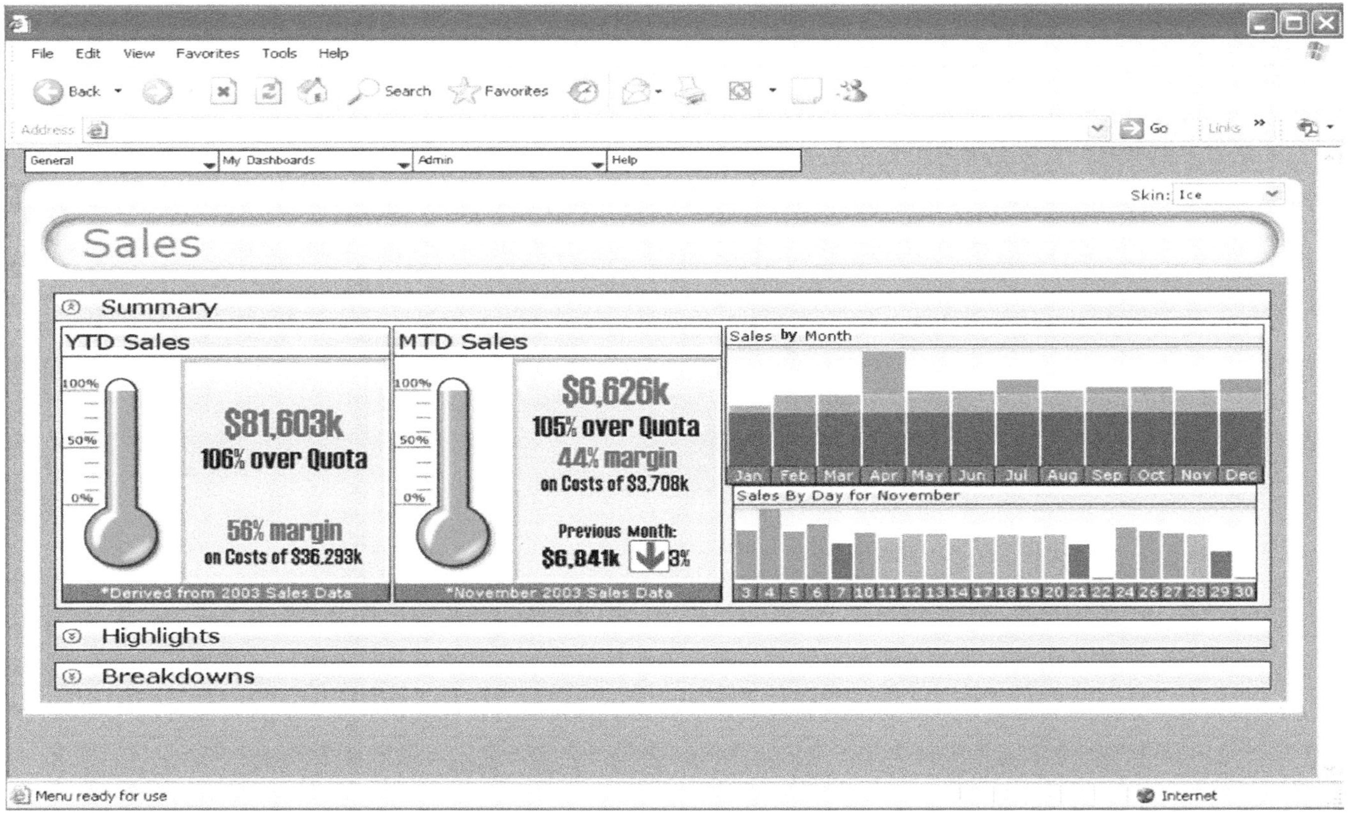

the information they need to make informed, reasoned decisions. The following is a set of high-level criteria for evaluating metrics management solutions.

Proactive Metrics Management

Look for the ability to trigger effective preventive and corrective action. The crucial differentiator for metrics management solutions is the level of *proactive* metrics management they support. Proactive metrics management is a function of a solution's ability both to trigger appropriate actions in systems with which it interfaces and to facilitate timely decisions by the persons responsible for key results areas (KRAs) in the organization.

It is important to understand here that more than one capability defines a product's level of support for proactive metrics management. Two important capabilities are:

1. Thresholds, alerts, and notifications: A proactive solution uses thresholds and alerts set from absolute values or trends (or a combination of these) to keep users informed.
2. Triggers and corrective/preventative action: A proactive solution is able to initiate actions when thresholds are compromised. For example, it can send an email alert or launch a script.

Technology Agnostic

Look for no restrictions on metrics data sources. TeamQuest Active Metrics can retrieve data from virtually any source, as shown in Figure 6.91. Very few organizations use a single technology to meet all their data and performance needs.

Typically, different technologies have been implemented because each is the best available for a specific area of the business: financial reporting, trouble-ticket management, etc. A proactive metrics management solution must be able to query seamlessly a wide array of different databases (and even spreadsheets) for key indicators and, if required, use indicators derived from different sources in the same dashboard.

Easy Availability

Look for web-based, easily customizable user interface (UI) design with multi-language support. Organizations today are rarely limited to a single geographic location and often work around the clock. A metrics management solution's dashboards must be available at all times and anywhere in the world, and because key metrics must be made available to many people, the solution should not require specialized viewing tools. Users should need nothing more than a web browser to view metrics dashboards over the Internet or an intranet.

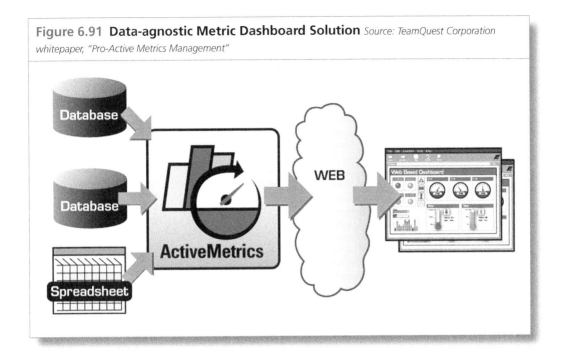

Figure 6.91 Data-agnostic Metric Dashboard Solution *Source: TeamQuest Corporation whitepaper, "Pro-Active Metrics Management"*

Security

Look for configurable, secure user and data source access. Since metrics tell critical information about the business, they usually convey sensitive information. Dashboards and the information they contain should therefore be easily protected without sacrificing their availability. Further, the metrics management solution should be able to access data that is protected behind various security barriers and firewalls without jeopardizing the integrity of these security measures.

Usability

Look for ease of use for both end users and dashboard developers.

As with anything else, a metrics management solution is of little value if it is not widely adopted by users. Users can be classed into two areas: end users and dashboard designers. Any solution must be easily usable by both.

Assuming that the metrics dashboards are easily accessible to those who need them anywhere and at any time, the design of the dashboards themselves and the metrics they present should be evaluated:

- Are the dashboards clear and easy to understand so that users grasp instantly what they are being shown and what it means for their areas of responsibility?
- Can thresholds, trends, and warnings be easily and clearly identified by the end user?

- Can legends to explain data be easily integrated into the dashboards?
- Are relationships between metrics easily established by the end users?
- Do the dashboards allow drill-down views for more detailed information, and other view changes requested by users?
- Does the solution offer clean and consistent display of metrics without visual noise?

Usability for dashboard designers may be less noticeable day to day, but it is critical, for it is the designers who will be responsible for the rapid development and deployment of dashboards to monitor KPIs, and hence for critical business information.

When evaluating solutions, consider the following:

- What skill sets do the dashboard designers require?
- Is training necessary, and if so, how much is required? A rule of thumb is that that no more than a few days' training should be needed for staff to become effective dashboard designers.
- Do the solution's dashboard design interfaces permit development of simple dashboards to start, with increasing complexity added as designers increase their expertise and decision makers develop their use of metrics to run the business?
- Does the solution allow users and designers to select dashboard look and feel, including branding the dashboards with your organization's logo and colors?

Deployment

Look for templates that facilitate rapid deployment without restricting future development.

Deployment brings together a host of other criteria that should be examined, and is often the litmus test of these: technology requirements, availability, security, and usability. (See Figure 6.92.)

TeamQuest Accelerators offer rapid deployment of dashboards for specific purposes.

Generally, the more effective the dashboards that can be designed, and the more rapidly that the metrics management solution can be deployed, the better the solution because its effect is immediate. Often, a proactive metrics management solution is brought into a new area of the business to facilitate improved decision making right from the start, or into an area that is struggling in order to effect profound changes. In either case, getting the solution up and running and the metrics dashboards to decision makers quickly is of primary importance. However, rapid deployment in itself is not sufficient reason to select a solution. Beware of instant solutions that limit the ability to grow and improve how metrics are monitored and managed in the future.

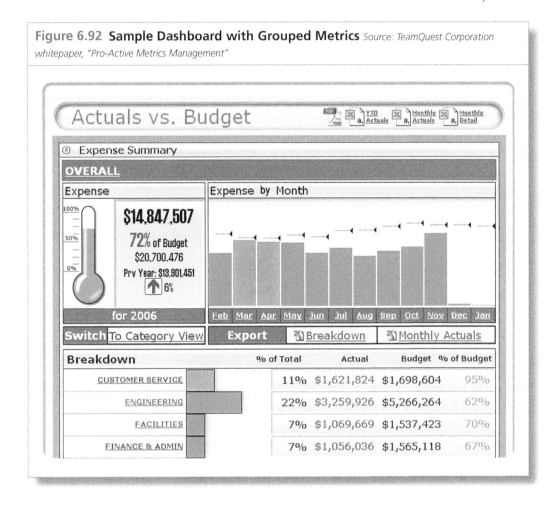

Figure 6.92 Sample Dashboard with Grouped Metrics *Source: TeamQuest Corporation whitepaper, "Pro-Active Metrics Management"*

Power and Flexibility

Look for the ability to deliver the metrics management capabilities that meet the needs of *your* business environment. The power and flexibility of a metrics management solution encompasses a wide range of capabilities. Questions to ask when evaluating solutions include:

- Can the dashboard pull data from diverse databases and technologies? If the proposed solution cannot do this, if it is limited to one or two proprietary database technologies, go no further.
- Can dashboard designers define the polling periods for queries, and can these be dynamically updated if required?
- Can metrics be based on aggregate queries; that is, can the metrics management solution pull together different types of data from different sources and make sense of it? For example, can it use one type of data for targets, another for progress, and yet another to show trends?

- Can the dashboards maintain histories based on user defined parameters?
- Does the solution offer different dashboard views for different users?
- Can metrics shown on dashboards be filtered, and can this filtering be configured by the dashboard designer?
- Do the dashboards manage thresholds, and what sort of actions can these thresholds trigger: emails, scripts, queries, etc.?
- Are metrics presented in context? Does the solution present numbers, or does it present meaningful information: quantities or percentages measured against targets, comparison of quantities for different periods, trends, etc.?

Context includes the date and time the metric was updated. In many cases, a metric has value only if the user knows when the data was retrieved.

Parameters can be used to filter data so that users see only the information they need to see. In Figure 6.93, data is filtered by region. All users see the same dashboard, but users in Region East see data for their region while users in Region West see information for their region. Whatever the case, the dashboard designer should spend time understanding not only the data behind the indicator but also

Figure 6.93 How Parameters Can Be Used to Simplify Dashboard Design and Implementation, and Improve Usability *Source: TeamQuest Corporation white paper, "Pro-Active Metrics Management"*

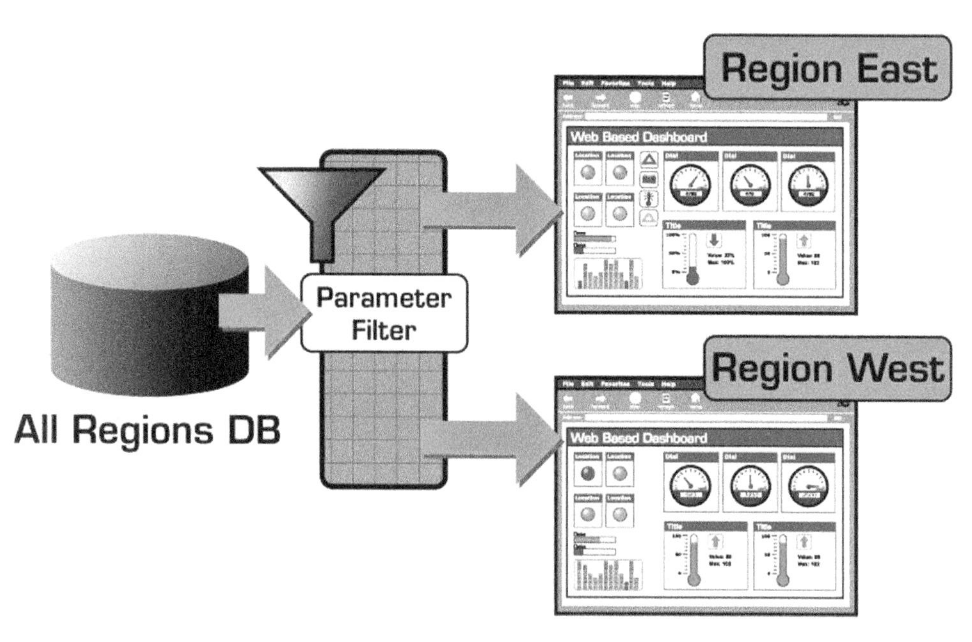

the data that creates the context for the indicator (such as revenue targets, time period coverage, maximum capacities, etc.). This rule should guide the dashboard designer through every step of design, from data gathering to detailed composition of displays.

The Future

Look for the capability to initiate and sustain positive transformations of the business.

It is wise to find out if the capabilities offered by a solution will restrict what can be done in the future. Does the solution lend itself easily to additions and improvements, both in the way data is collected for metrics and to the dashboards that users see?

Deployment of a truly proactive metrics management solution will almost immediately provide insights into how business decisions are made and will reveal opportunities for improving the decision-making process, for managing the metrics used in this process—and ultimately for transforming the business.

As shown in Figure 6.94, thresholds can be used with alerts to transform dashboards from passive monitoring devices into active vehicles, inducing corrective and, especially, preventive decisions and actions.

Conclusion

Managers and technical experts looking for a metrics management solution face a growing number of options.

When evaluating solutions for metrics management, prospective buyers should arm themselves with knowledge about what these

Figure 6.94 Simple Alert Triggered by a Threshold *Source: TeamQuest Corporation white paper, "Pro-Active Metrics Management"*

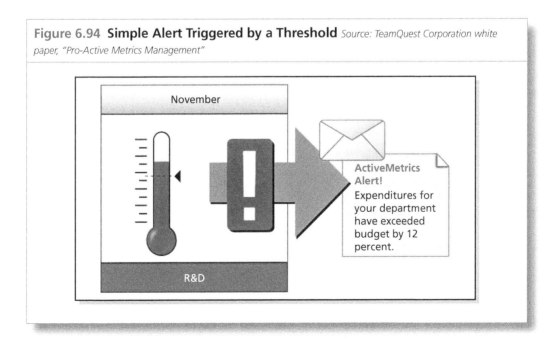

types of solutions can and cannot do, and with a list of evaluation criteria. And, finally, before buying they should ask two questions that encompass the others:

1. Will this solution simply record what happened in the past, or does it reveal what is going on now and, based on current trends, what is likely to happen in the future?
2. Is the solution truly proactive? Does it include displays, thresholds, and triggers not just to monitor what is going on, but also and critically to automatically initiate appropriate actions, and to facilitate reasoned, informed decisions?

6.21 A SIMPLE TEMPLATE

Table 6.7 shows a very simple template that can be used in the creation of a metric/KPI library. Rather than call it a KPI library, it may be referred to as a metric/KPI library because not all metrics are KPIs and KPIs on one project may serve as a simple metric on another project. Also, in the template there may be a reference to various parts of the *PMBOK" Guide*.[8]

The real purpose of the template is to record which metrics have been used successfully. We know that humans absorb information in a variety of ways and the metrics must always undergo continuous improvement efforts. Metrics that work well for one company may not work equally well for another company. The images, colors, and shading techniques may also have to be changed.

Organizations must be cautious not to go overboard in the number of metrics that will appear in the library. Too many metrics may be overwhelming to the users. Rad and Levin provide excellent checklists for assessing what may or may not be important as a metric on a given project or a stream of projects.[9] Items considered as critical for a given project or project stream may mandate that measurable metrics be defined.

6.22 SUMMARY OF DASHBOARD DESIGN REQUIREMENTS

We can now summarize the design requirements for dashboards using the rules for design.

---

8 PMBOK is a registered mark of the Project Management Institute, Inc.

9 Parviz Rad and Ginger Levin, *Metrics for Project Management* (Vienna, VA: Management concepts, 2006).

The information in Section 6.20 has been graciously provided by Hubert Lee, the founder of Dashboardspy.com. Hubert Lee is the founder of Dashboardspy.com. Known in BI circles as "The Dashboard Spy," Hubert Lee is an expert at business dashboard design and user experience. His collection of dashboard examples is among the largest in the world and can be viewed at http://dashboardspy.com.

TABLE 6.7   KPI Template

Description:_____
_____
_____

KPI Advantages:_____
_____

KPI Limitations:_____
_____

KPI Sponsor:_____
KPI Owner:_____

PMBOK® Guide Domain Area:

☐	Initiating
☐	Planning
☐	Executing
☐	Monitoring and Controlling
☐	Closing
☐	Professional Responsibility

PMBOK® Guide Area of Knowledge:

☐	Integration Mgt.
☐	Scope Mgt.
☐	Time Mgt.
☐	Cost Mgt.
☐	Quality Mgt.
☐	Human Resource Mgt.
☐	Communication Mgt.
☐	Risk Mgt.
☐	Quality Mgt.
☐	Stakeholder Mgt.

Relationship to CSF:_____
Objective/Target Limits:_____
KPI Start Date:_____

KPI End Date:_____

KPI Life Span:_____

Reporting Periodicity:_____

Graphic Display:

☐	Area Chart: Clustered	☐	Gauges
☐	Area Chart: Stacked	☐	Grids
☐	Area Chart: 100% Stacked	☐	Icons
☐	Bar Chart: Clustered	☐	Line Chart: Clustered
☐	Bar Chart: Stacked	☐	Line Chart: Stacked
☐	Bar Chart: 100% Stacked	☐	Line Chart: 100% Stacked
☐	Bubble Chart	☐	Pie Charts
☐	Column Chart: Clustered	☐	Radar Chart
☐	Column Chart: Stacked	☐	Table
☐	Column Chart: 100% Stacked	☐	Other:_____

Measurement Method:

☐ Calibration

☐ Confidence Limits

☐ Decision Models

☐ Decomposition Techniques

☐ Human Judgment

☐ Ordinal/Nominal Tables

☐ Ranges/Sets of Values

☐ Sampling Techniques

☐ Simulation_____

☐ Statistics_____

☐ Other:_____

The Importance of Design to Information Dashboards

The success of a project management dashboard (or any information dashboard for that matter) lies in the adoption and use of the dashboard by the end user as a truly helpful tool. Successful dashboards are those used every day to inform, manage, and optimize business and project processes. Failed dashboards are those that lie unused and forgotten, or provide no meaningful information.

So, the big question is this: On your dashboard, what makes the difference between success and failure? How will you ensure successful adoption of your dashboard by your users?

We all know of spreadsheets that have been derided and cast aside. They may contain all the data elements you need to monitor (and then some!), yet these spreadsheets are just not used as much as they should. Why is that? And why is it that if you pull the key metrics out of the same spreadsheets, lay them out on a properly designed dashboard, you suddenly have the attention of the entire user community? We'll examine several factors critical to the design of your dashboard.

The Rules for Color Usage on Your Dashboard

A truly effective dashboard makes good use of color to display information in an easily understood manner. Color theory and the cognitive effects of color are subjects close to the hearts of visual artists but seldom appreciated by creators of dashboards and other user interfaces. We must always be careful as the poor or careless use of color can mangle the real message of the data.

No discussion of color would be complete without a quick word on some fundamental warnings about the use of color in your dashboard projects. It is estimated that up to 8–12% of the male population suffers from some form of color blindness. Take a look at this series of graphics in Figure 6.95 that show how the colors of the rainbow appear to people with various forms of color blindness.

Note how similar yellow and green can be to some color-blind people and also how similar red and green can be to other color-blind people. Now, isn't that eye-opening?

The other warning about colors has to do with black-and-white versus color printers. As digital formats replace hard copy print quality is less of a concern than it once was, but some users may still print a paper copy to be studied later. Given the general decline in the importance of print in the workplace (not to mention the expense of color printing where print is still used), any printed copy is likely to be in grayscale. So for the time being at least, the effect of grayscale conversion must still be taken into consideration.

So, what do we do about these basic color challenges? Do these limitations mean we should not use color on our dashboards? Of course not. It does mean, however, that we must always use color in

Figure 6.95 Rainbow Colors and Their Perception *Source: Based on Wikipedia contributors.*

Note that the guidelines just provided refer to the sections of your dashboards that contain data. The rest of the dashboard (page background, navigation elements, branding areas, headers, and footers) may be designed to accommodate your usual graphic look and feel

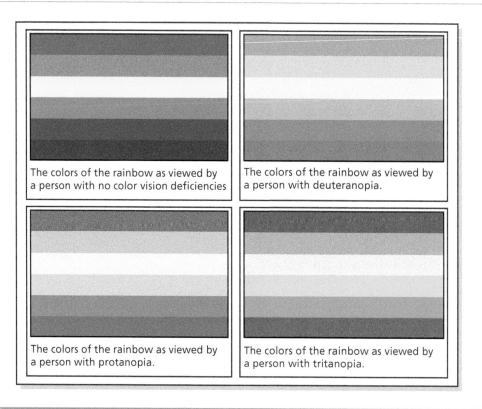

The colors of the rainbow as viewed by a person with no color vision deficiencies

The colors of the rainbow as viewed by a person with deuteranopia.

The colors of the rainbow as viewed by a person with protanopia.

The colors of the rainbow as viewed by a person with tritanopia.

conjunction with text labels. By that I mean explicitly including the relevant label right on or next to the graphic. For example, if you are showing a red/green/yellow status indicator graphic, put the text value right next to the graphic.

With that basic warning out of the way, let's look at some critical rules to follow concerning the use of color in your dashboards:

- Be aware of the background color of tables and graphs. Use a background color that contrasts sufficiently with the foreground objects. Also, use background color to group and unify different objects.
- Components of graphs and tables that are "nondata," that is, structural elements, should not call attention to themselves. Conversely, the data-centric elements of charts should be highlighted with color.

- When displaying a sequential or related group of metrics, use a small set of related hues and vary the intensity to correlate with the increasing data values if possible.
- Try to make the color usage meaningful. Use different colors to show different meanings.
- Use color consciously—check with yourself that each color (or the particular usage of colors) is meaningful.

The Rules for Graphic Design of Your Dashboard

The level of design appropriate to a dashboard (or any information visualization) is an age-old debate. Good information visualization practice and BI user interface design calls for effective use of screen real estate with clear and easily understood charts and visually simple charts such as sparklines and bullet graphs. Easy to read and interpret, these charts, combined with a monotone color scheme (so as to focus attention when needed with red/green/yellow alert indicators) have become a "best practice" design for data-heavy BI dashboards.

With its sparseness of color and avoidance of heavy design elements, this approach allows the user to concentrate on the information aspects. However, as dashboard project members often find when embarking on the design phase of the project, such a "sparse" interface often runs into the objections of the "eye-candy people"—often (but not always!) including the project sponsor, who is looking to impress users and peers with visual splash.

Many people want the "sizzle" of gaily colored gauges, complete with 3-D effects, gradients, and big splashy presentation. Maybe it "feels" like more of a user experience when you get the "wow" factor in there!

From a development standpoint, libraries of visual components make it very simple to throw all sorts of charts, gauges, dials, and widgets onto a dashboard. It makes it too easy to get carried away by the visual excitement. It becomes the responsibility of the dashboard designer to strike the proper balance between visual excitement and information visualization best practice.

One proven formula is to limit the efforts of the eager graphic designer to clearly defined areas. The following list is usually within the purview of the graphic designer.

Overall Layout of the Website or Software Application

- Header banner
- Logo
- Portlets (widgets)
- Title bars
- Tables
- Sidebars
- Navigation elements
- Footer

Reserve the design of the charts and other data visualization elements for the information visualization expert.

Other tips for the graphic design of your dashboard:

- Avoid the cliché of using a steering wheel—don't take the dashboard metaphor literally! An automobile's steering wheel has no place on your information dashboard.
- Bring a design professional into your team. It is uncommon to find design professionals already on your staff. Hire a proven consultant with a track record of designing dashboards. Find a visual designer with expertise in the layout and design of software applications and pair him/her with an expert information visualization professional. The look and feel of your dashboard will reflect the investment.
- Become aware of current trends in the visual design of software applications. Understand why some looks are considered "modern" and "cutting-edge" and adopt those practices to give your design a longer life.

The Rules for Placing the Dashboard in Front of Your Users—The Key to User Adoption

The most successful dashboards become invaluable tools that users rely on to facilitate their workflows, remind them of what they have to concentrate on, and guide them through metrics and KPI relevant to their changing situations.

To get your dashboard successfully adapted to such a level requires you to carefully think about how the users engage with your system in the first place. From where and exactly how do the users launch your dashboard? Typically, a dashboard is launched by typing the URL of the application into the address bar of a browser. While a common approach, more proactive steps can be taken that can lead to increased adoption by the user community. Here are some ideas:

- Identify launch points within existing company portals. Place branding elements (banner/logo) and provide a "sign in" button right in the sidebar of other applications or web sites.
- Speak with other application owners and insert launch points for your dashboard right into the other application. The idea is to place it at the appropriate portion of the user workflow.
- Consider using the "desktop widget" approach. Have you seen those mini-applications that your operating system displays on your desktop? Some dock on the side of the screen and pop out when you mouse over them. Others reside in full view and allow you to flip them over for more information. Create a widget for your dashboard and have it live on the desktop of your PC. Now all your users have to do is to log onto their PC and they have instant access to their dashboard. No logging in to a separate

application or website. This approach is particularly good for delivering messages and alerts to the users.

■ Give your dashboard email capability. Send users alerts and updates by email with links that they can click on to go right into their dashboards.

The goal is to make your dashboard, and the information that it delivers, absolutely unavoidable by the user. Don't be afraid to put your dashboard right in the face of your users. After all, that's what is being done by the automobile dashboard and the airplane cockpit. The information and the controls are literally in front of the users' faces.

The Rules for Accuracy of Information on Your Dashboard

This rule is easy to understand and absolutely inviolate: Your dashboard must never be seen as the cause of any problems with the data. The accuracy of the data on your dashboard can be questioned (unfortunately, that's sometimes the nature of data)—*but* the dashboard itself must be seen as a trusted reporting mechanism. If there is a problem with the data, the user should blame the team responsible for producing the data, but not the dashboard itself.

If there is a known issue (perhaps a technical problem or a stale data problem), the dashboard should display a warning message. If the dashboard does not show live, real-time data but snapshots captured from certain points in time, the tables and charts should be labeled to indicate that.

You must work hard at instilling faith and trust in your dashboard users. Transparency in process, and clear labeling and directions must be used in the dashboard design process for this to happen.

6.23 DASHBOARD LIMITATIONS

Dashboards are designed first and foremost with the viewer in mind. The intent of the project management dashboard is to convert data into a meaningful representation of the project's performance. But how does the viewer intend to use the information presented?

Viewers and stakeholders who are expected to make decisions in support of the project want sufficient data such that they can make "informed" decisions in a timely manner rather than just guesses. Having the correct information is critical. In this case, the dashboard may have to contain detailed information and the viewer may need to examine the information carefully. The amount of information that can be displayed accurately on a dashboard is limited due to space and readability requirements.

Figure 6.96 Simple Dashboard Icons

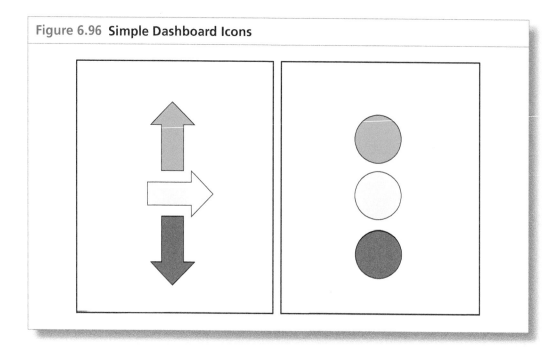

However, some stakeholders and viewers who are passively involved in the project and are not expected to make decisions are usually happy with a one-minute or at-a-glance dashboard. This type of dashboard has limitations on both the information presented and its intended use. The one-minute dashboard can be displayed using the three directional icons shown in Figure 6.96. In the figure, a green arrow always points upward and indicates a favorable condition. For example, if the one-minute dashboard is used to display how well we are managing the project according to the project's constraints, then the green arrow would indicate that we are well within the constraint's limits. A red arrow, which always points downward, indicates that we have an unfavorable condition and have exceeded the limits of the constraint. A yellow arrow might indicate that we are within but close to the threshold limits or tolerances for the constraint. Some people prefer to use the "traffic light" or circular icons rather than the arrows.

In Figure 6.97, we have a project that we assume has only five constraints. For each of the five constraints, we have identified an icon that shows how well we are performing according to the predetermined constraints. This type of one-minute dashboard provides only a cursory picture of the project's health and has limited value for decision making. If the viewers want additional information, they may need to use drill-down buttons for more detail. Dashboards cannot display all of the information needed in most circumstances.

Sometimes good intentions go astray. One company created both one-minute dashboards and detailed dashboards for its viewers.

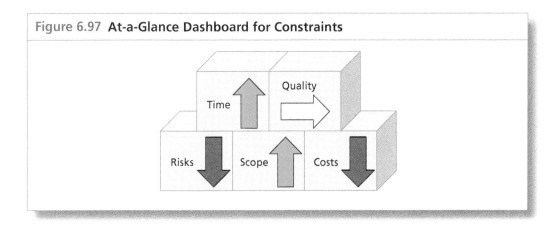

Figure 6.97 At-a-Glance Dashboard for Constraints

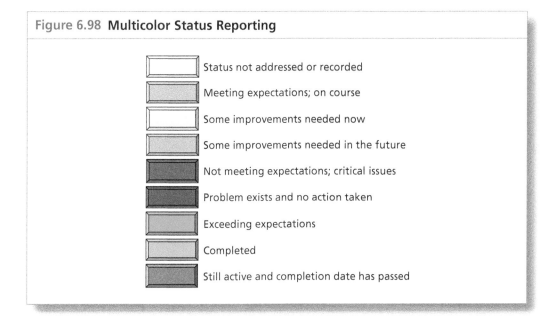

Figure 6.98 Multicolor Status Reporting

Status not addressed or recorded

Meeting expectations; on course

Some improvements needed now

Some improvements needed in the future

Not meeting expectations; critical issues

Problem exists and no action taken

Exceeding expectations

Completed

Still active and completion date has passed

In the one-minute dashboard, developers decided to go to nine colors, as shown in Figure 6.98, rather than just the three standard colors that had been used previously. This made the dashboard significantly more complex because there could be more than one color associated with each activity. For example, an activity could be both red and purple, indicating that the activity has some unresolved critical issues and is still active. It then took longer to read the dashboard because viewers had to continuously review the meaning of the colors.

Some of the viewers wanted to see the direction of the variances, specifically time and cost, and the company then added into the dashboard the color scheme shown in Figure 6.99. Although there is

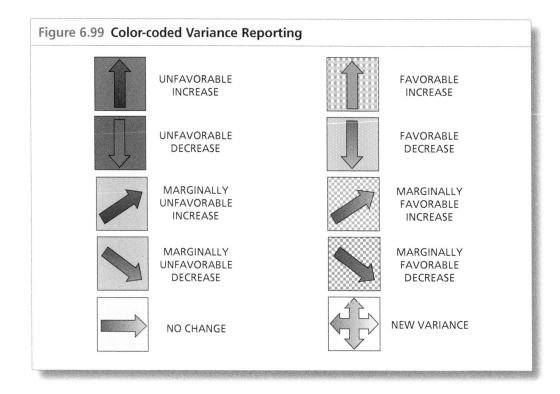

Figure 6.99 Color-coded Variance Reporting

some merit in using the style shown in this figure, it is inappropriate for the one-minute dashboard. Eventually, the one-minute dashboard was removed and replaced with detailed dashboards that displayed a rainbow of colors and gradients.

6.24 THE DASHBOARD PILOT RUN

Although the project manager, the project team, and the dashboard designer have a reasonably good understanding of the information that appears in the dashboard, the same cannot be said for the viewers of the dashboard. If possible, given the project's schedule and the size of the project, setting up a dashboard pilot run may be beneficial. It verifies that the client and the stakeholders:

- Understand what they are viewing
- Arrive at the correct conclusions
- Have faith in the dashboard concept
- Are willing to use the information for decision making

During the first few weeks of the project, it may be necessary for the project manager and the dashboard's viewers to work together closely to make sure that all of the dashboard information is clearly understood. A great deal of time and money can be wasted if the viewers lose faith in the dashboard concept. Regaining lost faith may be impossible.

6.25 EVALUATING DASHBOARD VENDORS

With numerous companies in the marketplace selling dashboard services, it is important to what how to evaluate both the dashboard features and the accompanying software. Shadan Malik has identified 10 categories that must be considered when evaluating dashboard vendors:

- End user experience
- User management
- Drill-down
- Reporting
- Data connectivity
- Alerts
- Visualization
- Collaboration
- System requirements
- Image capturing and printout[11]

Malik has also identified rules for good dashboard software.[12] It is worth noting that dashboards software must also meet the standards of any good software, which includes the following:

- **Fast response:** Users should not experience an inordinate delay in retrieving their dashboards and associated reports.
- **Intuitive:** End users need not be required to go through a big learning curve or mandatory training.
- **Web based:** Users should be able to access the dashboards through the web, if they have proper access rights.
- **Secure:** System administrators may administer software security easily to reduce and track wrongful access. The software must also provide (when necessary) data encryption to secure sensitive data transmission across the web.
- **Scalable:** A large number of users may access the software without crashing the system or causing it to slow down below an acceptable performance benchmark. The quality assumes a reasonable hardware and network bandwidth.
- **Industry compliant:** The software should integrate with the standard databases of different vendors and work with different server standards (e.g., .Net, J2EE) and various operating systems (e.g., Unix, Windows, Linux).
- **Open technology:** The software should not have proprietary standards that would make it difficult or impossible to extend its reach within a complex IT environment.
- **Supportable:** It should be easy to manage a large deployment within the existing IT staff with limited training on dashboard

---

11 Shadan Malik, *Enterprise Dashboards* (Hoboken, NJ: John Wiley & Sons, 2005), p. 147.
12 Ibid., pp. 9–10.

software. In other words, the software should not be so complex that it requires long-term contracting or the hiring of another expert to support its deployment, assuming that the organization has a reasonably qualified IT staff.

■ **Cost effective:** The total cost of ownership should be well below the monetary benefit it provides to justify a strong ROI. Therefore, the licensing cost, implementation cost, and support cost should be within a range that provides strong ROI and organizational benefits after deployment.

The main cost factors that should be considered for a successful dashboard solution are:

■ Software cost
■ Annual support cost
■ Additional hardware cost
■ Initial deployment cost
■ User training cost
■ Ongoing support personnel cost[13]

Software purchases require and understanding of the cost savings and business opportunities. This is shown in Table 6.8.

6.26 NEW DASHBOARD APPLICATIONS

Because dashboards can be updated in real time, new dashboard applications have emerged. Companies are using dashboards as part of capacity planning optimization analyses. This is similar to what-if scenarios used in project scheduling.

TABLE 6.8   Potential Cost Savings and Business Opportunities

COST SAVINGS	ADDITIONAL BUSINESS OPPORTUNITY
• Reduction or elimination of efforts for consolidating disparate reports	• Better decision making with more current or live information
• Reduction of time wasted in reviewing overwhelming amount of data and reports	• Better business insight because of improved data visibility through enhanced visualization
• Reduction of time spent coordinating and monitoring complex processes	• Proactive and timely decision making with exception management and alerts
• Reduction of effort enforcing regulatory compliances	• Greater democratization of information, empowering the front line in the organization
• Elimination of redundancies within the organization for processing of similar data	• Better customer service and enhanced value delivered to customers and/or vendors

Source: *Shadan Malik*, Enterprise Dashboards *(Hoboken, NJ: John Wiley & Sons, 2005), p. 172.*

---

13 Ibid., p. 168.

With optimization dashboards, a company can look at various ways of assigning resources to either a single project or a multitude of projects. For each change in the way that the resources are allocated, the dashboard will display the impact on the schedules of each project and possibly profitability. These applications are now being used as part of capacity-planning analyses during the portfolio project selection and project prioritization processes. Unfortunately, optimization dashboards focus on the higher levels of the work breakdown structures. As technology advances, these techniques may become commonplace for functional managers as well to perform functional resources optimization.

7
DASHBOARD APPLICATIONS

CHAPTER OVERVIEW

Not all dashboards have the same intended use. Some dashboards are for internal use only, whereas others serve as a means to communicate with customers. A dashboard's use cannot be discerned by looking at it. This chapter contains dashboards from several companies that have recognized the benefits of their usage.

CHAPTER OBJECTIVES

- To understand the various uses of dashboards
- To identify the type of material in each dashboard
- To understand the application of the dashboards

KEY WORDS

- Dashboard design
- Dashboard use
- Dashboard contents

7.0 INTRODUCTION

In recent years, there has been a rapid growth in the number of companies developing expertise in dashboard design techniques and graphical displays of business intelligence. These companies often have the knowledge and capability to construct effective dashboards for clients through a much less costly approach than if the company had to do it by itself and develop its own expertise. While these companies understand the necessity to custom-design the dashboards to fit the needs of the client, they also try to prevent a "reinvention of the wheel" by first looking at what dashboards worked well for other clients and seeing if any or all of the information in or design of the dashboard can be applied to new clients. The same holds true for the metrics used in the dashboards.

Sometimes the same metric can be represented differently on two different dashboards, especially if the information is for different clients or a different audience within the same company.

The remaining sections of the chapter include examples of dashboards that have been graciously provided by various companies.

Project Management Metrics, KPIs, and Dashboards: A Guide to Measuring and Monitoring Project Performance, Fourth Edition. Harold Kerzner.
© 2023 John Wiley & Sons, Inc. Published 2023 by John Wiley & Sons, Inc.
Companion Website: www.wiley.com/go/kerznermetrics4e

7.1 DASHBOARDS IN ACTION: DUNDAS DATA VISUALIZATION

Dundas Data Visualization is a leading, global provider of Business Intelligence (BI) and Data Visualization solutions. Dundas offers easy to use self-service, single BI experience allowing users to connect, interact and visualize powerful dashboards, reports and advanced data analytics for any data, on any device. Their flexible BI platform is fully supported by a consultative and best practice solutions approach. For over 20 years, Dundas has been helping organizations discover deeper insights faster, make better decisions and achieve greater success. Examples of Dundas dashboards are shown in Figures 7.1 through 7.11.

7.2 DASHBOARDS IN ACTION: PIE

Consider the following scenario, which appears to be happening in a multitude of companies. Senior management is actively involved in the selection of projects that will go into the portfolio. Once the projects are selected, however, management accesses one summary dashboard to review, but cannot easily find appropriate detailed information that could influence their decisions at the moment. Although the summary has merit, a drill-down ability for easy access to more critical information can make a big difference.

There are many project and portfolio management systems, but not very many that elegantly tie the executive dashboard down to the front-line team member execution the way PieMatrix's "Pie" solution does. Not only does the Pie software provide this drill-down capability, but it is done with a user-friendly look and feel that can be learned in minutes. Customers can turn complex projects into more manageable views that make it easy for executives and front-line people to make informed decisions in a timely manner.

Unlike many project systems that have complicated reporting or some that lack portfolio reporting, Pie provides a balance of simplicity with enough of the right high-level to drill down-level data to help executives know what's happening and how to make decisions. Also, Pie takes a process focus on project management. Pie's major focus is repeatable-type projects that can leverage a hybrid of the

Material in Section 7.1 has been provided by Dundas Data Visualization, Inc. © 2017 Dundas Data Visualization. All rights reserved.

Sections 7.2 and 7.3 have been graciously provided by Paul Dandurand, chief executive officer, PieMatrix Inc.; Office (802) 222-6094; Mobile 802-578-5653; paul. dandurand@piematrix.com; www.piematrix.com; Patent pending #12/258,637. Reproduced by permission of PieMatrix Inc. © 2021 PieMatrix Inc. All rights reserved.

Figure 7.1 Financial and Nonfinancial Dashboard Metrics

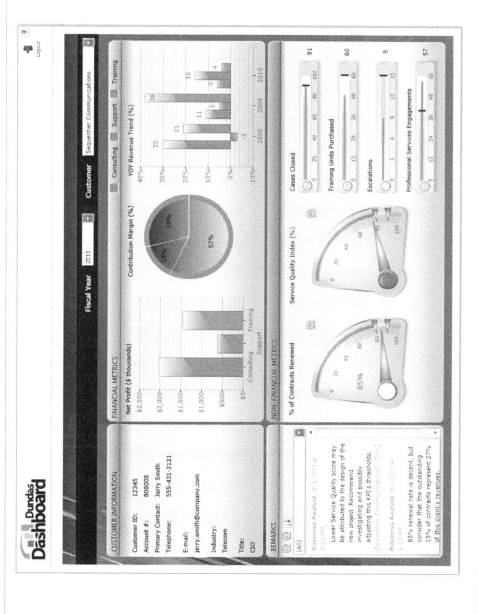

Figure 7.2 Overview Dashboard

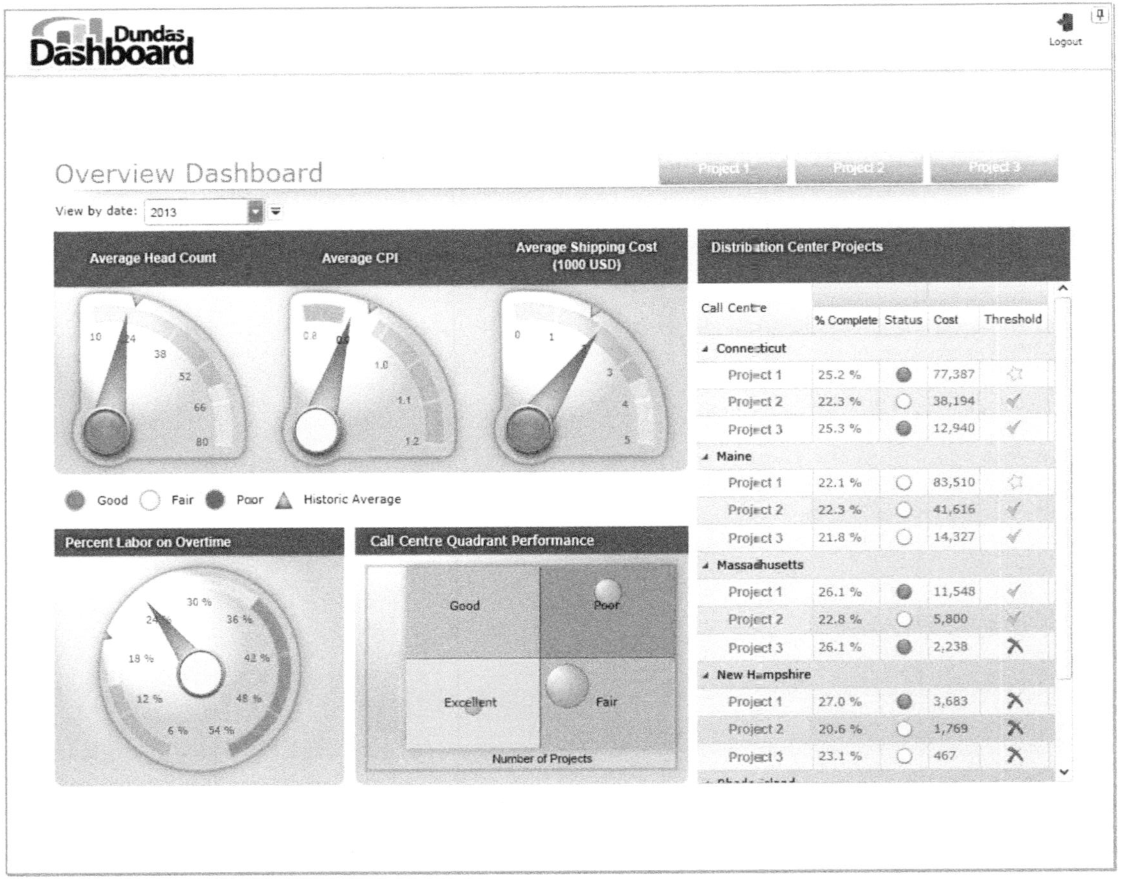

Figure 7.3 Executive Dashboard

Figure 7.4 Project Support Dashboard

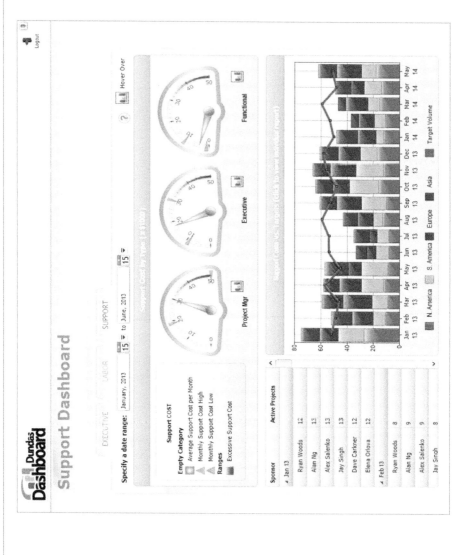

Figure 7.5 Business Intelligence Dashboard

Figure 7.6 IT Monitoring Dashboard

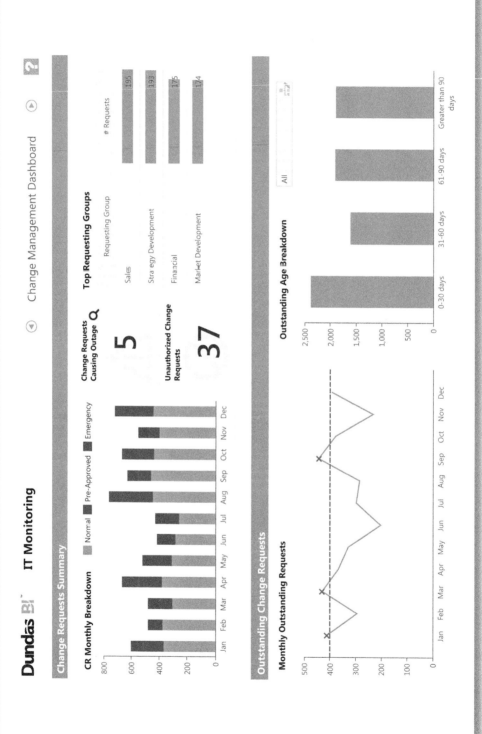

Figure 7.7 Wireless Dashboard

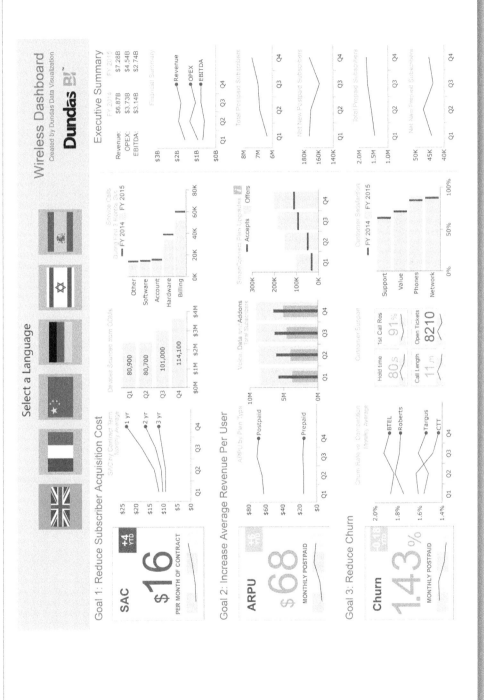

Figure 7.8 Hospital Performance Dashboard

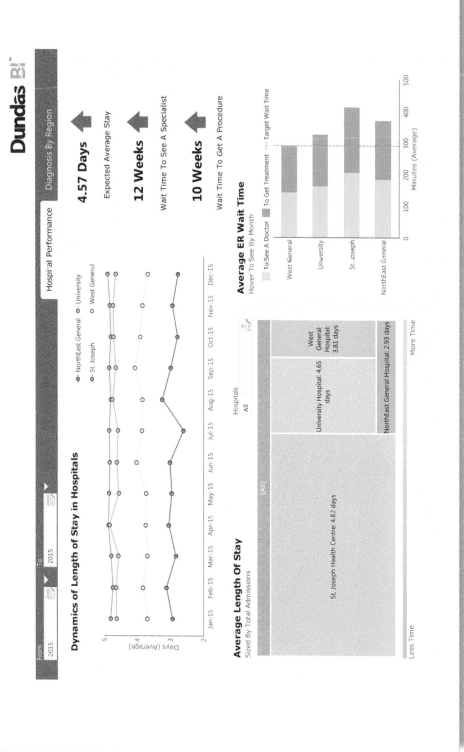

Figure 7.9 Business Intelligence Dashboard

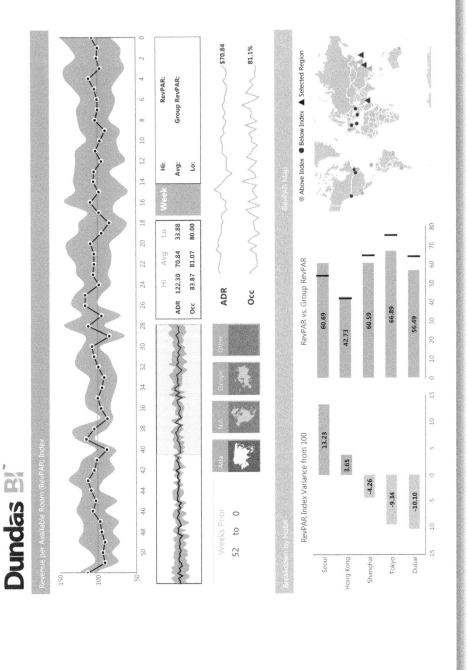

Figure 7.10 Insurance Call Center Dashboard

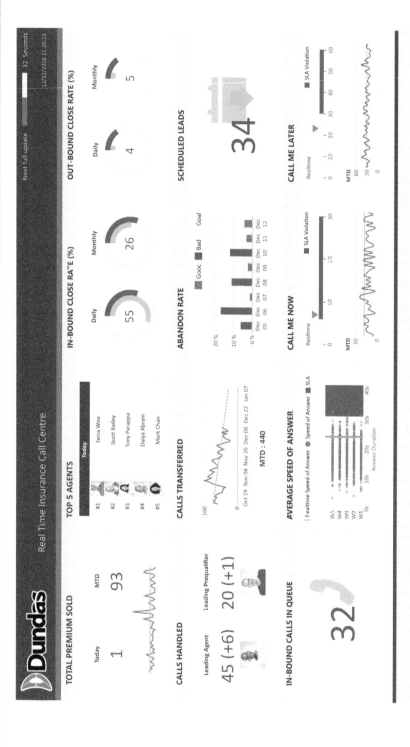

Figure 7.11 Business Intelligence Dashboard

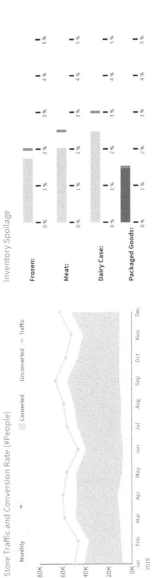

customer's flexible frameworks and methods (recipes) including agile approaches. This means that the data being displayed is not just data about how things are doing, but also how things are done consistently right. Predictability of success becomes more controlled, and project managers may have more confidence knowing that the execution is being done with the best selected frameworks designed for great project results.

7.3 PIE OVERVIEW

PieMatrix produces a visual project and process application called "Pie." Pie has three key differentiation points. One is a visual design made for fast user adoption. Second is its focus on enabling flexible process frameworks with how-to knowledge to help reduce risk and drive continuous improvement. Third is combining agile and waterfall methods into a hybrid approach within the same project.

Pie is made for any business initiative across industries. Pie customers range from government sectors to private industries such as consulting, healthcare, and new product development. The customer base includes small to medium sized organizations.

Pie is made for process owners, project managers, team members, stakeholders, and executives. Although it has many different functions for designing and executing projects, the following pages will focus mainly on the reporting visuals.

The following sections introduce the different reporting capabilities. All dashboard and report data are constantly updated in real time. In addition to these reports, you can export project data to CSV files which can then be imported into business analytics and business intelligent (BI) tools.

More information about Pie can be found at https://www.pie.me

Pie Dashboard

The Pie application is uniquely simple to use. The application's beautiful graphical design lends well with Pie's dashboards and reports. The value of being easy-to-use is that it will get used by executives, stakeholders, project managers, and team members. Pie stands uniquely in the market with simplicity, and some have called the Dashboard executive eye-candy.

Pie's Dashboard has three views.

- Charts (Figure 7.12)
- Table (Figure 7.13)
- Issues Log (Figure 7.14)

The Charts view (Figure 7.12) displays key indicator panels. The Portfolio Progress panel displays your top five projects, which you can define. This panel shows the project's status including the number

Figure 7.12 Dashboard Charts

Figure 7.13 Dashboard Table

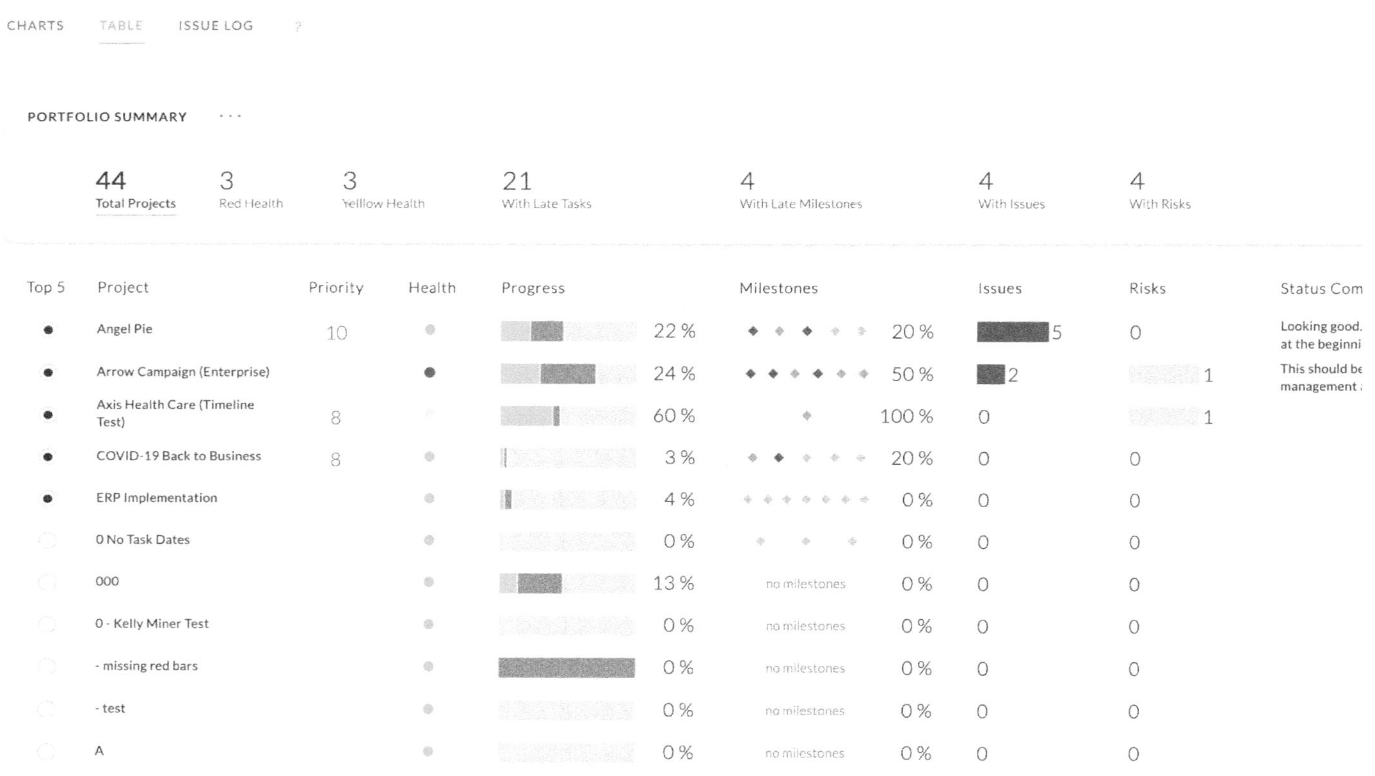

CHARTS
TABLE
ISSUE LOG
PORTFOLIO SUMMARY
44 Total Projects
3 Red Health
3 Yellow Health
21 With Late Tasks
4 With Late Milestones
4 With Issues
4 With Risks
Top 5
Project
Priority
Health
Progress
Milestones
Issues
Risks
Status Com
Angel Pie
10
22 %
20 %
5
0
Looking good.
at the beginni
Arrow Campaign (Enterprise)
24 %
50 %
2
1
This should be
management
Axis Health Care (Timeline Test)
8
60 %
100 %
0
1
COVID-19 Back to Business
8
3 %
20 %
0
0
ERP Implementation
4 %
0 %
0
0
0 No Task Dates
0 %
0 %
0
0
000
13 %
no milestones
0 %
0
0
0 - Kelly Miner Test
0 %
no milestones
0 %
0
0
- missing red bars
0 %
no milestones
0 %
0
0
- test
0 %
no milestones
0 %
0
0
A
0 %
no milestones
0 %
0
0

Figure 7.14 Dashboard Issues Log

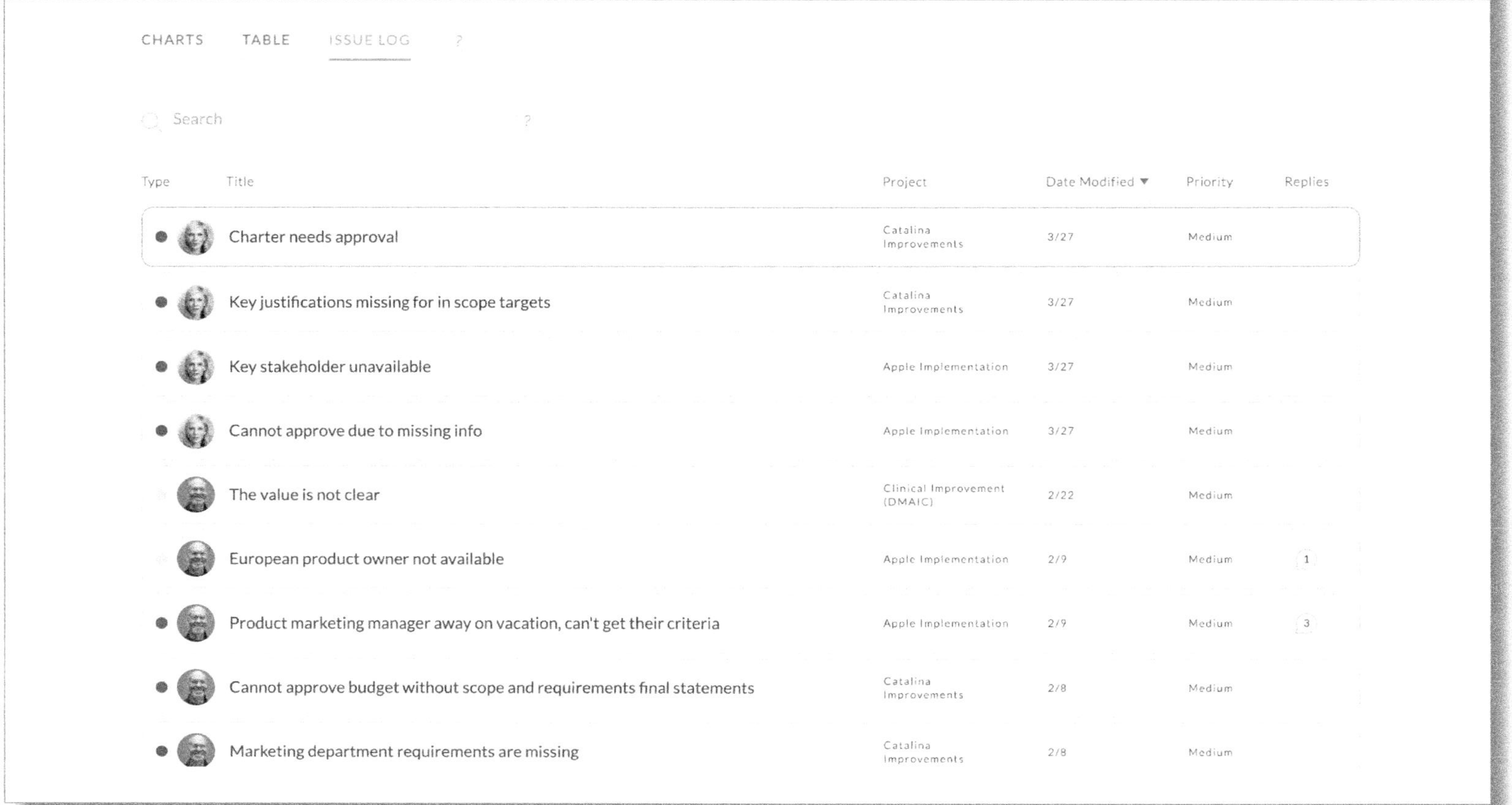

CHARTS   TABLE   ISSUE LOG   ?

Q Search   ?

Type	Title	Project	Date Modified ▼	Priority	Replies
●	Charter needs approval	Catalina Improvements	3/27	Medium	
●	Key justifications missing for in scope targets	Catalina Improvements	3/27	Medium	
●	Key stakeholder unavailable	Apple Implementation	3/27	Medium	
●	Cannot approve due to missing info	Apple Implementation	3/27	Medium	
●	The value is not clear	Clinical Improvement (DMAIC)	2/22	Medium	
●	European product owner not available	Apple Implementation	2/9	Medium	1
●	Product marketing manager away on vacation, can't get their criteria	Apple Implementation	2/9	Medium	3
●	Cannot approve budget without scope and requirements final statements	Catalina Improvements	2/8	Medium	
●	Marketing department requirements are missing	Catalina Improvements	2/8	Medium	

of tasks that are done, late, not done, and percentage done. The Portfolio Risks & Issues panel displays the count for the open issues and risks. The Portfolio Milestones panel displays the top five projects' milestones as interactive diamond icons. Hover over a milestone diamond to pop up its name, due date, and a link that jumps you to the task in the project view. The other panels include personal views for the individual team member to keep track of their own work and their open issues and risks. A project manager can select different team members from these panels to see how they are doing.

The Table tab view (Figure 7.13) displays a useful list of all projects the user has permission to see. This table can be sorted by column header and filtered by summary count headers. The Health column displays the overall health of a project, and the Comment column makes it easy for the executive to post comments to ask questions or provide support. You can export the table list as a PDF file for sharing outside of Pie.

The Issues Log tab view (Figure 7.14) displays open issue and risk posts across all projects the user has permission to see. Clicking on the issue (red dotted) or risk (yellow dotted) bar expands the item to show its details. The executive can easily post their own comments inside the bar to help with making decisions on solving the issue or risk. The list can be sorted and filtered in many ways.

Pie brings the dashboard to life with visualization. It also connects the executive with the project managers and teams with interactive collaboration features and drilldowns. The other report tabs as also display similar interactive benefits helping management and teams to collaborate for better project results.

Pie Portfolio Timeline

As the Dashboard provides simple visuals, the Portfolio Timeline report (Figure 7.15) is also designed for simplicity. This report displays the list of all projects the user has permission to see. The visualization is similar to a high-level Gantt view, but with a friendlier look and feel.

Each project has a bubble bar to its right showing the length of the project relative to the timeline dates. The green fill represents the amount of completed tasks. The top of the page has a zoom bar making it easy for the viewer to choose a timeline or to drag the zoom bar to a preferred timeline.

Hovering over a project name or bar reveals a popup with more information about that project including a link to open that project into its project's timeline page.

Pie Project Timeline

When the user drills down on a project from the Portfolio Timeline, Pie will automatically open the project into its Project Timeline view (Figure 7.16). The project example in Figure 7.16 has its project

Figure 7.15 Portfolio Timeline

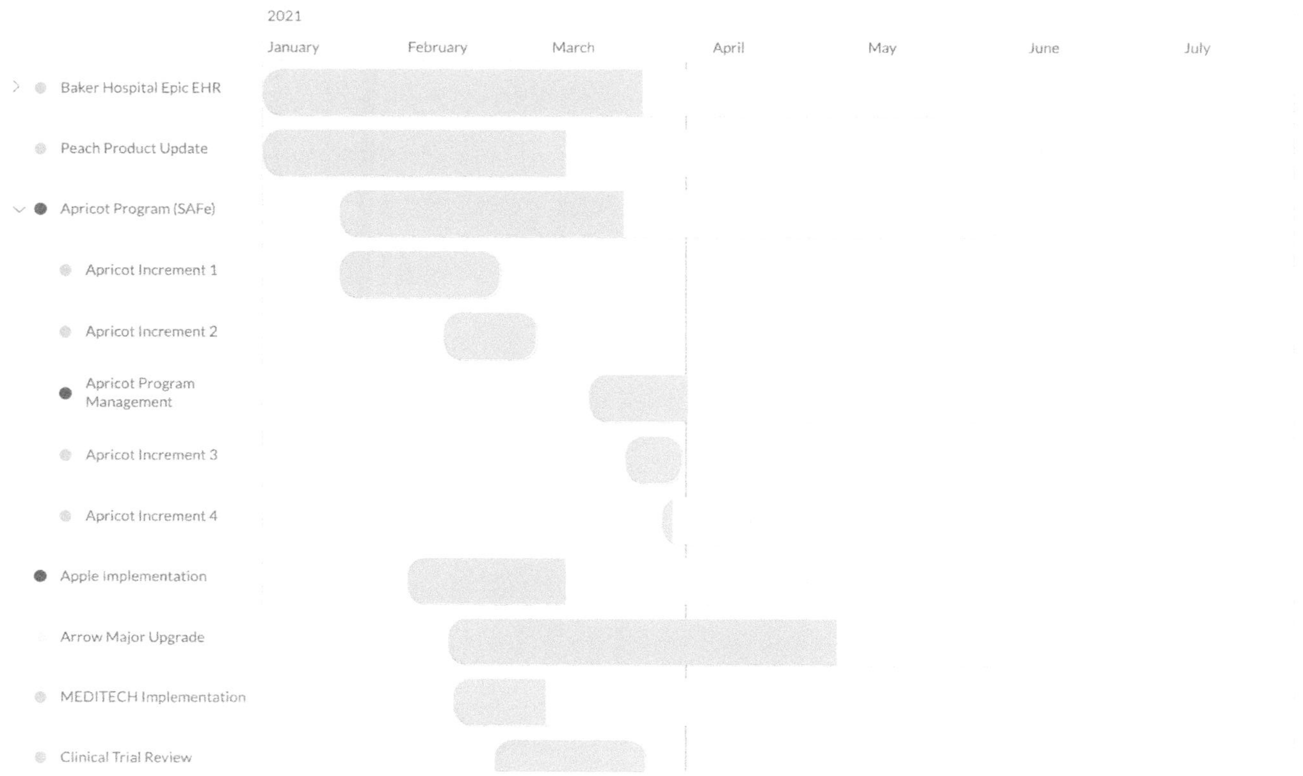

2021

	January	February	March	April	May	June	July

- Baker Hospital Epic EHR
- Peach Product Update
- Apricot Program (SAFe)
- Apricot Increment 1
- Apricot Increment 2
- Apricot Program Management
- Apricot Increment 3
- Apricot Increment 4
- Apple Implementation
- Arrow Major Upgrade
- MEDITECH Implementation
- Clinical Trial Review

Figure 7.16 Project Timeline

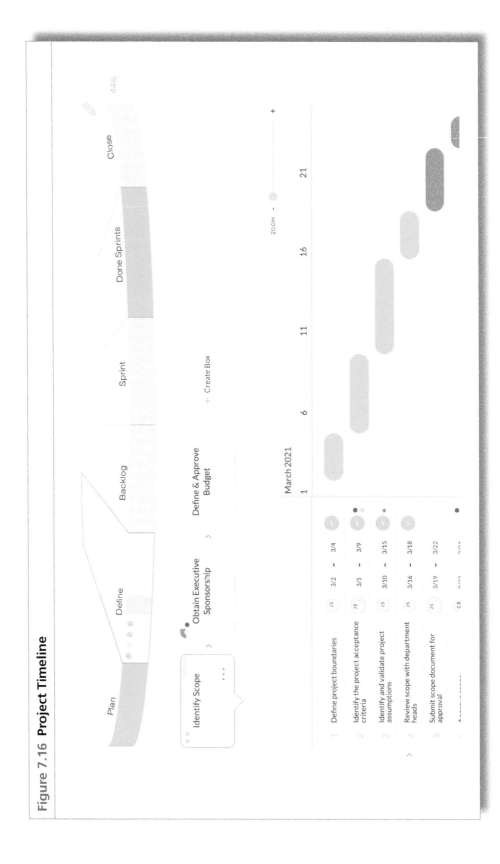

phase Define selected. This displays its content, which are three process boxes. The first box Identify Scope is selected to display its task list in a visual timeline (Gantt) format. The user can edit the tasks directly in this view with one or two clicks or drag and drop for re-ordering.

As the project manager works on their project, they can quickly tab over to this Project Timeline page for a friendly Gantt-like visual. They can view dates, progress, lateness, and assignments. The manager can easily filter this page to show the timeline of any of the project team members.

The Project Timeline and the associated Project Process views and are pages where team members focus on day-to-day work. In this example, the project has been created from a "recipe," which is like an advanced project template with all of the process and accountability content ready to go for the next project. This recipe project framework feature is created in a library of flexible frameworks where project managers can choose from a buffet of ready-to-execute recipes. See the Pie Recipes section below for more information.

Pie People Timeline (Resource Planning)

The Report section has a People Timeline view (Figure 7.17) that displays team members' workload across a timeline. This is a resource planning and forecasting report to help identify each person's workload based on task assignments and their durations on each project. This is a powerful report with real-time data and visualization. Like the other reports, you can filter the list based on workspaces and project tags.

The People Timeline as shown in Figure 7.17 shows the timeline bars with colors. The length of the bar displays the project's start date to end date span. Green color represents the individual's completed work, red is late, and blue is future work not yet started. A project manager or planning director can review future workloads to help with forecasting people resource needs. Each team member bar can be clicked to expand to see their assigned projects. Then you can click and expand their project bar to see what roles they are holding for that project and how the role assignments are aligned with the timeline.

This report includes a separate Roles tab to display roles assigned to projects and their timeline associations. Roles are like people titles, such as "project manager," "architect," and "business analyst." The Roles tab is similar to the People tab where you can view the past, current, and future workload forecast of roles. The color code works the same for the Roles view as they do for the People view. You can expand each role bar and see the assigned people to that role. Then if you expand its people bar, you can see their assigned projects.

Figure 7.17 Project Timeline

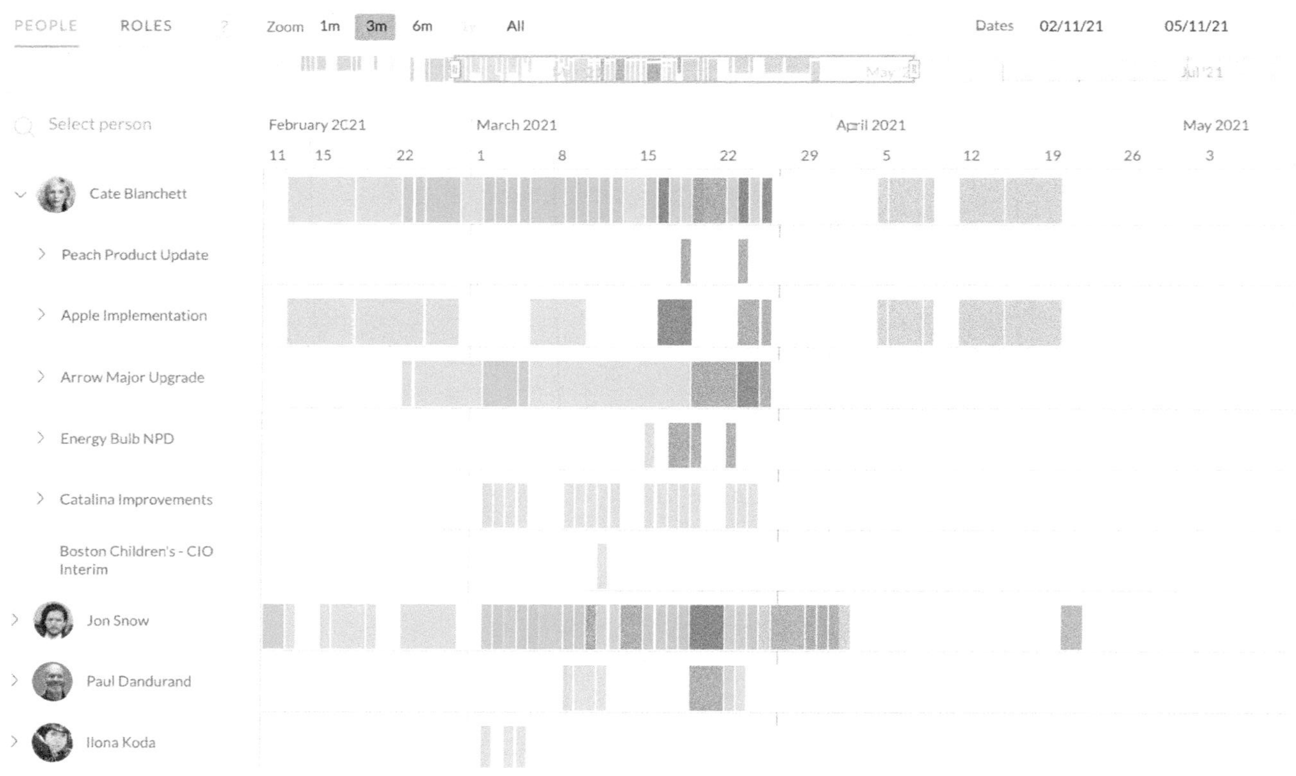

Pie Project List

The Project List page (Figure 7.18) is the portfolio view of active projects the user has permissions to see. The unique feature of this page is the visualization of projects and their phases as pie "slices." A manager or executive can quickly understand the progress of projects at each project phase. This page is also the main page for all listed projects, where you would create a new project or open an existing project to get work done.

The reporting features of this view include the visual colors of green for done and red for late. In the Figure 7.18 example, you can see diamond-shaped icons indicating milestones. You can simply switch the icons to display issues and risk icons. These icons provide mouse-hover popups for more information and quick drilldown capabilities.

The Project List page can also display complex project programs with sub-projects. This feature is called "stacking" projects (Figure 7.18). A value of stacking projects is to see a visual progress of all sub-projects under a project program. The stack feature is also available in the Report Portfolio Timeline view. The example Figure 7.19 shows a stack expanded showing you its child projects and their progress.

Pie on My Plate

Pie users don't need to open a project to see this week's work. Instead, they can go to their On My Plate page. This allows them to see their assigned workload across all of their projects for the next seven days including what's late. They can manage and edit their assigned tasks from this page, such as updating progress, adding posts, and dragging or snoozing the work to another day as needed. They can even add private personal to-dos.

For project managers, they can turn the On My Plate page to the On My Team's Plate page (Figure 7.20). This view will display all tasks their teams are working on for the week. The value is to get a sense of what the load is for the week and what are the tasks that are running late. Project managers will often bring up this report during team meetings. As they review and discuss the week's workload, they can update the tasks in real time with comments, changed dates, or any other updates.

The On My Plate section in Pie also has a Timesheet page (Figure 7.21) where team members can report their time. They can report hours worked either at the project level or task level. Pie was designed to make time tracking simple and visual. The example in Figure 7.21 displays a team member's page with hours allocated on different days. The clean visual display makes it easy for the team member to see the level of work on each day. They can also add keyword tags to each timesheet line item. Tags can be added to show what time entries are for billable or non-billable work. They can even track vacation days. Project managers can export the data into a CSV or Excel file for reporting or for exporting into other time tracking tools.

Figure 7.18 Project List

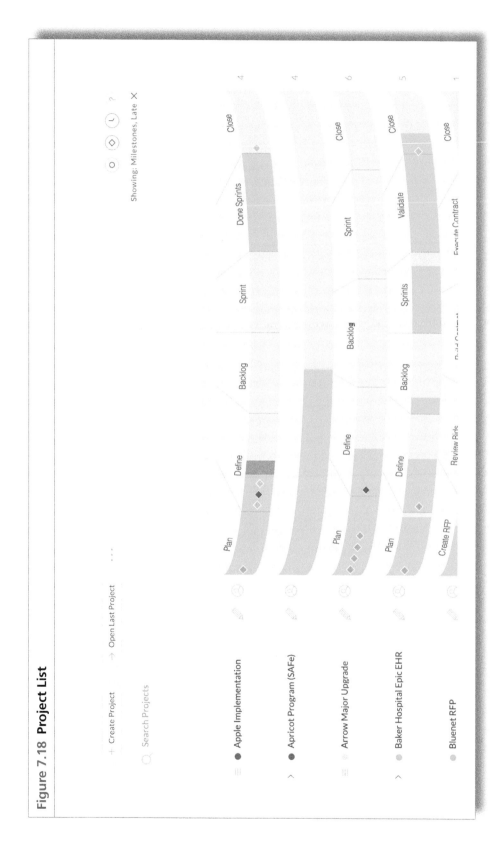

Figure 7.19 Project List—Stack

Figure 7.20 On My Team's Plate

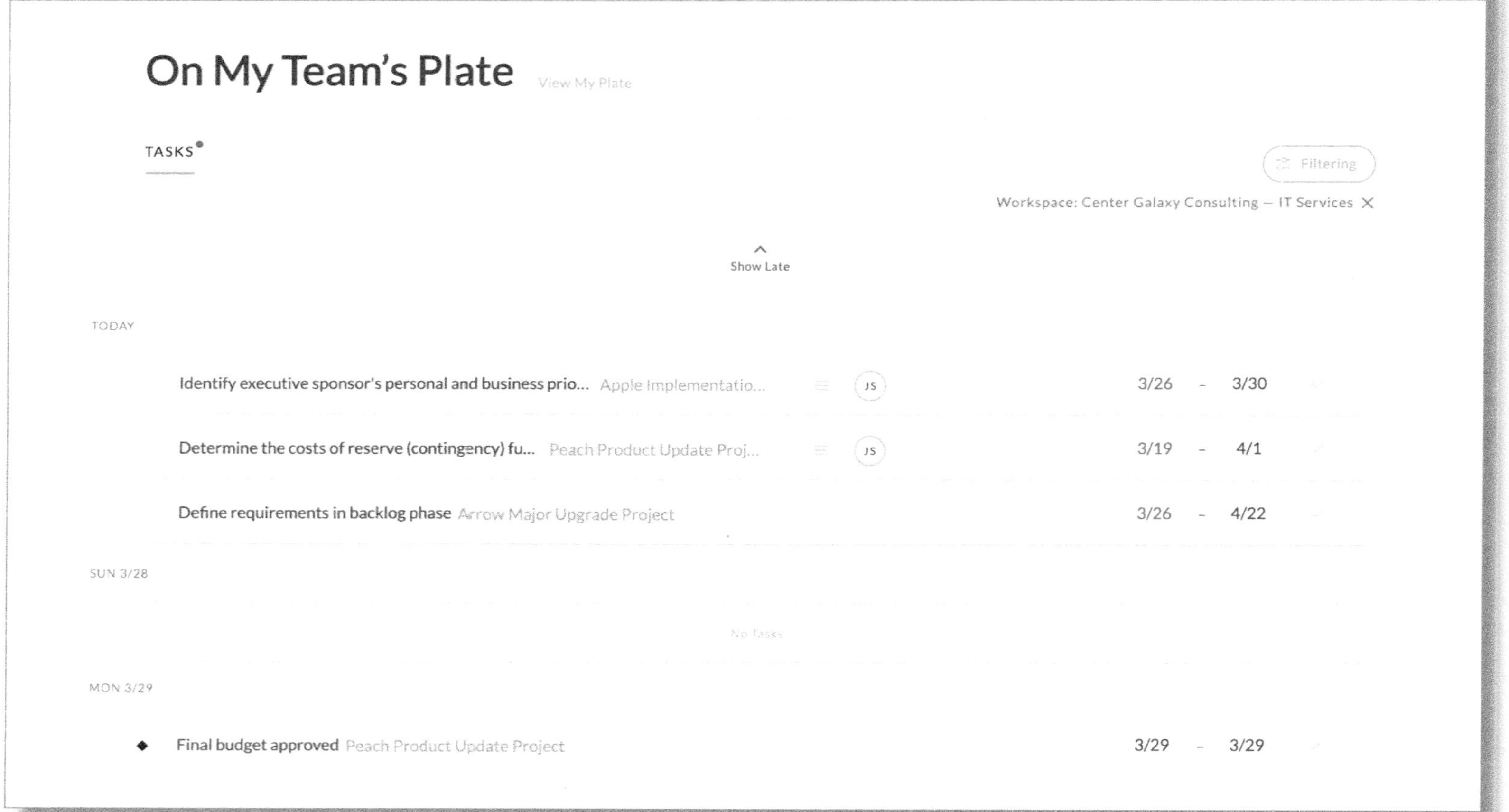

On My Team's Plate
View My Plate

TASKS

Workspace: Center Galaxy Consulting — IT Services ✕

(⠿ Filtering)

⌄ Show Late

TODAY

Identify executive sponsor's personal and business prio... Apple Implementatio... ≡ (JS)   3/26 – 3/30

Determine the costs of reserve (contingency) fu... Peach Product Update Proj... ≡ (JS)   3/19 – 4/1

Define requirements in backlog phase Arrow Major Upgrade Project   3/26 – 4/22

SUN 3/28

No Tasks

MON 3/29

◆ Final budget approved Peach Product Update Project   3/29 – 3/29

Figure 7.21 Timesheet

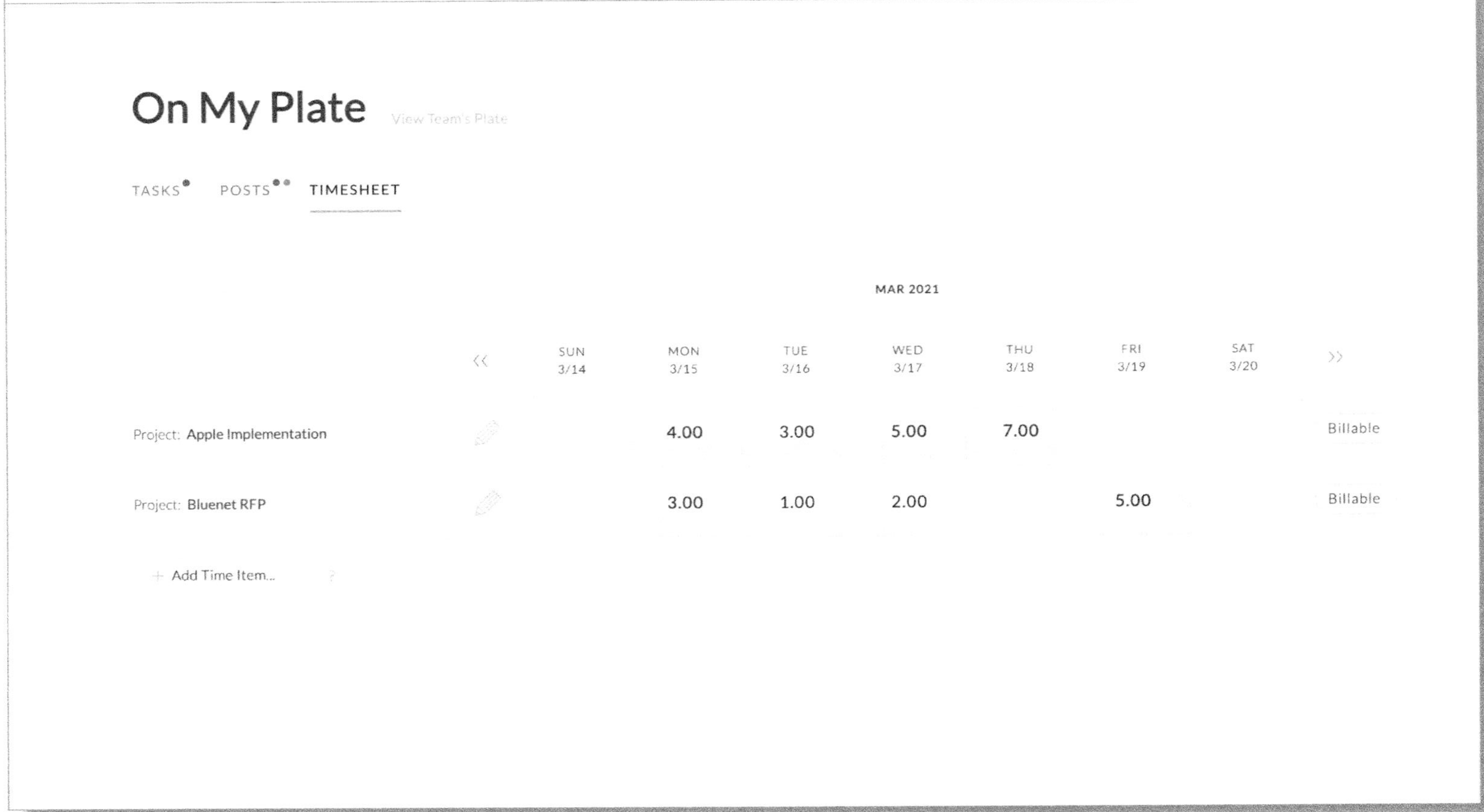

On My Plate View Team's Plate

TASKS • POSTS • • TIMESHEET

		SUN 3/14	MON 3/15	TUE 3/16	WED 3/17	THU 3/18	FRI 3/19	SAT 3/20	
	<<				MAR 2021				>>
Project: Apple Implementation			4.00	3.00	5.00	7.00			Billable
Project: Bluenet RFP			3.00	1.00	2.00		5.00		Billable
+ Add Time Item...									

Figure 7.22 **Recipes** *Source: Reproduced by permission of PieMatrix, Inc. All rights reserved.*

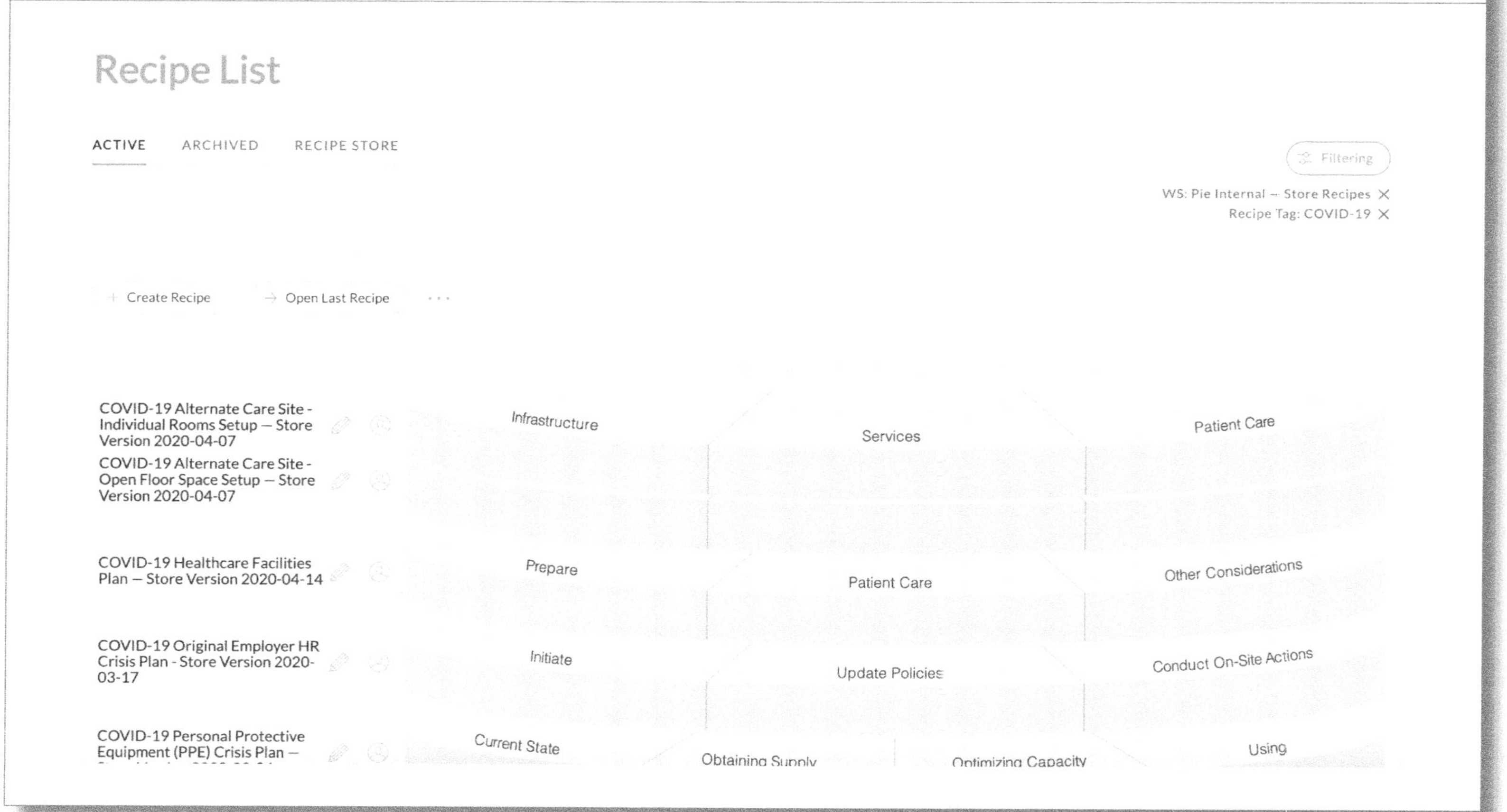

Recipe List
ACTIVE
ARCHIVED
RECIPE STORE
Filtering
WS: Pie Internal – Store Recipes
Recipe Tag: COVID-19
Create Recipe
Open Last Recipe
COVID-19 Alternate Care Site - Individual Rooms Setup – Store Version 2020-04-07
Infrastructure
Services
Patient Care
COVID-19 Alternate Care Site - Open Floor Space Setup – Store Version 2020-04-07
COVID-19 Healthcare Facilities Plan – Store Version 2020-04-14
Prepare
Patient Care
Other Considerations
COVID-19 Original Employer HR Crisis Plan - Store Version 2020-03-17
Initiate
Update Policies
Conduct On-Site Actions
COVID-19 Personal Protective Equipment (PPE) Crisis Plan –
Current State
Obtaining Supply
Optimizing Capacity
Using

Pie Recipes—Flexible Frameworks

Since Pie is process-focused, it has an entire section built for creating and managing repeatable best practices, processes, procedures, or flexible frameworks. These are called "Recipes," which are similar to project templates. The Recipe List page (Figure 7.22) displays a list ready-to-use project recipes. A process owner or a project manager can create new recipes from scratch or from an existing project. The example in Figure 7.22 shows a list of COVID-19 recipes made with content from the CDC website pages. These recipes can be imported for free from the Pie Store at https://my.pie.me/store.

You can make these standards available for project teams and you can also keep these standards fresh with new updates in real time. The value is continuous improvements, which leads to better project results.

For information on Pie, go to https://www.pie.me/store.

7.4 DASHBOARDS IN ACTION: INTERNATIONAL INSTITUTE FOR LEARNING

The International Institute for Learning (IIL) is one of the world's largest project management educational and consulting services providers. With a multitude of clients scattered across the globe and each at a possible different level of maturity in project management, IIL is often called upon to create different sets of dashboards for their clients.

Dashboards do not need to be highly complex. Even the simplest form of dashboard can provide effective information for decision makers. Figures 7.23 through 7.28 illustrate that simple dashboards

Figure 7.23 Impact upon Strategic Objectives. *Source: International Institute for Learning, Inc.*

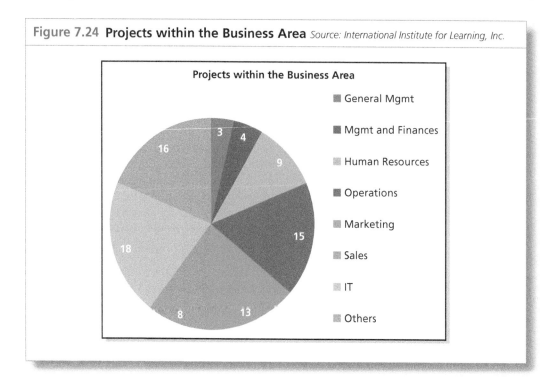

Figure 7.24 **Projects within the Business Area** *Source: International Institute for Learning, Inc.*

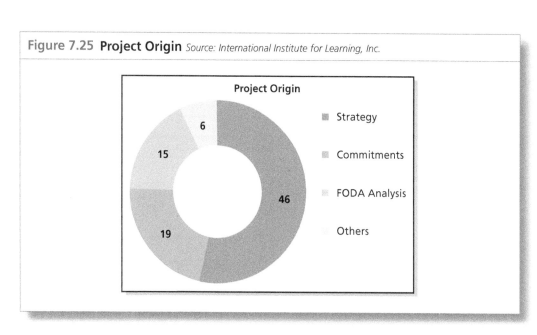

Figure 7.25 **Project Origin** *Source: International Institute for Learning, Inc.*

can often be as effective or even more effective than complex dashboards. Loading up a dashboard with unnecessary bells and whistles does not increase the quality of the information.

Figure 7.26 Project Status within the Business Area *Source: International Institute for Learning, Inc.*

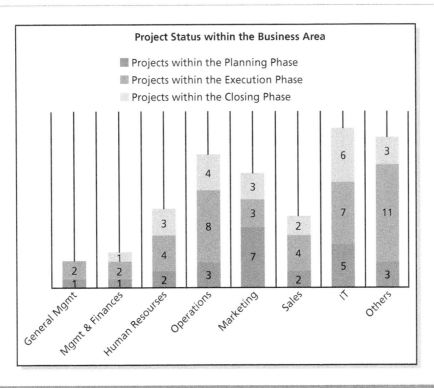

Figure 7.27 Projects by Year of Approval *Source: International Institute for Learning, Inc.*

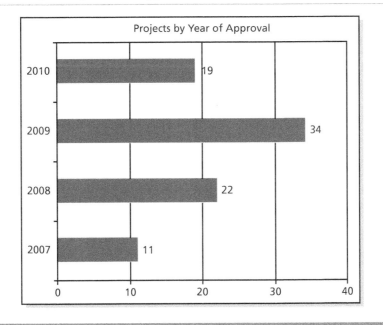

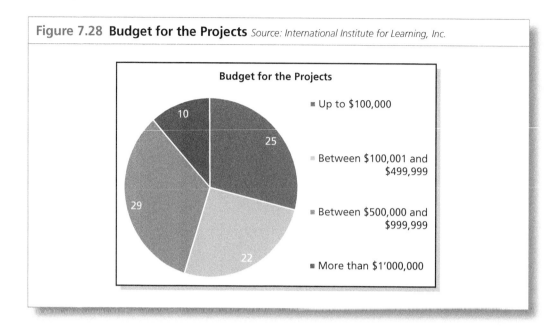

Figure 7.28 Budget for the Projects *Source: International Institute for Learning, Inc.*

In the last line of Figure 7.28, $1'000,000 looks strange, but it is correct. In the international format, which is used by many countries, the apostrophe is used to separate millions and comma is used for thousands.

THE PORTFOLIO MANAGEMENT PMO AND METRICS

CHAPTER OVERVIEW

To avoid metrics mania and save time and headaches, companies have given the responsibility of metrics management to various traditional or corporate project management offices (PMOs). However, as companies recognize the need for specialized PMOs, such as the portfolio management PMO, specialized metrics may be needed. The PMO generally establishes the metrics needed to validate the performance of the overall portfolio of projects as well as strategic metrics that are needed for executive decision making.

CHAPTER OBJECTIVES

- To understand the responsibilities of a portfolio management PMO
- To understand the differences between traditional metrics and value-based metrics
- To understand the types of metrics needed by a portfolio management PMO
- To understand the need for crisis dashboards

KEY WORDS

- Crisis dashboards
- Portfolio management PMO
- Value-based metrics
- Intangible value metrics
- Strategic value metrics

8.0 INTRODUCTION

Advances in project metrics have been rapid, but advances in portfolio metrics have been slow because not all companies maintain a project management office (PMO) dedicated to portfolio management activities. Some companies maintain just a single PMO. Although PMOs are often created to provide an independent view of project performance, metrics must also be established to measure PMO success as well as portfolio success.

Today, it is becoming more common to have a PMO dedicated to the management of the portfolio of projects. This can lead to

Project Management Metrics, KPIs, and Dashboards: A Guide to Measuring and Monitoring Project Performance, Fourth Edition. Harold Kerzner.
© 2023 John Wiley & Sons, Inc. Published 2023 by John Wiley & Sons, Inc.
Companion Website: www.wiley.com/go/kerznermetrics4e

changes in the role of the project manager, the metrics used, and the dashboard displays.

8.1 CRITICAL QUESTIONS

Three important questions that must be addressed by the portfolio management PMO:

1. *Is the company working on the right projects?* (i.e., Do the projects support strategic initiatives, and are they aligned with strategic objectives?)
2. *Is the company working on enough of the right projects?* (i.e., Is there the right mix of projects to maximize investment value? Shareholder value?)
3. *Is the company doing the right projects right?* (i.e., When will the project be finished, and at what cost?)

All three questions mention the word "right." Today, this word has a value meaning or at least implies value. Simply stated, why select a project as part of the portfolio if the intent is not to create business value? And if the project is completed within time and cost, is it a success if business value was not created? In the future, "value" will become increasingly important, and the definition of project success might be "achieving the desired business value within the competing constraints."

The importance of value should now be clear. Success must be defined in terms of the value that was expected to be delivered. Business cases today define the benefits to be achieved and how they will be measured. Value is what the benefits are actually worth to the business at the completion of the project. Metrics must be designed to reflect this value.

8.2 VALUE CATEGORIES

As shown in Figure 8.1, projects can be selected as part of the portfolio based on the type of value they are expected to deliver. The quadrants in the figure are generic, and each company can have its own categories of value based on how the company performs strategic planning.

Defining success on a project has never been an easy task. The focus has always been the triple constraints. Today it is acknowledged that there are four cornerstones for success, where success is defined in terms of value that is expected. Projects must be selected that can maximize the value based on the constraints placed on the available resources. To do this, project managers must be able to quantify value. Possible categories of value include:

- **Internal value:** These projects are designed to improve the efficiency and effectiveness of the firm. A side benefit might be the

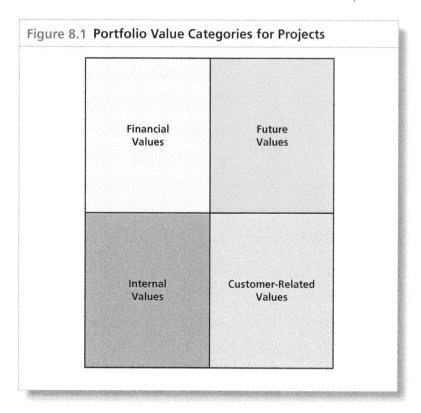

Figure 8.1 **Portfolio Value Categories for Projects**

building of relationships throughout the firm. The value obtained from these projects could be lowering costs, controlling scope changes, reducing waste, and shortening the time to market for new products. These projects can also be undertaken to improve the enterprise project management methodology, in which case people with process skills would be needed.

- **Customer-related value:** The near-term value in these projects is that they improve customer relations. It is not uncommon for near-term projects to drain cash rather than generate cash. The long-term value comes from future contracts to support cash flow. Resources needed on these projects are generally people who know the customer or may have worked on projects for the customer previously.

- **Future value:** These projects are designed to create future value through new products and services. In most companies, the best technically oriented people are assigned to these projects based upon the subcategories. These projects may be heavily oriented around research and development. Typical subcategories might be radical breakthrough, next generation, addition to the family, or add-ons and enhancements. Future value projects may require project managers with technical skills as well as business skills and a good understanding of business risk management.

TABLE 8.1	Typical Categories of Value and Tracking Metrics	
CATEGORY	**BENEFITS/VALUE**	**VALUE TRACKING METRICS**
Internal value	■ Adherence to constraints ■ Repetitive delivery ■ Control of scope changes ■ Control of action items ■ Reduction in waste ■ Efficiency	■ Time ■ Cost ■ Scope ■ Quality ■ Number of scope changes ■ Duration of open action items ■ Number of resources ■ Amount of waste ■ Efficiency
Financial value	■ Improvements in ROI, NPV, IRR, and payback period ■ Cash flow ■ Improvements in operating margins	■ Financial metrics ■ ROI calculators ■ Operating margins
Customer-related	■ Customer loyalty ■ Number of customers allowing you to use their name as a reference ■ Improvements in customer delivery ■ Customer satisfaction ratings	■ Loyalty/customer satisfaction surveys ■ Time to market ■ Quality
Future value	■ Reducing time to market ■ Image/reputation ■ Technical superiority ■ Creation of new technology or products	■ Time ■ Surveys on image and reputation ■ Number of new products ■ Number of patents ■ Number of retained customers ■ Number of new customers

Table 8.1 identifies the four broad categories from Figure 8.1 and the accompanying tracking metrics. Numerous benefits and metrics can be used for each category. Only a few appear here as examples.

8.3 PORTFOLIO METRICS

The value tracking metrics identified in Table 8.1 are design to track individual projects in each of the categories. These metrics are referred to as micro-level metrics. Specific metrics can be used to measure the overall effectiveness of a portfolio management PMO. Table 8.2 shows the metrics that can be used to measure the overall value of project management, a traditional PMO, and a portfolio PMO. The metrics listed under project management and many of the metrics

TABLE 8.2 Metrics for Specific PMO Types		
PROJECT MANAGEMENT (MICRO-LEVEL METRICS)	**TRADITIONAL PMO METRICS (MACRO-LEVEL METRICS)**	**PORTFOLIO PMO METRICS (MACRO-LEVEL METRICS)**
■ Adherence to schedule baselines	■ Growth in customer satisfaction	■ Business portfolio profitability or ROI
■ Adherence to cost baselines	■ Number of projects at risk	■ Portfolio health
■ Adherence to scope baselines	■ Conformance to the methodology	■ Percentage of successful portfolio projects
■ Adherence to quality requirements	■ Ways to reduce the number of scope changes	■ Portfolio benefits realization
■ Effective utilization of resources	■ Growth in the yearly throughput of work	■ Portfolio value achieved
■ Customer satisfaction levels	■ Validation of timing and funding	■ Portfolio selection and mix of projects
■ Project performance	■ Ability to reduce project closure rates	■ Resource availability
■ Total number of deliverables produced	■ Capturing and maintaining a best practices library	■ Capacity available for the portfolio
		■ Utilization of people for portfolio projects
		■ Hours per portfolio project
		■ Staff shortage
		■ Strategic alignment
		■ Business performance enhancements
		■ Portfolio budget versus actual
		■ Portfolio deadline versus actual

under the traditional PMO are considered to be micro-level metrics that focus on tactical objectives. The metrics listed under the portfolio PMO are macro-level metrics. Both the traditional and the portfolio PMOs are generally considered as overhead and subject to possible downsizing unless the PMOs can show through metrics how the organization benefits from their existence.

The portfolio management PMO may very well be involved in the establishment of a portfolio of projects that includes all four quadrants. In addition, the PMO may track the life cycle phase of each project in the portfolio as well as the priority. This can appear in one dashboard image, as shown in Figure 8.2. Specialized project portfolio software exists that can accomplish significantly more than just the image in the figure.

Sometimes several metrics can be combined into one table on a dashboard screen to show the portfolio governance committee the status of selected projects. Such a table appears in Figure 8.3.

8.4 MEASUREMENT TECHNIQUES AND METRICS

With the growth in measurement techniques, companies now have a multitude of metrics to support the decisions they must make and to measure portfolio benefits and value. Although some of these measurement techniques are still in the infancy stages, the growth

Figure 8.2 High-Level Project Portfolio Status

Project Numbers

Waiting for Funding	Waiting for Final Approval	Approved but Not Started	Started and No Issues	Started and Small Issues	Started and Major Issues
Proj . #31 Proj . #30 Proj . #22	Proj . #17 Proj . #24 Proj . #15	Proj . #9 Proj . #7 Proj . #8 Proj . #4	Proj . #2 Proj . #5 Proj . #13 Proj . #12	Proj . #1	Proj . #6 Proj . #11

Priorities: (A) is the highest priority

#31 (A) #30 (B) #22 (B)	#17 (A) #24 (B) #15 (C)	#9 (A) #7 (B) #8 (A) #4 (C)	#2 (A) #5 (B) #13 (B) #12 (C)	#1 (A)	#6 (A) #11 (C)

Figure 8.3 Grouping of Projects

Project	Business Owner	Project Manager	Priority	Status	Cost	Time	Risk	Business Value	Resources
1	Ralph	Bob	High	●	At Risk	On Track	High	High	Sufficient
2	Carol	Anne	Medium	◐	On Track	On Track	Med	High	Sufficient
3	Ruth	Frank	High	○	On Track	At Risk	High	Medium	Borderline
4	Paul	Joan	Medium	◉	On Track	On Track	Med	Low	Sufficient
5	Rich	Gary	Low	○	On Track	On Track	Low	Low	Borderline
6	Betty	Louis	Medium	◉	On Track	On Track	Low	Medium	Sufficient
7	Fran	Chris	High	○	At Risk	At Risk	Med	High	Shortage
8	Joe	Jean	Low	●	On Track	On Track	Low	Medium	Shortage

rate is expected to be rapid. The purpose of the portfolio metrics is to address concerns about the percentage of projects:

- On time and within budget
- With missed milestones
- On hold, canceled (before and/or after approval), or that have failed
- Aligned to strategic objectives
- That have undergone scope reduction
- That required rework
- That are used to run the business, grow the business, and to innovate

Portfolio metrics also address:

- How resources are being utilized across the portfolio
- How much time is spent approving a project
- How much time is spent approving the features/deliverables of a project
- How much time is spent developing the benefits realization plan and the business case

The actions over the concerns may require re-baselining the portfolio. This might include:

- Terminating or removing weak investments
- Recommending scope changes to some of the existing projects
- Cutting costs
- Accelerating some schedules
- Consolidating some projects
- Changing project personnel

Re-baselining often is necessary if there are (1) too many projects in the queue and not enough resources, and (2) critical resources being consumed on non-value-added projects.

8.5 THE GROWTH OF PROJECT METRICS

As stated in previous chapters, project managers are realizing that they are no longer just managing a traditional project to create an outcome or deliverable. Instead, project managers are managing part of the business and are being treated as business managers. The traditional metrics that we use, such as time, cost, and scope, may be insufficient for assisting in making some project decisions and may not capture the true business value of the deliverable of a project.

Today, we live in a world where digitalization is becoming increasingly important. Also, significant advances are being made in measurement techniques that are being applied to projects. As such,

Figure 8.4 Metrics Growth

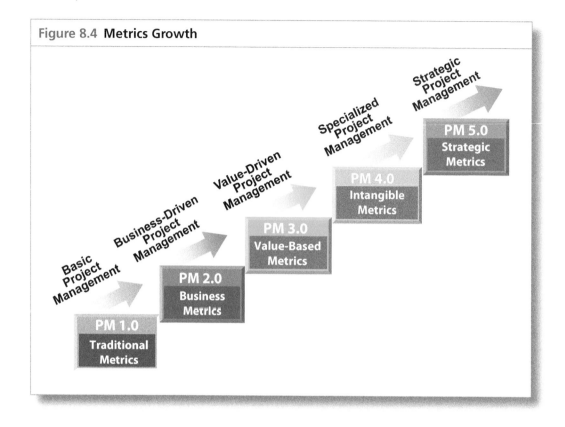

there has been a significant growth in types of metrics as seen in Figure 8.4.

The five levels identified in Figure 8.4 can be described as follows:

- *Basic Project Management (PM 1.0)*: This is traditional or operational project management which focuses on well-defined requirements and primarily time, cost, and scope metrics only. Once the deliverables are created, the project manager hands the project over to someone else and moves on to another assignment.
- *Business-Driven Project Management (PM 2.0)*: The project manager is viewed as a business manager and participates in some business-related decisions that impact the projects. The PM must capture and report certain business metrics so that management can make timely decisions based upon facts and evidence rather than guesses.

The business metrics referred to here are usually metrics related to the customers of the project's deliverables and may include the number of new customers, number of customers lost, and maintenance and support information for a targeted customer base. The business metrics are not necessarily strategic metrics but are used to assist the project sponsors and portfolio managers in the decisions they must make for the projects where they are providing governance.

- *Value-Driven Project Management (PM 3.0)*: Projects are now aligned to strategic business objectives where success is measured by the business benefits and value created from the deliverables. The PM must identify, capture, and report benefits and value metrics, and provide this information to executives and managers of strategic business portfolios of projects.
- *Specialized Project Management (PM 4.0)*: The outcome of some projects requires the measurement of intangibles rather than just tangible outcomes. Intangible metrics may focus on continuous improvement efforts for better management or governance of projects and programs.
- *Strategic Project Management (PM 5.0)*: As business managers, some project managers may be asked to manage strategic projects, such as those involving innovation and new business models, and must therefore monitor and report strategic business metrics in addition to other metrics. These metrics are needed to support the information portfolio PMOs must report such as the information in column 3 in Table 8.2.

8.6 METRICS FOR MEASURING INTANGIBLES

Effective project management education, especially at the executive level, is now addressing long-term benefits and value resulting from not only the outcome of projects, but also from the way that projects were managed. Many of the benefits and value are the result of intangible asset creation that was previously difficult to measure. Fortunately, measurement techniques have advanced to the point where we believe that we can measure anything. Projects now have both financial and nonfinancial metrics, and many of the nonfinancial metrics are regarded as intangible metrics. An example of an intangible metric might be the effectiveness of governance as shown in Figure 8.5.

The value of intangibles can have a greater impact on long-term considerations rather than short-term factors. Management support for the value measurements of intangibles can also prevent short-term financial considerations from dominating project decision-making. Measuring intangibles is dependent upon management's commitment to the measurement techniques used. Measuring intangibles does improve performance provided we have valid measurements free of manipulation.

The following are often regarded as intangible project management assets and can be measured:

- Project management governance (Did we have proper governance, was it effective and did the governance personnel understand their roles and responsibilities?)

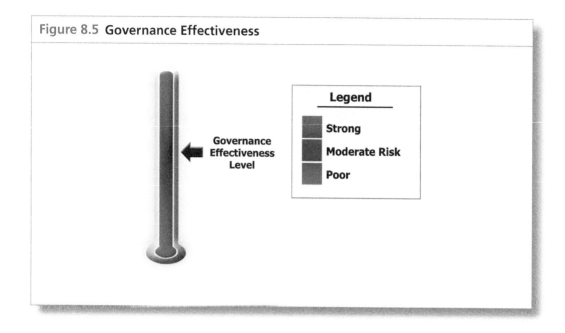

Figure 8.5 **Governance Effectiveness**

- Project management leadership (Did the project manager provide effective leadership?)
- Commitment (Is top management committed to continuous improvements in project management?)
- Lessons learned and best practices (Did we capture lessons learned and best practices?)
- Knowledge management (Are the best practices and lessons learned part of our knowledge management system?)
- Intellectual property rights (Is project management creating patents and other forms of intellectual property rights?)
- Working conditions (Are the people on the project teams satisfied with the working conditions?)
- Teamwork and trust (Are the people on the project teams working together as a team and do they trust one another's decisions?)

Businesses need to understand and create new sets of metrics and key performance indicators (KPIs) that can measure the intangibles and how they impact future decision-making and performance.

Even though most executives seem to understand the benefits of measuring intangibles there is still resistance such as:

- Intangibles are long-term measurements, and most companies focus on the short-term results.
- Companies argue that intangibles do not impact the bottom line.
- Companies are fearful of what the results will show.
- Companies argue that they lack the capability to measure intangibles.

The traditional metrics that we use, such as time, cost, and scope, may not capture the true business value of the deliverable of a project if the result is an intangible asset. Today, we live in a world where digitalization is becoming increasingly important. Also, significant advances are being made in measurement techniques that can be applied to projects. As such, the outcomes resulting from intangibles are becoming more important than the tangibles for the business to succeed and develop a sustainable competitive advantage.

There are many types of intangibles that benefit businesses and will be used as companies plan for the future. Examples include:

- Goodwill
- Customer satisfaction
- Our relationships with our customers
- Our relationship with our suppliers and distributors
- Brand image and reputation
- Patents, trademarks, and other intellectual property
- Our business processes
- Executive governance
- The company's culture and mindset
- Human capital, including retained knowledge and the ability to work together
- Strategic decision-making
- Strategy execution

Businesses need to understand and create new sets of metrics and key performance indicators (KPIs) that can measure the intangibles and how they impact future decision-making and performance.

For years, we had questions about intangibles that we had difficulty in answering:

- Are all corporate assets tangible, intangible, or both?
- Can we define intangible assets?
- Can the intangible assets be expressed in financial terms and their impact on the corporate balance sheet?
- Can intangible assets be measured?
- Are they value-added and can we establish intangible project management metrics?
- Do they impact the performance of the organization in the future?

Intangible assets today are more than just goodwill or intellectual property. They also include maximizing human performance. Intangible assets include such items as corporate culture, intellectual capital and the accompanying knowledge management system, executive and project leadership and governance, employee talent, employee satisfaction, trust and credibility, and workforce innovation capability. Understanding and measuring intangible asset value improves performance. While many of these may be hard to measure, they are not immeasurable.

8.7 THE NEED FOR STRATEGIC METRICS

Because of advances in metric measurement techniques, models have been developed by which we can show the alignment of projects to strategic business objectives. One such model appears in Figure 8.6. Years ago, the only metrics we would use were time, cost, and scope. Today, we can include metrics related to both strategic value and business value. This allows us to evaluate the health of the entire portfolio of projects as well as individual projects.

Since all metrics have established targets, we can award points for each metric based upon how close we come to the targets. The metrics selected help support strategic decisions that may be necessary. Figure 8.7 shows that the project identified in Figure 8.6 has received thus far 80 points out of a possible 100 points. Figure 8.8 shows the alignment of projects in the portfolio to strategic objectives. If the total score in Figure 8.7 is between 0 and 50 points, we would assume that the project is not contributing to strategic objectives at this time, and this would be shown as a zero or blank cell in Figure 8.8. Scores between 51 and 75 points would indicate a "partial" contribution to the objectives and shown as a one in Figure 8.8. Scores between 76 and 100 points would indicate fulfilling the objectives and shown as a two in Figure 8.8. Periodically we can summarize the results in Figure 8.8 to show management Figure 8.9 which illustrates our ability to create the desired benefits and final value.

Previously, we stated that project management has become a strategic competency and that most projects are aligned to strategic

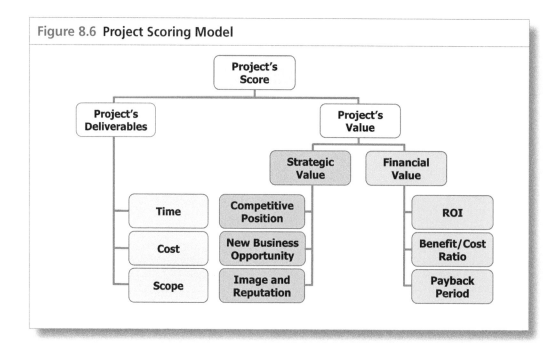

Figure 8.6 **Project Scoring Model**

Figure 8.7 Project Scoring Model with Points Assigned

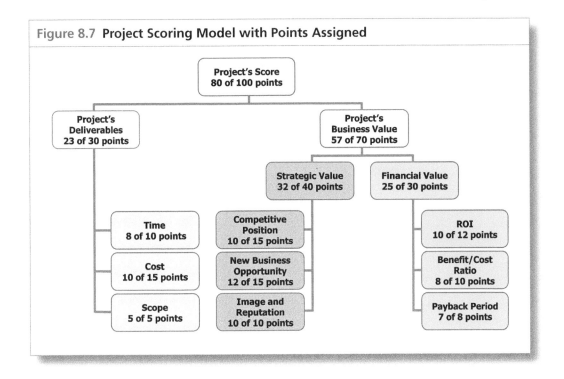

Figure 8.8 Matching Projects to Strategic Business Objectives

Strategic Objectives:	Project 1	Project 2	Project 3	Project 4	Project 5	Project 6	Project 7	Project 8	Scores
Projects									
Technical Superiority	2		1			2		1	6
Reduced Operating Costs				2	2				4
Reduced Time-To-Market	1		1	2	1	1		2	8
Increase Business Profits			2	1	1	1		2	7
Add Manufacturing Capacity	1		2	2		1		1	7
Column Scores	4	0	6	7	4	5	0	6	

	No Contribution
1	Supports Objective
2	Fulfills Objective

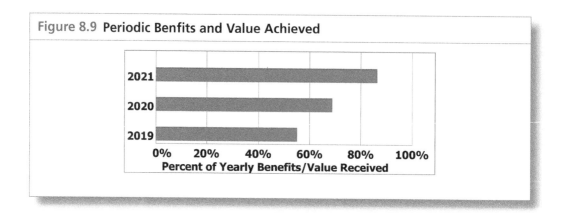

Figure 8.9 **Periodic Benfits and Value Achieved**

business objectives. To select and evaluate projects, there must exist strategic metrics other than value-driven and intangible metrics.

The strategic business metrics must be able to be combined to answer questions that executives might ask. The list below identifies metrics that executives need to make decisions concerning business and portfolio health.

- Business profitability
- Portfolio health
- Portfolio benefits realization
- Portfolio value achieved
- Portfolio mixes of projects
- Resource availability
- Capacity utilization
- Strategic alignment of projects
- Overall business performance

Project managers will be required to provide project performance information that supports the use of these metrics. The project-based strategic business metrics must be able to be combined to create this list of metrics that executives need for business decision-making and strategic planning. Typical questions that senior management must address, which requires use of these metrics, include:

- Do we have any weak investments that need to be cancelled or replaced?
- Must any of the programs and/or projects be consolidated?
- Must any of the projects be accelerated or decelerated?
- How well are the projects aligned to strategic objectives?
- Does the portfolio have to be rebalanced?
- Is value being created?
- Do we know the risks and how risks are being mitigated?
- Can we predict future corporate performance?

Will we need to perform portfolio resource re-optimization?

8.8 CRISIS DASHBOARDS

Projects in today's environments are significantly more complex than projects managed in the past. Governance is performed by a governance committee rather than just a project sponsor. Each stakeholder or member of the governance committee may very well require different metrics and key performance indicators (KPIs). If each stakeholder wishes to view 20 to 30 metrics, the costs of metric measurement and reporting can be significant and can defeat the purpose of going to paperless project management.

The solution to effective communications with stakeholders and governance groups is to show them that they can most likely get all of the critical data they need for informed decision making with 6 to 10 metrics or KPIs that can be displayed on one computer screen. This is not always the case, and drill-down to other screens may be necessary. But, in general, one computer screen shot should be sufficient.

If an out-of-tolerance condition or crisis situation exists with any of the metrics or KPIs on the dashboard screen, then the situation should be readily apparent to the viewer. But what if the crisis occurs due to metrics that do not appear on the screen? In this case, the viewer will be immediately directed to a crisis dashboard, which shows all of the metrics that are out of tolerance. The metrics will remain on the crisis dashboard until such time that the crisis or out-of-tolerance conditions are corrected. Each stakeholder will now see the regular screen shot and then be instructed to look at the crisis screen shot.

Defining a Crisis

A crisis can be defined as any event, whether expected or not, that can lead to an unstable or dangerous situation affecting the outcome of the project. Crises imply negative consequences that can harm the organization, its stakeholders, and the general public and can cause changes to the firm's business strategy, how it interfaces with the enterprise environmental factors, its social consciousness, and the way it maintains customer satisfaction. A crisis does not necessarily mean that the project will fail nor does it mean that the project should be terminated. The crisis simply could be that the project's outcome will not occur as expected.

Some crises may appear gradually and can be preceded by early warning signs. These can be referred to as smoldering crises. The intent of metrics and dashboards is to identify trends that could indicate that a crisis is coming and provide the project manager with sufficient time to develop contingency plans. The earlier the project manager knows about the impending crisis, the more options available as a remedy.

Another type of crisis is that which occurs abruptly with little or no warning. These are referred to as sudden crises. Examples that could impact projects might be elections or political uncertainty in the host country, natural disasters, or the resignation of an employee with critical skills. Metrics and dashboards cannot be created for every possible crisis that could exist on a project. Sudden crises cannot be prevented.

Not all out-of-tolerance conditions are crises. For example, being significantly behind schedule on a software project may be seen as a problem but not necessarily a crisis. But if the construction of a manufacturing plant is behind schedule and plant workers have already been hired to begin work on a certain date or the delay in the plant will activate penalty clauses for late delivery of manufactured items for a client, then the situation could be a crisis. Sometimes exceeding a target favorably also triggers a crisis. As an example, a manufacturing company had a requirement to deliver 10 and only 10 units to a client each month. The company manufactured 15 units each month but could only ship 10 per month to the client. Unfortunately, the company did not have storage facilities for the extra units produced, and a crisis occurred.

How do project managers determine whether the out-of-tolerance condition is just a problem or a crisis that needs to appear on the crisis dashboard? The answer is in the potential damage that can occur. If any of the items on the next list could occur, then the situation would most likely be treated as a crisis and necessitate a dashboard display:

- There is a significant threat to:
- The outcome of the project.
- The organization as a whole, its stakeholders, and possibly the general public.
- The firm's business model and strategy.
- Worker health and safety
- There is a possibility for loss of life.
- Redesigning existing systems is now necessary.
- Organizational change will be necessary.
- The firm's image or reputation will be damaged.
- Degradation in customer satisfaction could result in a present and future loss of significant revenue.

It is important to understand the difference between risk management and crisis management.

In contrast to risk management, which involves assessing potential threats and finding the best ways to avoid those threats, crisis management involves dealing with threats before, during, and after they have occurred. [That is, crisis management is proactive, not merely reactive.] It is a discipline within the broader context of management consisting of skills and techniques required to identify, assess, understand, and cope with a serious situation, especially from the moment it first occurs to the point that recovery procedures start.[1]

---

1 Wikipedia contributors, "Crisis Management," *Wikipedia, The Free Encyclopedia,* https://en.wikipedia.org/w/index.php?title=Crisis_management&oldid=774782017.

Crises often require that immediate decisions be made. Effective decision making requires information. If one metric appears to be in crisis mode and shows up on the crisis dashboard, viewers may find it necessary to look at several other metrics that may not be in a crisis mode and may not appear on the crisis dashboard but are possible causes of the crisis. Looking at metrics on dashboards is a lot easier than reading reports.

The difference between a problem and a crisis is like beauty; it is in the eyes of the beholder. What one stakeholder sees as a problem another stakeholder may see it as a crisis. Table 8.3 shows how difficult it is to make the differentiation.

TABLE 8.3   Differentiating between a Problem and a Crisis

METRIC/KPI	PROBLEM	CRISIS
Time	The project will be late but still acceptable to the client.	The project will be late and the client is considering cancellation.
Cost	Costs are being overrun, but the client can provide additional funding.	Costs are being overrun and no additional funding is available. Cancellation is highly probable.
Quality	The customer is unhappy with the quality but can live with it.	Quality of the deliverables is unacceptable, personal injury is possible, the client may cancel the contract, and no further work may come from this client.
Resources	The project is either understaffed or the resources assigned have marginal skills to do the job. A schedule delay is probable.	The quality or lack of resources will cause a serious delay in the schedule, and the quality of workmanship may be unacceptable such that the project may be canceled.
Scope	Numerous scope changes cause changes to the baselines. Delays and cost overruns are happening but are acceptable to the client for now.	The number of scope changes has led the client to believe that the planning is not correct and more scope changes will occur. The benefits of the project no longer outweigh the cost, and project termination is likely.
Action items	The client is unhappy with the amount of time taken to close out action items, but the impact on the project is small.	The client is unhappy with the amount of time taken to close out action items, and the impact on the project is significant. Governance decisions are being delayed because of the open action items, and the impact on the project may be severe.
Risks	Significant risk levels exist, but the team may be able to mitigate some of the risks.	The potential damage that can occur because of the severity of the risks is unacceptable to the client.
Assumptions and constraints	New assumptions and constraints have appeared and may adversely affect the project.	New assumptions and constraints have appeared such that significant project replanning will be necessary. The value of the project may no longer be there.
Enterprise environmental factors	The enterprise environmental factors have changes and may adversely affect the project.	The new enterprise environmental factors will greatly reduce the value and expected benefits of the project.

These conclusions about crisis dashboards can be drawn:

- The definition of what is or is not a crisis is not always clear to the viewers.
- Not all problems are crises.
- Sometimes unfavorable trends are treated as a crisis and appear on crisis dashboards.
- The crisis dashboard may contain a mixture of crisis metrics and metrics that are treated just as problems.
- The metrics that appear on a traditional dashboard reporting system may have to be redrawn when placed on a crisis dashboard to ensure that the metrics are easily understood.

Crisis metrics generally imply that either this situation must be monitored closely or that some decisions must be made. But project managers must be careful not to overreact.

INDEX

Project Management Metrics, KPIs, and Dashboards: A Guide to Measuring and Monitoring Project Performance, Fourth Edition. Harold Kerzner.
© 2023 John Wiley & Sons, Inc. Published 2023 by John Wiley & Sons, Inc.
Companion Website: www.wiley.com/go/kerznermetrics4e